Advances in Functional Analysis and Operator Theory

Contemporary Mathematics

798

Advances in Functional Analysis and Operator Theory

AMS-EMS-SMF Special Session
Advances in Functional Analysis and Operator Theory
July 18–22, 2022
Université de Grenoble-Alpes, Grenoble, France

Marat V. Markin
Igor V. Nikolaev
Carsten Trunk
Editors

AMS American Mathematical Society

2020 *Mathematics Subject Classification.* Primary 11G10, 34A09, 34B24, 37A55, 46L05, 46L10, 46L70, 46L85.

Library of Congress Cataloging-in-Publication Data

Names: AMS-EMS-SMF Special Session on Advances in Functional Analysis and Operator Theory (2022 : Grenoble, France) | Markin, Marat V., editor.

Title: Advances in functional analysis and operator theory : AMS-EMS-SMF Special Session on Advances in Functional Analysis and Operator Theory, July 18-22, 2022, Université de Grenoble-Alpes, Grenoble, France / Marat V. Markin, Igor V. Nikolaev, Carsten Trunk, editors.

Description: Providence, Rhode Island : American Mathematical Society, [2024] | Series: Contemporary mathematics, 0271-4132 ; volume 798 | Includes bibliographical references.

Identifiers: LCCN 2023042586 | ISBN 9781470473051 (paperback) | 9781470476113 (ebook)

Subjects: LCSH: Functional analysis–Congresses. | Operator theory–Congresses. | AMS: Number theory – Arithmetic algebraic geometry (Diophantine geometry) – Abelian varieties of dimension | Ordinary differential equations – General theory – Implicit equations, differential-algebraic equations. | Ordinary differential equations – Boundary value problems – Sturm-Liouville theory. | Dynamical systems and ergodic theory – Ergodic theory – Relations with the theory of -algebras. | Functional analysis – Selfadjoint operator algebras (algebras, von Neumann algebras, etc.) – General theory of -algebras. | Functional analysis – Selfadjoint operator algebras (algebras, von Neumann algebras, etc.) – General theory of von Neumann algebras. | Functional analysis – Selfadjoint operator algebras (algebras, von Neumann algebras, etc.) – Nonassociative selfadjoint operator algebras. | Functional analysis – Selfadjoint operator algebras (algebras, von Neumann algebras, etc.) – Noncommutative topology.

Classification: LCC QA319 .A47 2022 | DDC 515/.7–dc23/eng/20231229

LC record available at https://lccn.loc.gov/2023042586

---

10 9 8 7 6 5 4 3 2 1 29 28 27 26 25 24

Contents

Preface

This volume contains the proceedings of the Special Session on Advances in Functional Analysis and Operator Theory, held July 18–22, 2022, as part of the 2nd Joint Congress of Mathematics co-organized by the American Mathematical Society, the European Mathematical Society, and the Société Mathématique de France at Université Grenoble Alpes Campus in Grenoble, France.

The papers reflect the modern interplay between differential equations, functional analysis, operator algebras, and their applications from the dynamics to quantum groups to number theory. The topics discussed are: the Sturm-Liouville and boundary value problems, axioms of quantum mechanics, C^*-algebras and symbolic dynamics, Schlesinger systems, von-Neumann algebras and low-dimensional topology, quantum permutation groups, the Jordan algebras, the Villadsen idempotents, the Kadison-Singer transforms and the Shafarevich-Tate groups of abelian varieties.

Contemporary Mathematics
Volume **798**, 2024
https://doi.org/10.1090/conm/798/15978

Minimality conditions for Sturm-Liouville problems with a boundary condition depending affinely or quadratically on an eigenparameter

Nazim Kerimov and Yagub Aliyev

ABSTRACT. In the paper we study Sturm-Liouville problems with a boundary condition depending affinely or quadratically on an eigenparameter. The necessary and sufficient conditions for minimality and completeness of the chosen system of root functions of the corresponding operator were given in two forms, one with the use of special associated functions and another one with the direct use of characteristic functions. This direct method was known for the affine case and was extensively discussed in the literature. The aim of the present paper is to develop this direct method for the quadratic case and to consider the affine and quadratic cases together in a unified way.

1. Introduction

Consider the spectral problem

$$-y'' + q(x)y = \lambda y,\ 0 < x < 1, \tag{1}$$

$$y'(0)\sin\beta = y(0)\cos\beta,\ 0 \le \beta < \pi, \tag{2}$$

$$y'(1) = (a\lambda^2 + b\lambda + c)y(1), \tag{3}$$

where λ is the spectral parameter, $q(x)$ is a real valued and continuous function on the interval $[0, 1]$, and a, b, c are real. We will focus on two cases Quadratic $a \neq 0$, and Affine $a = 0, b < 0$. In the papers by Binding, Browne, Watson, and Code [**7**], [**9**] the existence and asymptotics of the eigenvalues of (1)-(3) were studied. It was proved that the eigenvalues of (1)-(3) form an infinite sequence, accumulating only at $+\infty$, and the following cases are possible:

(a) All the eigenvalues are real and simple;

(b) All the eigenvalues are simple and all, except a conjugate pair of non-real, are real;

(c) All the eigenvalues are real and all, except one double, are simple;

(d) All the eigenvalues are real and all, except one triple, are simple.

The eigenvalues λ_n $(n \ge 0)$ will be considered to be listed according to non-decreasing real part and repeated according to algebraic multiplicity. The asymptotic formulas of eigenvalues for more general polynomial case were studied in [**8**].

2020 *Mathematics Subject Classification.* Primary 34B24; Secondary 34L10.

Key words and phrases. Sturm-Liouville, eigenparameter-dependent boundary conditions, minimal system, root functions.

For the affine decreasing case the asymptotic formula is the following

$$\lambda_n = \begin{cases} (n-1/2)^2 \pi^2 + O(1) & \text{if } \beta \neq 0, \\ n^2\pi^2 + O(1) & \text{if } \beta = 0. \end{cases}$$

For the quadratic case the formula is

$$\lambda_n = \begin{cases} (n-3/2)^2 \pi^2 + O(1) & \text{if } \beta \neq 0, \\ (n-1)^2\pi^2 + O(1) & \text{if } \beta = 0. \end{cases}$$

We are interested in the basis properties in $L_p(0,1)$ $(1 < p < \infty)$ of the root function system of the boundary value problem (1)-(3). In order to obtain a basis consisted of eigenfunctions one needs to eliminate one or two of the eigenfunctions. This phenomenon of defect in the system of eigenfunctions is well known and was discussed in many studies [**12**], [**17**], [**21**], [**22**], [**23**]. As in [**1**] and [**2**], our main objective will be the minimality conditions for the system of root functions (cf. [**18**]). But these minimality conditions can be easily extended to basis properties in $L_p(0,1)$ $(1 < p < \infty)$ (see e.g. [**4**]).

The following special problem appears in the studies of the torsional vibrations of a shaft with a disk at one end [**5**], and the vibrations of a homogeneous string with a weight at one end [**24**]:

$$-y'' = \lambda y, \ 0 < x < 1,$$

$$y(0) = 0, \ y'(1) = b\lambda y(1), \ b > 0.$$

Similar problems appear in the study of the surface viscosity of nematic liquid crystals in contact with solid surfaces [**6**], in the study of propogation of heat in a rod with concentrated heat capacity at one end [**17**]. All the eigenvalues of this special boundary value problem are real and simple. But if $b < 0$ or if linear function in the boundary condition $(b\lambda + c)$ is replaced by a more general quadratic function $(a\lambda^2 + b\lambda + c)$ then not allways all the eigenvalues are real and when all are real then not always all are simple. There are possibilities of a conjugate pair of non-real eigenvalues and a double or a triple eigenvalue for these more general problems. In these cases, there are many choices of the eliminated root functions (eigenfunctions or associated functions). In some cases, there are singularities when the system of root functions with some eliminated functions is not minimal or not complete. These cases can be described by some necessary and sufficient conditions. In the paper two ways of describing these singular cases will be discussed. One of them is indirect and uses specially defined associated functions. The second way is more direct and therefore more suitable for calculations. The focus of the present paper will be the description of these direct methods. The comparison of the methods was done on several examples at the end of the paper.

2. Terminology and notations

The notations and the preliminary results in this section were given in [**2**], [**4**] and some of them are included here just for completeness. We define $y(x, \lambda)$ to be the non-zero solution of (1), (2), analytic in $\lambda \in \mathbf{C}$, and we write the characteristic equation as

$$\omega(\lambda) = y'(1,\lambda) - (a\lambda^2 + b\lambda + c)y(1,\lambda).$$

The eigenvalue λ_k is multiple if $\omega'(\lambda_k) = 0$, in particular, we say that λ_k is a double eigenvalue if in addition $\omega''(\lambda_k) \neq 0$, and a triple eigenvalue if $\omega''(\lambda_k) = 0 \neq \omega'''(\lambda_k)$. Let y_n be an eigenfunction corresponding to eigenvalue λ_n. Note that

$y(x,\lambda) \to y(x,\lambda_n) = y_n$, uniformly in $x \in [0,1]$, as $\lambda \to \lambda_n$. If λ_k is a multiple eigenvalue ($\lambda_k = \lambda_{k+1}$) then the first order associated function y_{k+1} is defined by

$$-y''_{k+1} + q(x)y_{k+1} = \lambda_k y_{k+1} + y_k,$$

$$y'_{k+1}(0)\sin\beta = y_{k+1}(0)\cos\beta,$$

$$y'_{k+1}(1) = (a\lambda_k^2 + b\lambda_k + c)y_{k+1}(1) + (2a\lambda_k + b)y_k(1).$$

Here and in the following the defintion of the associated function from [**19**] was used. Since $\omega(\lambda_k) = \omega'(\lambda_k) = 0$ then $y(x,\lambda) \to y_k$, $y_\lambda(x,\lambda) \to \widetilde{y}_{k+1}$, uniformly according to $x \in [0,1]$, as $\lambda \to \lambda_k$, where $\widetilde{y}_{k+1}$ is one of the first order associated functions, $\widetilde{y}_{k+1} = y_{k+1} + \widetilde{c}y_k$, and $\widetilde{c} = (\widetilde{y}_{k+1}(1) - y_{k+1}(1))/y_k(1)$.

If λ_k is a triple eigenvalue ($\lambda_k = \lambda_{k+1} = \lambda_{k+2}$) then together with the first order associated function y_{k+1} there exists the second order associated function y_{k+2}, for which the following relations hold:

$$-y''_{k+2} + q(x)y_{k+2} = \lambda_k y_{k+2} + y_{k+1},$$

$$y'_{k+2}(0)\sin\beta = y_{k+2}(0)\cos\beta,$$

$$y'_{k+2}(1) = (a\lambda_k^2 + b\lambda_k + c)y_{k+2}(1) + (2a\lambda_k + b)y_{k+1}(1) + ay_k(1).$$

If λ_k is a triple eigenvalue $\omega''(\lambda_k) = 0$ then $y_{\lambda\lambda} \to 2\widetilde{y}_{k+2}$, uniformly in $x \in [0,1]$, as $\lambda \to \lambda_k$, where $\widetilde{y}_{k+2}$ is one of the second order associated functions corresponding to the first order associated function $\widetilde{y}_{k+1}$, $\widetilde{y}_{k+2} = y_{k+2} + \widetilde{c}y_{k+1} + \widetilde{d}y_k$, and $\widetilde{d} = (\widetilde{y}_{k+2}(1) - y_{k+2}(1) - \widetilde{c}y_{k+1}(1))/y_k(1)$.

Definition 1. *If λ_k is a double eigenvalue then let $y^*_{k+1} = y_{k+1} + c_1 y_k$, where*

$$c_1 = -\frac{\omega'''(\lambda_k)}{3\omega''(\lambda_k)} - \frac{\widehat{y}_{k+1}(1)}{y_k(1)},$$

and $\widehat{y}_{k+1} = y_{k+1} - \widetilde{c}y_k$.

Definition 2. *If λ_k is a triple eigenvalue then let $y^{\#}_{k+1} = y_{k+1} + c_2 y_k$ and $y^{\#}_{k+2} = y_{k+2} + c_2 y_{k+1} + d_1 y_k$, where*

$$c_2 = -\frac{\omega^{IV}(\lambda_k)}{4\omega'''(\lambda_k)} - \frac{\widehat{y}_{k+1}(1)}{y_k(1)},$$

$$d_1 = -\frac{\omega^{V}(\lambda_k)}{20\omega'''(\lambda_k)} - \frac{\widehat{y}_{k+1}(1)\omega^{IV}(\lambda_k)}{4y_k(1)\omega'''(\lambda_k)} - \frac{\widehat{y}_{k+2}(1)\omega'''(\lambda_k)}{y_k(1)\omega'''(\lambda_k)} + c_2^2,$$

and $\widehat{y}_{k+2} = y_{k+2} - \widetilde{c}\widetilde{y}_{k+1} - \widetilde{d}y_k$.

3. Affine case

Theorem 1 below was given in [**2**] and mentioned in [**12**]. Let $a = 0$. The case $b \geq 0$ is well known and was studied in many papers before (see e.g. [**11**], [**13**]). So, we assume in this section that $b < 0$.

Theorem 1. Let $a = 0$ and $b < 0$.

1) In the case **(c)** the system $\{y_n\}$ ($n = 0,\ 1,\ldots;\ n \neq k$) forms a minimal system in $L_2(0,1)$ if and only if $y^*_{k+1}(1) \neq 0$, which is equivalent to

$$\widetilde{c} \neq \frac{1}{3} \cdot \frac{\omega'''(\lambda_k)}{\omega''(\lambda_k)}.$$

2) In the case **(d)** the system $\{y_n\}$ $(n = 0,\ 1, \ldots;\ n \neq k+1)$ forms a minimal system in $L_2(0,1)$ if and only if $y^{\#}_{k+1}(1) \neq 0$, which is equivalent to

$$\widetilde{c} \neq \frac{1}{4} \cdot \frac{\omega^{IV}(\lambda_k)}{\omega'''(\lambda_k)}.$$

3) In the case **(d)** the system $\{y_n\}$ $(n = 0,\ 1, \ldots;\ n \neq k)$, forms a minimal system in $L_2(0,1)$ if and only if $y^{\#}_{k+2}(1) \neq 0$, which is equivalent to

$$\widetilde{d} \neq \frac{1}{4} \cdot \frac{\omega^{IV}(\lambda_k)}{\omega'''(\lambda_k)} \cdot \left(\widetilde{c} - \frac{1}{4} \cdot \frac{\omega^{IV}(\lambda_k)}{\omega'''(\lambda_k)}\right) + \frac{1}{20} \cdot \frac{\omega^{V}(\lambda_k)}{\omega'''(\lambda_k)}.$$

In each case the minimality conditions are also conditions to be a basis in $L_p(0,1)$ $(1 < p < \infty)$ for the corresponding systems (see [**2**]).

Similar problem was also studied in [**1**]. In more general form these basis properties (minimality and completeness) of differential operators with eigenparameter dependent boundary conditions were studied in [**23**]. In particular, the results mentioned in Theorem 3 of [**22**] for the cases **(c)** and **(d)** above are in perfect agreement with Theorem 1 of the current paper.

4. Quadratic case

In the current section, more general results will be proved for the quadratic case and Theorem 1 above is a consequence of Theorem 2 (part 2) and Theorem 3 below.

Theorem 2. Let $a \neq 0$.

1) In the case **(c)** the system $\{y_n\}$ $(n = 0,\ 1, \ldots;\ n \neq k, j)$, where $j \neq k, k+1$ is arbitrary non-negative integer, forms a minimal system in $L_2(0,1)$ if and only if $y^{*}_{k+1}(1)(\lambda_j - \lambda_k) \neq y_k(1)$, which is equivalent to

$$\widetilde{c} \neq \frac{1}{\lambda_j - \lambda_k} + \frac{1}{3} \cdot \frac{\omega'''(\lambda_k)}{\omega''(\lambda_k)}.$$

In the case **(d)** the system

2) $\{y_n\}$ $(n = 0,\ 1, \ldots;\ n \neq k, k+2)$ forms a minimal system in $L_2(0,1)$ if and only if $y^{\#}_{k+1}(1) \neq 0$, which is equivalent to

$$\widetilde{c} \neq \frac{1}{4} \cdot \frac{\omega^{IV}(\lambda_k)}{\omega'''(\lambda_k)};$$

3) $\{y_n\}$ $(n = 0,\ 1, \ldots;\ n \neq k, k+1)$ forms a minimal system in $L_2(0,1)$ if and only if $y^{\#}_{k+1}(1)^2 \neq y_k(1) y^{\#}_{k+2}(1)$, which is equivalent to

$$\widetilde{d} \neq \widetilde{c} \cdot \left(\widetilde{c} - \frac{1}{4} \cdot \frac{\omega^{IV}(\lambda_k)}{\omega'''(\lambda_k)}\right) + \frac{1}{20} \cdot \frac{\omega^{V}(\lambda_k)}{\omega'''(\lambda_k)};$$

4) $\{y_n\}$ $(n = 0,\ 1, \ldots;\ n \neq k+1, j)$, where $j \neq k, k+1, k+2$ is arbitrary non-negative integer, forms a minimal system in $L_2(0,1)$ if and only if $y^{\#}_{k+1}(1)(\lambda_j - \lambda_k) \neq y_k(1)$, which is equivalent to

$$\widetilde{c} \neq \frac{1}{\lambda_j - \lambda_k} + \frac{1}{4} \cdot \frac{\omega^{IV}(\lambda_k)}{\omega'''(\lambda_k)};$$

5) $\{y_n\}$ $(n = 0,\ 1, \ldots;\ n \neq k, j)$, where $j \neq k, k+1, k+2$ is arbitrary non-negative integer, forms a minimal system in $L_2(0,1)$ if and only if $y^{\#}_{k+2}(1)(\lambda_j - \lambda_k) \neq y^{\#}_{k+1}(1)$, which is equivalent to

$$\widetilde{d} \neq \left(\frac{1}{\lambda_j - \lambda_k} + \frac{1}{4} \cdot \frac{\omega^{IV}(\lambda_k)}{\omega'''(\lambda_k)}\right) \cdot \left(\widetilde{c} - \frac{1}{4} \cdot \frac{\omega^{IV}(\lambda_k)}{\omega'''(\lambda_k)}\right) + \frac{1}{20} \cdot \frac{\omega^{V}(\lambda_k)}{\omega'''(\lambda_k)}.$$

Proof. The part of the theorem about minimality in $L_2(0,1)$ was proved in [**4**] (See also [**16**]). So, we will prove only the equivalency of the two inequalities in each case.

1) Let us write the first inequality as $\frac{y^*_{k+1}(1)}{y_k(1)} \neq \frac{1}{\lambda_j - \lambda_k}$. By Definition 1 (Lemma 3.1 in [4]), $y^*_{k+1} = y_{k+1} + c_1 y_k$, where

$$c_1 = -\frac{\omega'''(\lambda_k)}{3\omega''(\lambda_k)} - \frac{y_{k+1}(1)}{y_k(1)} + \widetilde{c}.$$

So, we obtain $\frac{y_{k+1}(1)}{y_k(1)} + c_1 = -\frac{\omega'''(\lambda_k)}{3\omega''(\lambda_k)} + \widetilde{c} \neq \frac{1}{\lambda_j - \lambda_k}$, which completes the proof for this case.

2) By Definition 2 (Lemma 3.2 in [**4**]), $y^{\#}_{k+1} = y_{k+1} + c_2 y_k$, where

$$c_2 = -\frac{\omega^{IV}(\lambda_k)}{4\omega'''(\lambda_k)} - \frac{y_{k+1}(1)}{y_k(1)} + \widetilde{c}.$$

Then the inequality $y^{\#}_{k+1}(1) \neq 0$ is equivalent to

$$\frac{y_{k+1}(1)}{y_k(1)} + c_2 = -\frac{\omega^{IV}(\lambda_k)}{4\omega'''(\lambda_k)} + \widetilde{c} \neq 0,$$

which completes the proof for this case.

3) By Definition 2 (Lemma 3.3 in [**4**]),

$$d_1 - c_2^2 = -\frac{\omega^{V}(\lambda_k)}{20\omega'''(\lambda_k)} - \frac{1}{4}\left(\frac{y_{k+1}(1)}{y_k(1)} - \widetilde{c}\right)\frac{\omega^{IV}(\lambda_k)}{\omega'''(\lambda_k)} - \frac{y_{k+2}(1)}{y_k(1)} + \widetilde{c}\frac{y_{k+1}(1)}{y_k(1)} + \widetilde{d} - \widetilde{c}^2.$$

Let us write the inequality as $\frac{y^{\#}_{k+1}(1)}{y_k(1)} \neq \frac{y^{\#}_{k+2}(1)}{y^{\#}_{k+1}(1)}$. So, we can write

$$\frac{y_{k+1}(1)}{y_k(1)} + c_2 \neq \frac{y_{k+2}(1) + (d_1 - c_2^2)y_k(1)}{y_{k+1}(1) + c_2 y_k(1)} + c_2,$$

or

$$\frac{y_{k+1}(1)}{y_k(1)} \neq \frac{\frac{y_{k+2}(1)}{y_k(1)} + (d_1 - c_2^2)}{\frac{y_{k+1}(1)}{y_k(1)} + c_2}.$$

Noting the expressions for c_2 and $d_1 - c_2^2$, we obtain

$$\frac{y_{k+1}(1)}{y_k(1)} \neq \frac{-\frac{\omega^{V}(\lambda_k)}{20\omega'''(\lambda_k)} - \frac{1}{4}\left(\frac{y_{k+1}(1)}{y_k(1)} - \widetilde{c}\right)\frac{\omega^{IV}(\lambda_k)}{\omega'''(\lambda_k)} + \widetilde{c}\frac{y_{k+1}(1)}{y_k(1)} + \widetilde{d} - \widetilde{c}^2}{-\frac{\omega^{IV}(\lambda_k)}{4\omega'''(\lambda_k)} + \widetilde{c}}.$$

Reorganizing the numerator of the right hand side, we obtain

$$\frac{y_{k+1}(1)}{y_k(1)} \neq \frac{-\frac{\omega^{V}(\lambda_k)}{20\omega'''(\lambda_k)} + \widetilde{d}}{-\frac{\omega^{IV}(\lambda_k)}{4\omega'''(\lambda_k)} + \widetilde{c}} + \left(\frac{y_{k+1}(1)}{y_k(1)} - \widetilde{c}\right),$$

which completes the proof for this case.

4) Let us write the inequality as $\frac{y_{k+1}^{\#}(1)}{y_k(1)} \neq \frac{1}{\lambda_j-\lambda_k}$. Using the method of the previous cases, we can write

$$\frac{y_{k+1}(1)}{y_k(1)} + c_2 \neq \frac{1}{\lambda_j - \lambda_k},$$

which simplifies to

$$-\frac{\omega^{IV}(\lambda_k)}{4\omega'''(\lambda_k)} + \widetilde{c} \neq \frac{1}{\lambda_j - \lambda_k}.$$

5) Let us write the inequality as $\frac{y_{k+2}^{\#}(1)}{y_{k+1}^{\#}(1)} \neq \frac{1}{\lambda_j-\lambda_k}$. We already simplified the left side of this inequality in the proof of the case **3**. Using it we obtain

$$\frac{y_{k+2}(1) + (d_1 - c_2^2)y_k(1)}{y_{k+1}(1) + c_2 y_k(1)} + c_2 \neq \frac{1}{\lambda_j - \lambda_k}.$$

Using the formula for c_2, we obtain

$$\frac{-\frac{\omega^V(\lambda_k)}{20\omega'''(\lambda_k)} + \widetilde{d}}{-\frac{\omega^{IV}(\lambda_k)}{4\omega'''(\lambda_k)} + \widetilde{c}} + \left(\frac{y_{k+1}(1)}{y_k(1)} - \widetilde{c}\right) - \frac{\omega^{IV}(\lambda_k)}{4\omega'''(\lambda_k)} - \frac{y_{k+1}(1)}{y_k(1)} + \widetilde{c} \neq \frac{1}{\lambda_j - \lambda_k},$$

or just

$$\frac{-\frac{\omega^V(\lambda_k)}{20\omega'''(\lambda_k)} + \widetilde{d}}{-\frac{\omega^{IV}(\lambda_k)}{4\omega'''(\lambda_k)} + \widetilde{c}} - \frac{\omega^{IV}(\lambda_k)}{4\omega'''(\lambda_k)} \neq \frac{1}{\lambda_j - \lambda_k},$$

which is equivalent to the required inequality.

Theorem 3. 1) In the case **(c)**, $y_{k+1}^{*}(1) \neq 0$ if and only if

$$\widetilde{c} \neq \frac{1}{3} \cdot \frac{\omega'''(\lambda_k)}{\omega''(\lambda_k)}.$$

2) In the case **(d)**, $y_{k+2}^{\#}(1) \neq 0$ if and only if

$$\widetilde{d} \neq \frac{1}{4} \cdot \frac{\omega^{IV}(\lambda_k)}{\omega'''(\lambda_k)} \cdot \left(\widetilde{c} - \frac{1}{4} \cdot \frac{\omega^{IV}(\lambda_k)}{\omega'''(\lambda_k)}\right) + \frac{1}{20} \cdot \frac{\omega^V(\lambda_k)}{\omega'''(\lambda_k)}.$$

Proof. **1)** Let us write the inequality as $\frac{y_{k+1}(1)}{y_k(1)} + c_1 \neq 0$. Using the formula for c_1, we obtain $\frac{y_{k+1}(1)}{y_k(1)} + c_1 = -\frac{\omega'''(\lambda_k)}{3\omega''(\lambda_k)} + \widetilde{c} \neq 0$, which completes the proof for this case.

2) Let us write the inequality as $\frac{y_{k+2}(1)}{y_k(1)} + c_2\left(\frac{y_{k+1}(1)}{y_k(1)} + c_2\right) + d_1 - c_2^2 \neq 0$. Using the formula for c_2 and d_1, we obtain

$$\frac{y_{k+2}(1)}{y_k(1)} + c_2\left(-\frac{\omega^{IV}(\lambda_k)}{4\omega'''(\lambda_k)} + \widetilde{c}\right)$$

$$+\left(-\frac{\omega^V(\lambda_k)}{20\omega'''(\lambda_k)} + \left(\frac{y_{k+1}(1)}{y_k(1)} - \widetilde{c}\right)\left(-\frac{\omega^{IV}(\lambda_k)}{4\omega'''(\lambda_k)} + \widetilde{c}\right) - \frac{y_{k+2}(1)}{y_k(1)} + \widetilde{d}\right) \neq 0$$

or

$$c_2\left(-\frac{\omega^{IV}(\lambda_k)}{4\omega'''(\lambda_k)} + \widetilde{c}\right)$$

$$-\frac{\omega^V(\lambda_k)}{20\omega'''(\lambda_k)} + \left(-\frac{\omega^{IV}(\lambda_k)}{4\omega'''(\lambda_k)} - c_2\right)\left(-\frac{\omega^{IV}(\lambda_k)}{4\omega'''(\lambda_k)} + \widetilde{c}\right) + \widetilde{d} \neq 0.$$

Finally, after cancelling the terms with c_2, we obtain

$$-\frac{\omega^V(\lambda_k)}{20\omega'''(\lambda_k)}+\frac{\omega^{IV}(\lambda_k)}{4\omega'''(\lambda_k)}\left(-\frac{\omega^{IV}(\lambda_k)}{4\omega'''(\lambda_k)}+\widetilde{c}\right)+\widetilde{d}\neq 0,$$

which proves the theorem.

Remark: As in the previous section, one can extend these minimality conditions to basis properties in $L_p(0,1)$ $(1<p<\infty)$ using the asymptotic formula for the eigenvalues and the fact that the corresponding systems of eigenfunctions are quadratically close to some trigonometric systems (see e.g. [**4**]).

5. Examples

In [**1**], [**2**], [**3**] many examples of the affine case were given. In those examples both ways of describing the necessary and sufficient conditions were presented. In all of these examples the two ways gave coinciding answers. So, the examples about the affine case will not be discussed here. Now we will use the new results in the current paper to do the same for the quadratic case. The examples below were previously studied only with one method. For completeness, we included those studies here and compared them with the new alternative method.

First example of a double eigenvalue. The following example from [**4**] was also discussed in [**14**], [**15**], [**16**]. Consider the spectral problem

$$-y''=\lambda y,\ 0<x<1,$$

$$y'(0)=0,\ y'(1)=\left(\frac{\lambda^2}{\pi^2}-\lambda\right)y(1).$$

For this problem

$$\omega(\lambda)=-\sqrt{\lambda}\sin\sqrt{\lambda}-\left(\frac{\lambda^2}{\pi^2}-\lambda\right)\cos\sqrt{\lambda}.$$

and

$$\omega(0)=\omega'(0)=0,\quad \omega''(0)=-\frac{2}{3}-\frac{2}{\pi^2},\quad \omega'''(0)=\frac{1}{5}+\frac{3}{\pi^2},$$

$$\omega(\pi^2)=0\neq\omega'(\pi^2).$$

In [**4**] we applied part 1 of Theorem 4.1 to this example. It was shown that the system

$$\left\{-\frac{x^2}{2}+c,\quad \cos\sqrt{\lambda_i}x\quad (i=3,4,5,\ldots)\right\},$$

is minimal in $L_2(0,1)$ if and only if

$$c\neq\frac{\pi^2+15}{10(\pi^2+3)}-\frac{1}{\pi^2}.$$

Let us now solve the same problem using the new method. By Theorem 2, the condition $y_1^*(1)(\lambda_2-\lambda_0)\neq y_0(1)$ is equivalent to

$$\widetilde{c}\neq\frac{1}{\lambda_2-\lambda_0}+\frac{\omega'''(0)}{3\omega''(0)}=\frac{1}{\pi^2-0}+\frac{1}{3}\cdot\frac{\frac{1}{5}+\frac{3}{\pi^2}}{-\frac{2}{3}-\frac{2}{\pi^2}}=$$

$$=\frac{1}{\pi^2}-\frac{\pi^2+15}{10(\pi^2+3)}.$$

Since $c=-\widetilde{c}$, we obtain again

$$c\neq\frac{\pi^2+15}{10(\pi^2+3)}-\frac{1}{\pi^2}.$$

Second example of a double eigenvalue. The following example is taken from [**3**]. Consider the spectral problem

$$-y''=\lambda y,\ 0<x<1,$$

$$y(0)=0,\ y'(1)=\left(\frac{\lambda^2}{4\pi^2}-\frac{5}{8}\lambda+\frac{9\pi^2}{64}\right)y(1).$$

For this problem

$$y(x,\lambda)=\sin\sqrt{\lambda}x,\quad y'(x,\lambda)=\sqrt{\lambda}\cos\sqrt{\lambda}x,$$

and

$$\omega(\lambda)=\sqrt{\lambda}\cos\sqrt{\lambda}-\left(\frac{\lambda^2}{4\pi^2}-\frac{5}{8}\lambda+\frac{9\pi^2}{64}\right)\sin\sqrt{\lambda}.$$

Here $\lambda=0$ is not an eigenvalue, $\lambda_0=\lambda_1=\frac{\pi^2}{4}$ is the double eigenvalue, $\lambda_2=\frac{9\pi^2}{4}$ and all other eigenvalues $\lambda_3<\lambda_4<\dots$ are simple. Note that $y_0=\sin\frac{\pi}{2}x$, $y_2=\sin\frac{3\pi}{2}x$, $y_i=\sin\sqrt{\lambda_i}x$ $(i\geq 3)$, and

$$\widetilde{y}_1(x)=\lim_{\lambda\to\frac{\pi^2}{4}}y_\lambda(x,\lambda)=\lim_{\lambda\to\frac{\pi^2}{4}}\left(\frac{x\cos\sqrt{\lambda}x}{2\sqrt{\lambda}}\right)=\frac{x}{\pi}\cdot\cos\frac{\pi}{2}x.$$

So, we define the general first order associated function by $y_1=\frac{x}{\pi}\cdot\cos\frac{\pi}{2}x+c\cdot\sin\frac{\pi}{2}x$, that is $c=-\widetilde{c}$ (c is a constant). Let $y_1^*=\frac{x}{\pi}\cdot\cos\frac{\pi}{2}x+c'\cdot\sin\frac{\pi}{2}x$. From Lemma 5 it follows that $c'=\frac{4}{3\pi^2}-\frac{2}{9}-c$, so

$$y_1^*=\frac{x}{\pi}\cdot\cos\frac{\pi}{2}x+\left(\frac{4}{3\pi^2}-\frac{2}{9}-c\right)\cdot\sin\frac{\pi}{2}x.$$

The system

$$\left\{\frac{x}{\pi}\cdot\cos\frac{\pi}{2}x+c\cdot\sin\frac{\pi}{2}x,\quad \sin\sqrt{\lambda_i}x\quad(i=3,4,5,\dots)\right\},$$

that is the system of root functions without the removed functions y_0 and y_2, is minimal in $L_2(0,1)$ if and only if $y_1^*(1)(\lambda_2-\lambda_0)\neq y_0(1)$, or more explicitly, if $c\neq\frac{5}{6\pi^2}-\frac{2}{9}$. If $c=\frac{5}{6\pi^2}-\frac{2}{9}$ then the function $g(x)=2\pi x\cdot\cos\frac{\pi}{2}x+\sin\frac{\pi}{2}x+\sin\frac{3\pi}{2}x$, is orthogonal to all the elements of the system. By Theorem 2, the condition $y_1^*(1)(\lambda_2-\lambda_0)\neq y_0(1)$ is equivalent to

$$\widetilde{c}\neq\frac{1}{\lambda_2-\lambda_0}+\frac{\omega'''(\frac{\pi^2}{4})}{3\omega''(\frac{\pi^2}{4})}=\frac{1}{\frac{9\pi^2}{4}-\frac{\pi^2}{4}}+\frac{1}{3}\cdot\frac{\frac{6-\pi^2}{\pi^4}}{-\frac{3}{2\pi^2}}=-\frac{5}{6\pi^2}+\frac{2}{9}.$$

Since $c=-\widetilde{c}$, we obtain again $c\neq\frac{5}{6\pi^2}-\frac{2}{9}$.

First example of a triple eigenvalue. The following example is taken from [**4**] and it was also discussed in [**20**]. Consider the problem

$$-y''=\lambda y,\ 0<x<1,$$

$$y'(0)=0,\ y'(1)=\left(-\frac{\lambda^2}{3}-\lambda\right)y(1).$$

For this problem

$$\omega(\lambda) = -\sqrt{\lambda}\sin\sqrt{\lambda} + \left(\frac{\lambda^2}{3} + \lambda\right)\cos\sqrt{\lambda},$$

$$\omega(0) = \omega'(0) = \omega''(0) = 0, \; \omega'''(0) = -\frac{4}{5},$$

$$\omega^{IV}(0) = \frac{32}{105}, \; \omega^{V}(0) = -\frac{10}{189}.$$

We define the first and second order associated functions by $y_1 = -\frac{x^2}{2} + c$, $y_2 = \frac{x^4}{24} + c\left(-\frac{x^2}{2} + c\right) + d$, that is $c = -\widetilde{c}$, $d = -\widetilde{d}$ (c, d are constants). In [**4**] we have also shown that

$$y_1^{\#} = -\frac{x^2}{2} + \frac{25}{42} - c, \; y_2^{\#} = \frac{x^4}{24} - \left(\frac{25}{42} - c\right)\frac{x^2}{2} + \frac{1385}{5292} - \frac{25}{42}c - d.$$

In [**4**] we applied part 2 of Theorem 4.1 to this example and proved that the system

$$\left\{-\frac{x^2}{2} + c, \quad \cos\sqrt{\lambda_i}x \;\; (i = 3, 4, 5, \ldots)\right\},$$

is minimal in $L_2(0,1)$ if and only if $y_1^{\#}(1) \neq 0$ or more explicitly if $c \neq \frac{2}{21}$. By Theorem 2, $y_1^{\#}(1) \neq 0$ if and only if

$$\widetilde{c} \neq \frac{\omega^{IV}(0)}{4\omega'''(0)} = \frac{1}{4} \cdot \frac{32/105}{-4/5} = -\frac{2}{21},$$

which gives the same result. Similarly, in [**4**] we applied part 3 of Theorem 4.1 to the system

$$\left\{\frac{x^4}{24} + c\left(-\frac{x^2}{2} + c\right) + d, \quad \cos\sqrt{\lambda_i}x \;\; (i = 3, 4, 5, \ldots)\right\},$$

and proved that it is minimal in $L_2(0,1)$ if and only if $y_1^{\#}(1)^2 \neq y_0(1)y_2^{\#}(1)$ or more explicitly if $d \neq -\frac{5}{1512} + \frac{2}{21}c - c^2$.

By Theorem 2, $y_1^{\#}(1)^2 \neq y_0(1)y_2^{\#}(1)$ if and only if

$$\widetilde{d} \neq \widetilde{c}^2 - \frac{\widetilde{c}}{4} \cdot \frac{\omega^{IV}(0)}{\omega'''(0)} + \frac{1}{20} \cdot \frac{\omega^{V}(0)}{\omega'''(0)} =$$

$$= \widetilde{c}^2 - \frac{\widetilde{c}}{4} \cdot \frac{32/105}{-4/5} + \frac{1}{20} \cdot \frac{-10/189}{-4/5} = \widetilde{c}^2 + \frac{2}{21} \cdot \widetilde{c} + \frac{5}{1512},$$

which gives the same result.

Second example of a triple eigenvalue. The following example is taken from [**3**]. Consider the problem

$$-y'' = \lambda y, \; 0 < x < 1,$$

$$y(0) = 0, \; y'(1) = \left(-\frac{\lambda^2}{45} - \frac{\lambda}{3} + 1\right) y(1).$$

For this problem $y(x,\lambda) = \frac{\sin\sqrt{\lambda}x}{\sqrt{\lambda}}$, $y'(x,\lambda) = \cos\sqrt{\lambda}x$,

$$\omega(\lambda) = \cos\sqrt{\lambda} - \left(-\frac{\lambda^2}{45} - \frac{\lambda}{3} + 1\right)\frac{\sin\sqrt{\lambda}}{\sqrt{\lambda}},$$

$$\omega(0) = \omega'(0) = \omega''(0) = 0, \; \omega'''(0) = -\frac{4}{315},$$

$$\omega^{IV}(0) = \frac{16}{4725}, \ \omega^{V}(0) = -\frac{2}{4455}.$$

So, $\lambda_0 = \lambda_1 = \lambda_2 = 0$ is a triple eigenvalue, and $\lambda_3 < \lambda_4 < \dots$ are the positive solutions of the equation $\omega(\lambda) = 0$. Also, $y_0 = x, \ y_i = \sin\sqrt{\lambda_i}x \ \ (i \geq 3), \ y_1(x) = -\frac{x^3}{6} + cx$, and $y_2(x) = \frac{x^5}{120} - c\frac{x^3}{6} + dx$. We also find $c_2 = -2c + \frac{7}{30}, d_1 = \frac{329}{9900} - \frac{7}{15}c + 3c^2 - 2d$. Therefore, $y_1^{\#}(x) = -\frac{x^3}{6} + (-c + \frac{7}{30})x, y_2^{\#}(x) = \frac{x^5}{120} + (-c + \frac{7}{30})(-\frac{x^3}{6} - cx) + (\frac{329}{9900} - d)x$.

By part 2 of Theorem 4.1 in [**4**], the system

$$\left\{-\frac{x^3}{6} + cx, \ \ \sin\sqrt{\lambda_i}x \ \ (i = 3, 4, 5, \dots)\right\},$$

is minimal in $L_2(0,1)$ if and only if $y_1^{\#}(1) \neq 0$ or more explicitly if $c \neq \frac{1}{15}$. If $c = \frac{1}{15}$ then the function $\frac{x^5}{120} - \frac{x^3}{36} + \frac{7}{360}$ is orthogonal to all the elements of the system.

By Theorem 2, $y_1^{\#}(1) \neq 0$ if and only if

$$\widetilde{c} \neq \frac{\omega^{IV}(0)}{4\omega'''(0)} = \frac{1}{4} \cdot \frac{16/4725}{-4/315} = -\frac{1}{15}.$$

By part 3 of Theorem 4.1 in [**4**] the system

$$\left\{\frac{x^5}{120} + c\left(-\frac{x^3}{3} + c\right) + d, \ \ \sin\sqrt{\lambda_i}x \ \ (i = 3, 4, 5, \dots)\right\},$$

is minimal in $L_2(0,1)$ if and only if $y_1^{\#}(1)^2 \neq y_0(1)y_2^{\#}(1)$ or more explicitly if $d \neq -\frac{7}{3960} + \frac{1}{15}c - c^2$. If $d = -\frac{7}{3960} + \frac{1}{15}c - c^2$ then the same function $\frac{x^5}{120} - \frac{x^3}{36} + \frac{7}{360}$ is orthogonal to all the elements of the system.

By Theorem 2, $y_1^{\#}(1)^2 \neq y_0(1)y_2^{\#}(1)$ if and only if

$$\widetilde{d} \neq \widetilde{c}^2 - \frac{\widetilde{c}}{4} \cdot \frac{\omega^{IV}(0)}{\omega'''(0)} + \frac{1}{20} \cdot \frac{\omega^{V}(0)}{\omega'''(0)} =$$

$$= \widetilde{c}^2 - \frac{\widetilde{c}}{4} \cdot \frac{16/4725}{-4/315} + \frac{1}{20} \cdot \frac{-2/4455}{-4/315} = \widetilde{c}^2 + \frac{1}{15} \cdot \widetilde{c} + \frac{7}{3960},$$

which coincides with the previous result.

As these examples suggest, the alternative method, developed in the current paper, is simpler and gives the inequalities for minimality directly.

References

[1] Y. N. Aliyev, *Minimality properties of Sturm-Liouville problems with increasing affine boundary conditions*, Operator theory, functional analysis and applications, Oper. Theory Adv. Appl., vol. 282, Birkhäuser/Springer, Cham, [2021] ©2021, pp. 33–49, DOI 10.1007/978-3-030-51945-2_3. MR4248011

[2] Y. N. Aliyev, *Minimality of the system of root functions of Sturm-Liouville problems with decreasing affine boundary conditions*, Colloq. Math. **109** (2007), no. 1, 147–162, DOI 10.4064/cm109-1-12. MR2308832

[3] Y.N. Aliyev, *On the singularities of the root space of Sturm-Liouville boundary value problem with a boundary condition quadratically dependent on spectral parameter* (in Russian), Journal of Qafqaz University **29** (2010), 80–87.

[4] Y. N. Aliyev and N. B. Kerimov, *The basis property of Sturm-Liouville problems with boundary conditions depending quadratically on the eigenparameter* (English, with English and Arabic summaries), Arab. J. Sci. Eng. Sect. A Sci. **33** (2008), no. 1, 123–136. MR2414478

[5] I. G. Aramanovič and V. I. Levin, *Uravneniya matematicheskoĭ fiziki* (Russian), Second unrevised edition, Izdat. "Nauka", Moscow, 1969. MR259310

[6] G. Barbero; I. Dahl; L. Komitov,*Continuum description of the interfacial layer of nematic liquid crystals in contact with solid surfaces*, J. Chem. Phys., **130**, 174902 (2009) 1-10. https://doi.org/10.1063/1.3126657

[7] P. A. Binding, P. J. Browne, W. J. Code, and B. A. Watson, *Transformation of Sturm-Liouville problems with decreasing affine boundary conditions*, Proc. Edinb. Math. Soc. (2) **47** (2004), no. 3, 533–552, DOI 10.1017/S0013091504000197. MR2096617

[8] P. A. Binding, P. J. Browne, and B. A. Watson, *Equivalence of inverse Sturm-Liouville problems with boundary conditions rationally dependent on the eigenparameter*, J. Math. Anal. Appl. **291** (2004), no. 1, 246–261, DOI 10.1016/j.jmaa.2003.11.025. MR2034071

[9] W. J. Code and P. J. Browne, *Sturm-Liouville problems with boundary conditions depending quadratically on the eigenparameter*, J. Math. Anal. Appl. **309** (2005), no. 2, 729–742, DOI 10.1016/j.jmaa.2004.11.067. MR2154149

[10] S. Goktas, N. B. Kerimov, and E. A. Maris, *On the uniform convergence of spectral expansions for a spectral problem with a boundary condition rationally depending on the eigenparameter*, J. Korean Math. Soc. **54** (2017), no. 4, 1175–1187, DOI 10.4134/JKMS.j160414. MR3668863

[11] N. B. Kerimov and Y. N. Aliyev, *The basis property in L_p of the boundary value problem rationally dependent on the eigenparameter*, Studia Math. **174** (2006), no. 2, 201–212, DOI 10.4064/sm174-2-6. MR2238462

[12] N. B. Kerimov, S. Goktas, and E. A. Maris, *Uniform convergence of the spectral expansions in terms of root functions for a spectral problem*, Electron. J. Differential Equations (2016), Paper No. 80, 14 pp. MR3489988

[13] N. B. Kerimov and V. S. Mirzoev, *On the basis properties of a spectral problem with a spectral parameter in the boundary condition* (Russian, with Russian summary), Sibirsk. Mat. Zh. **44** (2003), no. 5, 1041–1045, DOI 10.1023/A:1025932618953; English transl., Siberian Math. J. **44** (2003), no. 5, 813–816. MR2019557

[14] C.-S. Liu, *The Lie-group shooting method for computing the generalized Sturm-Liouville problems*, CMES Comput. Model. Eng. Sci. **56** (2010), no. 1, 85–112. MR2666710

[15] C.-S. Liu, *Computing the eigenvalues of the generalized Sturm-Liouville problems based on the Lie-group $SL(2,\mathbb{R})$*, J. Comput. Appl. Math. **236** (2012), no. 17, 4547–4560, DOI 10.1016/j.cam.2012.05.006. MR2942448

[16] E. A. Maris and S. Goktas, *A study on the uniform convergence of spectral expansions for continuous functions on a Sturm-Liouville problem*, Miskolc Math. Notes **20** (2019), no. 2, 1063–1081. MR4066370

[17] E.I. Moiseev and N.Yu. Kapustin, *On the singularities of the root space of one spectral problem with a spectral parameter in the boundary condition*, Doklady Mathematics **66**(1) (2002) 14–18.

[18] M. Möller, *Minimality of eigenfunctions and associated functions of ordinary differential operators*, Adv. Oper. Theory **5** (2020), no. 3, 1014–1025, DOI 10.1007/s43036-020-00065-7. MR4126816

[19] M. A. Naimark, *Linear differential operators. Part I: Elementary theory of linear differential operators*, Frederick Ungar Publishing Co., New York, 1967. MR216050

[20] S. Yu. Reutskiy, *The method of external excitation for solving generalized Sturm-Liouville problems*, J. Comput. Appl. Math. **233** (2010), no. 9, 2374–2386, DOI 10.1016/j.cam.2009.10.022. MR2577774

[21] E. M. Russakovskiĭ, *Operator treatment of a boundary value problem with a spectral parameter that occurs rationally in the boundary conditions* (Russian), Teor. Funktsiĭ Funktsional. Anal. i Prilozhen. **30** (1978), 120–128. MR496113

[22] A. A. Shkalikov, *Basis properties of root functions of differential operators with spectral parameter in the boundary conditions*, Differ. Equ. **55** (2019), no. 5, 631–643, DOI 10.1134/s0012266119050057. Translation of Differ. Uravn. **55** (2019), no. 5, 647–659. MR3976532

[23] A. A. Shkalikov, *Boundary value problems for ordinary differential equations with a parameter in the boundary conditions* (Russian, with English summary), Trudy Sem. Petrovsk. **9** (1983), 190–229. MR731903

[24] A. N. Tihonov and A. A. Samarskiĭ, *Uravneniya matematičeskoĭ fiziki [The equations of mathematical physics]* (Russian), Gosudarstv. Izdat. Tehn.-Teor. Lit., Moscow, 1953. 2d ed. MR64971

Khazar University, Mehseti Genjevi str.11, Baku, 1096, Azerbaijan
Email address: nazimkerimov@yahoo.com

ADA University, Ahmadbey Aghaoglu str. 61 Baku, 1008, Azerbaijan
Email address: yaliyev@ada.edu.az

Contemporary Mathematics
Volume **798**, 2024
https://doi.org/10.1090/conm/798/15979

On the solvability of boundary value problems for linear differential-algebraic equations with constant coefficients

Anar Assanova, Carsten Trunk, and Roza Uteshova

ABSTRACT. We study a two-point boundary value problem for a linear differential-algebraic equation with constant coefficients by using the method of parameterization. The parameter is set as the value of the continuously differentiable component of the solution at the left endpoint of the interval. Applying the Weierstrass canonical form to the matrix pair associated with the differential-algebraic equation, we obtain a criterion for the unique solvability of the problem.

1. Introduction

We consider the linear differential-algebraic equation with constant coefficients of the form

$$E\dot{x}(t) = Ax(t) + f(t), \quad t \in (0,T), \tag{1.1}$$

subject to the boundary condition

$$Bx(0) + Cx(T) = d. \tag{1.2}$$

Here $E, A, B, C \in \mathbb{R}^{n\times n}$, $d \in \mathbb{R}^n$, $T > 0$. We suppose that the matrix pair (E, A) is regular, i.e. $\det(\lambda E - A) \neq 0$ for some $\lambda \in \mathbb{C}$.

By a solution of the boundary value problem (1.1), (1.2) we mean a function $x \in C^1([0,T], \mathbb{R}^n)$ satisfying equation (1.1) and the boundary condition (1.2).

Differential-algebraic equations have become widespread over the last decades, being a tool for modeling and simulation of dynamical systems with constraints in numerous applications [**4, 10, 11, 19, 23, 29, 30**]. The theory of boundary value problems for differential-algebraic equations started to develop by applying modified versions of the shooting and collocation methods designed for boundary value problems for ordinary differential equations [**3, 6, 7, 12, 20–22, 25, 31**]. In [**1**], P. Amodio and F. Mazzia studied problem (1.1), (1.2) by the method of boundary values. R. März applied the method of projectors and methods of perturbation

---

2020 *Mathematics Subject Classification.* Primary 34A09, 34B05; Secondary 34B99, 15A22.

Key words and phrases. Differential-algebraic equation, boundary value problem, Weierstrass canonical form, parameterization method.

This paper is supported by European Union's Horizon 2020 research and innovation programme under the Marie Sklodowska-Curie grant agreement ID: 873071, project SOMPATY (Spectral Optimization: From Mathematics to Physics and Advanced Technology).

This research has been funded by the Science Committee of the Ministry of Science and Higher Education of the Republik of Kazakhstan (Grant no. AP19675193).

theory [**26**–**28**] to problem (1.1), (1.2). The monograph by R. Lamour, R. März, and C. Tischendorf [**23**], devoted to the projector study of differential-algebraic equations, provides a detailed review in this field.

A series of papers by C. Trunk et al. [**8**,**9**,**17**,**24**] investigate possibilities of generalization and extension of the Kronecker canonical form to differential-algebraic equations with rectangular matrices. A number of methods and approaches have been developed for constructing their solutions.

However, the methods developed for solving problem (1.1), (1.2) may not always be applicable for a wide class of boundary conditions. As a consequence, this requires the development of new methods or modification of known methods of the theory of differential equations, which would be applicable to differential-algebraic equations and would be of constructive nature.

In this paper, we use the method of parameterization proposed by Dzhumabaev [**13**], which has proven to be an efficient constructive method allowing both to derive criteria for the unique solvability and obtain approximate solutions of various classes of boundary value problems [**2**,**5**,**14**–**16**]. This method was originally proposed for solving the linear boundary value problem (1.1), (1.2) provided $\det E \neq 0$. In this case, a criterion for the unique solvability was obtained in terms of coefficients and an algorithm for approximate solution was developed.

Our goal is to apply the method of parameterization to the boundary value problem (1.1), (1.2) in the case when the matrix E is not necessarily non-singular. We derive a criterion for the existence of a unique solution under certain assumptions on the matrices of the boundary condition.

2. Main results

We introduce the parameter $\mu \in \mathbb{R}^n$ defined as $\mu := x(0)$. By substituting

$$u(t) := x(t) - \mu,$$

equation (1.1) is transformed into the initial value problem with parameter

$$E\dot{u}(t) = A(u(t) + \mu) + f(t), \quad t \in (0, T), \tag{2.1}$$

$$u(0) = 0. \tag{2.2}$$

and the boundary condition (1.2) becomes

$$B(\mu + u(0)) + C(\mu + u(T)) = d. \tag{2.3}$$

The following statement shows the equivalence of the boundary value problem (1.1), (1.2) and the boundary value problem with parameter (2.1)-(2.3).

LEMMA 2.1. *If $x^* \in C^1([0,T], \mathbb{R}^n)$ is a solution of problem (1.1), (1.2), then the pair $(\mu^*, u^*) \in \mathbb{R}^n \times C^1([0,T], \mathbb{R}^n)$, where $\mu^* = x^*(0)$ and $u^*(t) = x^*(t) - \mu^*$, is a solution of problem (2.1)-(2.3).*

Conversely, if a pair $(\mu^{**}, u^{**}) \in \mathbb{R}^n \times C^1([0,T], \mathbb{R}^n)$ is a solution of problem (2.1)-(2.3), then the function $x^{**} \in C^1([0,T], \mathbb{R}^n)$, defined by $x^{**}(t) = \mu^{**} + u^{**}(t)$, is a solution of problem (1.1), (1.2).

Let P and Q be non-singular matrices which transform (2.1) and (2.2) to Weierstrass canonical form [**32**, **33**], i.e.,

$$PEQ = \begin{bmatrix} I_{n_1} & 0 \\ 0 & N \end{bmatrix}, \quad PAQ = \begin{bmatrix} J & 0 \\ 0 & I_{n_2} \end{bmatrix}, \quad Pf = \begin{bmatrix} \tilde{f}_1 \\ \tilde{f}_2 \end{bmatrix}, \tag{2.4}$$

where J is an $n_1 \times n_1$ matrix in real Jordan canonical form and N is an $n_2 \times n_2$ nilpotent matrix also in Jordan canonical form; $n_1 + n_2 = n$. Following [**19**], we call the index of nilpotency of N in (2.4) the index of the matrix pair (E, A), denoted by

$$\nu = \operatorname{ind}(E, A).$$

According to the space decomposition given by (2.4), we have the function

$$\tilde{u}(t) = (\tilde{u}_1(t), \tilde{u}_2(t))^T := Q^{-1}u(t)$$

and the vector

$$\tilde{\mu} = (\tilde{\mu}_1, \tilde{\mu}_2)^T := Q^{-1}\mu$$

with $\tilde{u}_j \in C^1([0,T], \mathbb{R}^{n_j})$ and $\tilde{\mu}_j \in \mathbb{R}^{n_j}$, $j = 1, 2$.

We then can rewrite problem (2.1), (2.2) in the following form:

$$\dot{\tilde{u}}_1(t) = J(\tilde{u}_1(t) + \tilde{\mu}_1) + \tilde{f}_1(t), \tag{2.5}$$

$$\tilde{u}_1(0) = 0, \tag{2.6}$$

$$N\dot{\tilde{u}}_2(t) = \tilde{u}_2(t) + \tilde{\mu}_2 + \tilde{f}_2(t), \tag{2.7}$$

$$\tilde{u}_2(0) = 0. \tag{2.8}$$

If we set $\tilde{B} := BQ$ and $\tilde{C} := CQ$, the boundary condition (2.3) transforms into

$$\tilde{B}(\tilde{\mu} + \tilde{u}(0)) + \tilde{C}(\tilde{\mu} + \tilde{u}(T)) = d. \tag{2.9}$$

By a solution of the boundary value problem (2.5)-(2.9) we mean a pair $(\tilde{\mu}, \tilde{u}) \in \mathbb{R}^n \times C^1([0,T], \mathbb{R}^n)$ with $\tilde{\mu} = (\tilde{\mu}_1, \tilde{\mu}_2)^T$ and $\tilde{u}(t) = (\tilde{u}_1(t), \tilde{u}_2(t))^T$ satisfying the initial value problems (2.5), (2.6) and (2.7), (2.8), and the boundary condition (2.9).

Before we start to study problem (2.5)-(2.9), let us note that we are able to write down explicitly the solutions of the initial value problems (2.5),(2.6) and (2.7),(2.8).

LEMMA 2.2. *For any $\tilde{\mu}_1 \in \mathbb{R}^{n_1}$, the initial value problem for the linear differential equation* (2.5),(2.6) *has the unique solution*

$$\tilde{u}_1(t) = J\int_0^t e^{(t-s)J}ds\tilde{\mu}_1 + \int_0^t e^{(t-s)J}\tilde{f}_1(s)ds. \tag{2.10}$$

PROOF. The general solution of the ordinary differential equation (2.5) has the form [**18**, p. 58]

$$\tilde{u}_1(t) = e^{tJ}\tilde{u}_1(0) + J\int_0^t e^{(t-s)J}ds\tilde{\mu}_1 + \int_0^t e^{(t-s)J}\tilde{f}_1(s)ds.$$

Applying the initial condition (2.6), we obtain (2.10). □

By Lemma 2.8 [**19**], equation (2.7) for fixed $\tilde{\mu}_2$ has the unique solution

$$\tilde{u}_2(t) = -\sum_{i=0}^{\nu-1} N^i[\tilde{\mu}_2 + \tilde{f}_2(t)]^{(i)} = -\tilde{\mu}_2 - \sum_{i=0}^{\nu-1} N^i\tilde{f}_2^{(i)}(t), \tag{2.11}$$

without specifying any initial values. Hence, taking into account (2.8), we obtain that the second component of the parameter $\tilde{\mu}$ is uniquely determined by

$$\tilde{\mu}_2 = -\sum_{i=0}^{\nu-1} N^i \tilde{f}_2^{(i)}(0). \tag{2.12}$$

Let us now turn to the boundary value problem with parameter (2.5)-(2.9). The second component $(\tilde{\mu}_2, \tilde{u}_2(t))$ of its solution $(\tilde{\mu}, \tilde{u}(t))$ is already known from (2.11) and (2.12). So we are interested in finding only the first component $(\tilde{\mu}_1, \tilde{u}_1(t))$ with $\tilde{\mu}_1$ and $\tilde{u}_1(t)$ satisfying (2.10). Hence, the appropriate number of imposed boundary conditions must coincide with the number n_1 of differential equations in (1.1). So, it is reasonable to assume that the matrices $\tilde{B}, \tilde{C} \in \mathbb{R}^{n\times n}$ and the right-hand side $d \in \mathbb{R}^n$ of the boundary condition (2.9) are of the form

$$\tilde{B} = \begin{bmatrix} \tilde{B}_1 & \tilde{B}_2 \\ 0 & 0 \end{bmatrix}, \quad \tilde{C} = \begin{bmatrix} \tilde{C}_1 & \tilde{C}_2 \\ 0 & 0 \end{bmatrix}, \quad d = \begin{bmatrix} d_1 \\ 0 \end{bmatrix}, \tag{2.13}$$

where $\tilde{B}_1, \tilde{C}_1 \in \mathbb{R}^{n_1\times n_1}$, $\tilde{B}_2, \tilde{C}_2 \in \mathbb{R}^{n_1\times n_2}$, $d_1 \in \mathbb{R}^{n_1}$.

Now, inserting (2.10) into (2.9) and taking into account Lemma 2.2 and (2.11), we obtain the following algebraic equation in $\tilde{\mu}_1$:

$$\tilde{D}\tilde{\mu}_1 = \tilde{d}, \tag{2.14}$$

where

$$\tilde{D} = \tilde{B}_1 + \tilde{C}_1 + \tilde{C}_1 J \int_0^T e^{(T-s)J} ds$$

and

$$\tilde{d} = d_1 - \tilde{C}_1 \int_0^T e^{(T-s)J} \tilde{f}_1(s)ds + \tilde{B}_2 \sum_{i=0}^{\nu-1} N^i \tilde{f}_2^{(i)}(0) + \tilde{C}_2 \sum_{i=0}^{\nu-1} N^i \tilde{f}_2^{(i)}(T).$$

The equation (2.14) has a unique solution if the matrix $\tilde{D}$ is non-singular. Let us assume, for instance, that $J = \operatorname{diag}(\lambda_1, \ldots, \lambda_{n_1})$. In this case ,

$$\tilde{D} = \tilde{B}_1 + \tilde{C}_1 \operatorname{diag}\left(1 + \lambda_1 \int_0^T e^{(T-s)\lambda_1} ds, \ldots, 1 + \lambda_{n_1} \int_0^T e^{(T-s)\lambda_{n_1}} ds\right),$$

and $\tilde{D}$ is non-singular if $\det(\tilde{B}_1 + \tilde{C}_1 e^{TJ}) \neq 0$.

Suppose that the matrix $\tilde{D}$ is non-singular. Then equation (2.14) has the unique solution $\tilde{\mu}_1 = \tilde{D}^{-1}\tilde{d}$. Substituting $\tilde{\mu}_1$ into (2.10), we obtain $\tilde{u}_1(t)$ and hence the first components of the unique solution $(\tilde{\mu}, \tilde{u}(t))$ of the boundary value problem with parameter (2.5)-(2.9):

$$(\tilde{\mu}_1, \tilde{u}_1(t)) = \Big(\tilde{D}^{-1}\tilde{d}, \int_0^t e^{(t-s)J} ds J \tilde{D}^{-1}\tilde{d} + \int_0^t e^{(t-s)J} \tilde{f}_1(s)ds\Big). \tag{2.15}$$

As previously stated, the second components of $(\tilde{\mu}, \tilde{u}(t))$ are determined by (2.12) and (2.11):

$$(\tilde{\mu}_2, \tilde{u}_2(t)) = \Big(-\sum_{i=0}^{\nu-1} N^i \tilde{f}_2^{(i)}(0), \ -\sum_{i=0}^{\nu-1} N^i [\tilde{f}_2^{(i)}(t) - \tilde{f}_2^{(i)}(0)]\Big). \tag{2.16}$$

Thus, taking into account the equivalence of the original problem (1.1), (1.2) and problems with parameter (2.1)-(2.3) and (2.5)-(2.9), we can summarize our results.

THEOREM 2.3. *Let (E, A) be a regular pair of square matrices and let P and Q be nonsingular matrices which transform (1.1) to Weierstrass canonical form (2.4). Furthermore, let $\nu = \operatorname{ind}(E, A)$ and $f \in C^\nu([0,T], \mathbb{R}^n)$. Then the boundary value problem with parameter (2.5)-(2.9), where $B = \tilde{B}Q^{-1}$, $C = \tilde{C}Q^{-1}$, and $d = (d_1, 0)^T$, has a unique solution if and only if:*

(i) *the matrix*

$$\tilde{D} = \tilde{B}_1 + \tilde{C}_1 + \tilde{C}_1 J \int_0^T e^{(T-s)J} ds$$

is non-singular;

(ii) $\tilde{\mu}_2 = -\sum_{i=0}^{\nu-1} N^i \tilde{f}_2^{(i)}(0)$.

The unique solution $(\tilde{\mu}, \tilde{u}(t) = \begin{bmatrix} (\tilde{\mu}_1, \tilde{u}_1(t)) \\ (\tilde{\mu}_2, \tilde{u}_2(t)) \end{bmatrix}$ of the boundary value problem with parameter (2.5)-(2.9) is determined by (2.15) and (2.16).

THEOREM 2.4. *Under the assumptions of Theorem 2.3 the boundary value problem (1.1), (1.2) with $B = \tilde{B}Q^{-1}$, $C = \tilde{C}Q^{-1}$, and $d = (d_1, 0)^T$ has a unique solution*

$$x(t) = Q(\tilde{\mu} + \tilde{u}(t)).$$

REMARK 2.5. We can also apply our method to initial value problems for linear differential-algebraic equations with constant coefficients. Indeed, if in the boundary condition (1.2) we replace B and C by the identity matrix I and zero matrix of order n, respectively, we get an initial condition

$$x(0) = d, \quad d \in \mathbb{R}^n. \tag{2.17}$$

Then, the parameter $\tilde{\mu}$ defined as $\tilde{\mu} = Q^{-1}\mu$, where $\mu = x(0)$, satisfies the equation

$$Q\tilde{\mu} = d.$$

The matrix Q is non-singular by assumption, so the initial value problem (1.1),(2.17) has a unique solution whenever

$$d_2 = \sum_{i=0}^{\nu-1} N^i \tilde{f}_2^{(i)}(0).$$

References

[1] Pierluigi Amodio and Francesca Mazzia, *Numerical solution of differential algebraic equations and computation of consistent initial/boundary conditions*, J. Comput. Appl. Math. **87** (1997), no. 1, 135–146, DOI 10.1016/S0377-0427(97)00178-7. MR1488825

[2] Anar T. Asanova and Dulat S. Dzhumabaev, *Well-posedness of nonlocal boundary value problems with integral condition for the system of hyperbolic equations*, J. Math. Anal. Appl. **402** (2013), no. 1, 167–178, DOI 10.1016/j.jmaa.2013.01.012. MR3023246

[3] Uri M. Ascher and Linda R. Petzold, *Projected collocation for higher-order higher-index differential-algebraic equations*, J. Comput. Appl. Math. **43** (1992), no. 1-2, 243–259, DOI 10.1016/0377-0427(92)90269-4. Orthogonal polynomials and numerical methods. MR1193304

[4] Uri M. Ascher and Linda R. Petzold, *Computer methods for ordinary differential equations and differential-algebraic equations*, Society for Industrial and Applied Mathematics (SIAM), Philadelphia, PA, 1998, DOI 10.1137/1.9781611971392. MR1638643

[5] Anar T. Assanova and Roza Uteshova, *Solution of a nonlocal problem for hyperbolic equations with piecewise constant argument of generalized type*. part 2, Chaos Solitons Fractals **165** (2022), no. part 2, Paper No. 112816, 7 pp., DOI 10.1016/j.chaos.2022.112816. MR4500657

[6] Yin Bai, *A perturbed collocation method for boundary-value problems in differential-algebraic equations*, Appl. Math. Comput. **45** (1991), no. 3, 269–291, DOI 10.1016/0096-3003(91)90084-Z. MR1126086

[7] Y. Bai, *A modified Lobatto collocation for linear boundary value problems of differential-algebraic equations* (English, with English and German summaries), Computing **49** (1992), no. 2, 139–150, DOI 10.1007/BF02238746. MR1197662

[8] Thomas Berger, Henk de Snoo, Carsten Trunk, and Henrik Winkler, *Linear relations and their singular chains* (English, with English and Ukrainian summaries), Methods Funct. Anal. Topology **27** (2021), no. 4, 287–301. MR4425707

[9] Thomas Berger, Carsten Trunk, and Henrik Winkler, *Linear relations and the Kronecker canonical form*, Linear Algebra Appl. **488** (2016), 13–44, DOI 10.1016/j.laa.2015.09.033. MR3419771

[10] A. A. Boichuk and A. M. Samoilenko, *Generalized inverse operators and Fredholm boundary-value problems*, VSP, Utrecht, 2004. Translated from the Russian by P. V. Malyshev and D. V. Malyshev, DOI 10.1515/9783110944679. MR2124880

[11] K. E. Brenan, S. L. Campbell, and L. R. Petzold, *Numerical solution of initial-value problems in differential-algebraic equations*, Classics in Applied Mathematics, vol. 14, Society for Industrial and Applied Mathematics (SIAM), Philadelphia, PA, 1996. Revised and corrected reprint of the 1989 original. MR1363258

[12] Kenneth D. Clark and Linda R. Petzold, *Numerical solution of boundary value problems in differential-algebraic systems*, SIAM J. Sci. Statist. Comput. **10** (1989), no. 5, 915–936, DOI 10.1137/0910053. MR1009546

[13] D. S. Dzhumabaev, *Conditions for the unique solvability of a linear boundary value problem for an ordinary differential equation* (Russian), Zh. Vychisl. Mat. i Mat. Fiz. **29** (1989), no. 1, 50–66, 157; English transl., U.S.S.R. Comput. Math. and Math. Phys. **29** (1989), no. 1, 34–46 (1990). MR984335

[14] D. S. Dzhumabaev, *On a method for solving a linear boundary value problem for an integrodifferential equation* (Russian, with Russian summary), Zh. Vychisl. Mat. Mat. Fiz. **50** (2010), no. 7, 1209–1221, DOI 10.1134/S0965542510070043; English transl., Comput. Math. Math. Phys. **50** (2010), no. 7, 1150–1161. MR2760445

[15] D. S. Dzhumabaev, *On one approach to solve the linear boundary value problems for Fredholm integro-differential equations*, J. Comput. Appl. Math. **294** (2016), 342–357, DOI 10.1016/j.cam.2015.08.023. MR3406986

[16] Dulat Dzhumabaev, *Computational methods of solving the boundary value problems for the loaded differential and Fredholm integro-differential equations*, Math. Methods Appl. Sci. **41** (2018), no. 4, 1439–1462, DOI 10.1002/mma.4674. MR3767243

[17] K. D. Clark and L. R. Petzold, *On characteristic invariants of matrix pencils and linear relations*, to appear in SIAM J. Matrix Anal. Appl. (2023).

[18] Philip Hartman, *Ordinary differential equations*, Classics in Applied Mathematics, vol. 38, Society for Industrial and Applied Mathematics (SIAM), Philadelphia, PA, 2002. Corrected reprint of the second (1982) edition [Birkhäuser, Boston, MA; MR0658490 (83e:34002)]; With a foreword by Peter Bates, DOI 10.1137/1.9780898719222. MR1929104

[19] Peter Kunkel and Volker Mehrmann, *Differential-algebraic equations*, EMS Textbooks in Mathematics, European Mathematical Society (EMS), Zürich, 2006. Analysis and numerical solution, DOI 10.4171/017. MR2225970

[20] Peter Kunkel and Ronald Stöver, *Symmetric collocation methods for linear differential-algebraic boundary value problems*, Numer. Math. **91** (2002), no. 3, 475–501, DOI 10.1007/s002110100315. MR1907868

[21] René Lamour, *A well-posed shooting method for transferable DAEs*, Numer. Math. **59** (1991), no. 8, 815–829, DOI 10.1007/BF01385812. MR1128035

[22] René Lamour, *A shooting method for fully implicit index-2 differential algebraic equations*, SIAM J. Sci. Comput. **18** (1997), no. 1, 94–114, DOI 10.1137/S1064827595287274. Dedicated to C. William Gear on the occasion of his 60th birthday. MR1433379

[23] René Lamour, Roswitha März, and Caren Tischendorf, *Differential-algebraic equations: a projector based analysis*, Differential-Algebraic Equations Forum, Springer, Heidelberg, 2013, DOI 10.1007/978-3-642-27555-5. MR3024597

[24] Leslie Leben, Francisco Martínez Pería, Friedrich Philipp, Carsten Trunk, and Henrik Winkler, *Finite rank perturbations of linear relations and matrix pencils*, Complex Anal. Oper. Theory **15** (2021), no. 2, Paper No. 37, 37 pp., DOI 10.1007/s11785-021-01082-x. MR4223454

[25] R. März, *On difference and shooting methods for boundary value problems in differential-algebraic equations* (English, with German and Russian summaries), Z. Angew. Math. Mech. **64** (1984), no. 11, 463–473, DOI 10.1002/zamm.19840641108. MR777496

[26] R. März, *Canonical projectors for linear differential algebraic equations*, Comput. Math. Appl. **31** (1996), no. 4-5, 121–135, DOI 10.1016/0898-1221(95)00224-3. Selected topics in numerical methods (Miskolc, 1994). MR1373460

[27] Roswitha März, *Solvability of linear differential algebraic equations with properly stated leading terms*, Results Math. **45** (2004), no. 1-2, 88–105, DOI 10.1007/BF03323000. MR2044016

[28] R. März, *Characterizing differential algebraic equations without the use of derivative arrays*, Comput. Math. Appl. **50** (2005), no. 7, 1141–1156, DOI 10.1016/j.camwa.2005.08.015. MR2167750

[29] Ricardo Riaza, *Differential-algebraic systems: Analytical aspects and circuit applications*, World Scientific Publishing Co. Pte. Ltd., Hackensack, NJ, 2008, DOI 10.1142/6746. MR2426820

[30] A. M. Samoilenko, M. I. Shkil, and V. P. Yakovets, *Linear systems of differential equations with degeneration*, Vyshcha Shkola, 2000.

[31] Ronald Stöver, *Collocation methods for solving linear differential-algebraic boundary value problems*, Numer. Math. **88** (2001), no. 4, 771–795, DOI 10.1007/PL00005458. MR1836879

[32] Peter Lancaster and Leiba Rodman, *Canonical forms for Hermitian matrix pairs under strict equivalence and congruence*, SIAM Rev. **47** (2005), no. 3, 407–443, DOI 10.1137/S003614450444556X. MR2178635

[33] Thomas Berger, Achim Ilchmann, and Stephan Trenn, *The quasi-Weierstraßform for regular matrix pencils*, Linear Algebra Appl. **436** (2012), no. 10, 4052–4069, DOI 10.1016/j.laa.2009.12.036. MR2914563

Institute of Mathematics and Mathematical Modeling, Pushkin Str. 125, 050010, Almaty, Kazakhstan

Email address: assanova@math.kz

Institut für Mathematik, Technische Universität Ilmenau, Postfach 100565, D-98684, Ilmenau, Germany

Email address: carsten.trunk@tu-ilmenau.de

Institute of Mathematics and Mathematical Modeling, Pushkin Str. 125, 050010, Almaty, Kazakhstan

Email address: r.uteshova@iitu.edu.kz

Contemporary Mathematics
Volume **798**, 2024
https://doi.org/10.1090/conm/798/15980

Invertible and noninvertible symbolic dynamics and their C*-algebras

Kevin Aguyar Brix

ABSTRACT. This paper surveys the recent advances in the interactions between symbolic dynamics and C*-algebras. We explain how conjugacies and orbit equivalences of both two-sided (invertible) and one-sided (noninvertible) symbolic systems may be encoded into C*-algebras, and how the dynamical systems can be recovered from structure-preserving *-isomorphisms of C*-algebras. We have included many illustrative examples as well as open problems.

CONTENTS

1. Introduction

This paper surveys some of the recent advances in the interactions between topological dynamical systems and C*-algebras with an emphasis on symbolic dynamics. The earliest examples of encoding dynamics into operators on Hilbert space include the group von Neumann algebra of Murray and von Neumann [**MvN43**], and von Neumann's ergodic theorem [**vN31**]; a myriad of examples later followed via the crossed product construction (broadly construed). It was Cuntz and Krieger [**CK80**] who established the first connections between C*-algebras and symbolic dynamics (viz. shifts of finite type), and this provided an abundance of simple and purely infinite C*-algebras. An immediate advantage of constructing C*-algebras (e.g. Cuntz–Krieger algebras) from dynamics is that C*-features such

2020 *Mathematics Subject Classification.* Primary 37A55; Secondary 37B10.
The author gratefully acknowledges a DFF-international postdoc grant (Case number 1025-00004B).

as ideal structure, KMS structure, or K-theory are intimately connected to the irreducible components, the entropy, and known invariants of the dynamical systems, respectively.

With this work, I hope to also show that this influence can go the other way: dynamics can benefit from operator algebras. This is perhaps most starkly seen in Giordano, Putnam, and Skau's classification of Cantor minimal systems up to orbit equivalence [**GPS95, GPS99**], their use of crossed product C*-algebras, and topological full groups (which now live a mature life of their own), but also in Krieger's dimension groups that serve to completely characterise shifts of finite type up to shift equivalence [**Kr80a, Kr80b**], as well as Huang's classification of two-component shifts of finite type up to flow equivalence using an idea of Cuntz from C*-algebras [**Hu94**].

Symbolic systems [**LM95**] are useful e.g. in the analysis of Axiom A diffeomorphisms [**Bo78**], mathematical linguistics (regular languages and sofic shifts), and coding theory (e.g. digitalising analogue signals). In this paper we focus on general shift spaces although there will be an emphasis on shifts of finite type because they are quite well understood and provide inspiration for the more general systems. We are interested in classification up to e.g. conjugacy or orbit equivalences, and how this is reflected in the C*-algebras. Invariably, this leads to a study of the fine-structure of C*-algebras (e.g. diagonal subalgebras and gauge actions) and structure-preserving *-isomorphisms between them. For graphs, this approach is prevalent in the study of how geometrical moves are reflected in *-isomorphisms of graph C*-algebras [**ERRS16, ER**].

During the time of writing this paper, I gave talks at Queen's University Belfast and University of Cardiff on related topics, and I thank Ying-Fen Lin and Ulrich Pennig for the kind invitations. The fact that so many subtly different equivalence relations on dynamical systems appear in this topic presents a challenge when giving a talk. It is my hope that this work may serve as a gentle introduction to the broader picture of how symbolic dynamical system and C*-algebras are related. Most of the material covered here is established in the literature, and plenty of references are provided that the interested reader may consult, although a few observations (especially in Section 6) seem to not have appeared before.

I have tried to keep the exposition short and, regrettably, there are many (interesting) aspects we cannot address here. This includes graphs (there is a plethora of literature on directed graphs and their C*-algebras, see [**Ra05**] and its references), Leavitt path algebras (the algebraic counterpart of graph C*-algebras), multidimensional dynamics, and Smale spaces. Topological full groups will briefly be mentioned in Section 8 on continuous orbit equivalence.

What we will be able to discuss are the topics listed in the table of content initiated by a preliminary section that covers the basics of symbolic dynamics and associated C*-algebras. Two things to emphasise. First, there is a tension between two-sided (invertible) systems, which seem to be of most interest to dynamicists, and one-sided (noninvertible) systems that seem more natural from a C*-algebraic perspective. For example, [**LM95**] only dedidates a few pages to one-sided shifts and the new addendum to the book does not discuss them at all; I hope to convey that one-sided dynamics can be just as interesting as its invertible sibling. Second, there is an unfortunate discrepancy in the notion of one-sided eventual conjugacy and this is addressed in Section 6.

2. Preliminaries

Let $\mathbb{Z}$ be the integers, let $\mathbb{N}$ be the nonnegative integers (including 0), and let $\mathbb{N}_+$ be the positive integers. An $\mathbb{N}$-matrix A is a matrix all of whose entries are in $\mathbb{N}$, and if A is square, we let $|A|$ denote the dimension of A. All matrices are finite-dimensional.

2.1. Symbolic dynamics. For a general and thorough introduction to classical theory of symbolic dynamics, we refer the reader to the excellent textbooks [**LM95**] and [**Ki98**]. Although the two-sided systems consisting of bi-infinite sequences seem to be more popular among dynamicists we shall here emphasise the one-sided systems consisting of right-infinite sequences as they fit nicer with the perspectives on C*-algebras.

Let $N \in \mathbb{N}_+$ and consider the set of *symbols* $\{0, \ldots, N-1\}$ as a discrete space. The *full one-sided N-shift* is the compact metric space $\{0, \ldots, N-1\}^{\mathbb{N}}$ (equipped with the product topology) of right-infinite sequences together with the *shift operation* σ given by $\sigma(x)_i = x_{i+1}$ for all $i \in \mathbb{N}$ and $x \in \{0, \ldots, N-1\}^{\mathbb{N}}$. A metric d on $\{0, \ldots, N-1\}^{\mathbb{N}}$ may be given as $d(x, y) = 2^{-\min\{k \in \mathbb{N} : x_k \neq y_k\}}$, for distinct points $x, y \in \{0, \ldots, N-1\}^{\mathbb{N}}$. The shift is a local homeomorphism (open and locally injective) and positively expansive in the sense that there is $\varepsilon > 0$ such that if $x, y \in \{0, \ldots, N-1\}^{\mathbb{N}}$ are distinct, then

$$d(\sigma^k(x), \sigma^k(y)) > \varepsilon.$$

for some $k \in \mathbb{N}$.

If $x \in \{0, \ldots, N-1\}^{\mathbb{N}}$, then we write $x = x_{[0,\infty)} = x_0 x_1 \cdots$ where $x_i \in \{0, \ldots, N-1\}$, and for integers $0 \leq i \leq j$, we have $x_{[i,j]} = x_i x_{i+1} \cdots x_j$ is a finite word in the symbols $\{0, \ldots, N-1\}$ (similarly, $x_{[i,j)}$ and $x_{(i,j]}$ will have the obvious meaning). We say the finite word $x_{[i,j}$ occurs in x, and we allow for a unique empty word ϵ that occurs in every sequence.

A *one-sided subshift* (or a *one-sided shift space*) is a closed subset X of $\{0, \ldots, N-1\}^{\mathbb{N}}$ that is shift-invariant in the sense that $\sigma(X) \subset X$ (we do not assume equality here). The restriction σ_X of the shift to X need not be open but it is always positively expansive, in particular it is locally injective. The language $L(X)$ of a subshift (σ, X) is all the finite words that occur in some $x \in X$, and this in fact determines the whole subshift, cf. [**LM95**, Proposition 1.3.4]. Moreover, a basis for the topology on X is given by the cylinder sets

$$Z(\mu) = \{x \in X : x_{[0,|\mu|)} = \mu\},$$

where μ is a word that occurs in X.

A *two-sided subshift* is a closed and shift-invariant subset $\underline{X} \subset \{0, \ldots, N-1\}^{\mathbb{Z}}$. We use the notation $\underline{X}$ to emphasise that it is the space of bi-infinite sequences although we use the same notation σ for the shift; it should always be clear from the context what we mean. The above notation also extends to the two-sided setting. There is a natural continuous shift-commuting surjection (a *factor map*) $\rho \colon \underline{X} \to \{0, \ldots, N-1\}^{\mathbb{N}}$ given by $\rho(x) = x_{[0,\infty)}$ for all $x \in \underline{X}$, and the image X is a one-sided subshift for which the shift is surjective. Conversely, if X is a one-sided shift for which σ_X is surjective, we can recover the two-sided subshift $(\sigma_X, \underline{X})$ as the *natural extension* given as the projective limit $\varprojlim_i (\sigma, X)$.

Two subshifts (σ_X, X) and (σ_Y, Y) are *conjugate* if there exists a homeomorphism $h \colon X \to Y$ satisfying $h \circ \sigma_X = \sigma_Y \circ h$.

DEFINITION 2.1. A *shift of finite type* is a subshift (σ, X) for which there exists a finite set of words (the forbidden words) such that X consists of all sequences that do not contain any forbidden words.

EXAMPLE 2.2. The full 2-shift may be represented by the directed graph

The space of right-infinite edge paths is $\{e, f\}^{\mathbb{N}}$ and we may equip this with the shift operation. The golden mean graph is given as

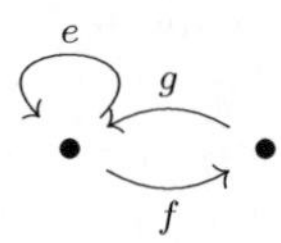

and its space of right-infinite paths is contained in $\{e, f, g\}^{\mathbb{N}}$, but the words $\{ff, fe, gg, eg\}$ are not allowed. This is a finite list of words that are forbidden, so the path space of the graph represents a shift of finite type.

The shift on the path space of a directed graph (in particular for a shift of finite type) is a local homeomorphism. Parry [**Pa66**, Theorem 1] shows that a subshift is a local homeomorphism if and only if it is of finite type.

REMARK 2.3. Any shift of finite type may be represented (up to conjugacy) by a finite directed graph, [**LM95**, Chapter 2]. As such the system is equivalently determined by the adjacency matrix of the graph, so we shall frequently denote a shift of finite type as X_A where A is the adjacency matrix of the graph that represents the system.

EXAMPLE 2.4. Consider the labelled graph

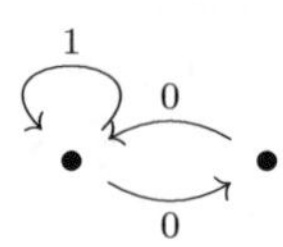

with the labelled path space contained in $\{0, 1\}^{\mathbb{N}}$. Any words of the form $10^{2n+1}1$ cannot occur in any right-infinite path on the labelled graph, and this set of forbidden words cannot be reduced to a finite set. The shift is therefore not of finite type. It is usually called the *even shift*; only an even number of 0s is allowed between any two 1s. Note that there is an obvious continuous and shift-commuting map (a factor map) from the path space of the golden mean graph of Example 2.2 to the even shift. This is an example of a sofic shift.

The even shift shows that shifts of finite type have the intrinsic flaw of not being closed under factor maps (*an unfair feature* in the words of Fischer). The smallest class containing shifts of finite type that is also closed under images is the sofic shifts which were coined and studied by Weiss in [**We73**].

DEFINITION 2.5. A subshift (σ_X, X) is *sofic* if it is the image of a continuous and shift-commuting map from a shift of finite type.

REMARK 2.6. Any sofic shift may be represented (up to conjugacy) by a finite *labelled* directed graph, [**LM95**, Chapter 3]. The underlying graph will then represent a shift of finite type from which there exists a factor map onto the sofic shift. Nasu [**Na86**] has described a representation matrix (in non-commutative variables) that describe the labelled graph and this is a very useful tool to generalise many perspectives from the finite type case. We shall not discuss this further here.

The representation of a sofic shift is not unique, but Fischer [**Fi75**] showed that any *irreducible* sofic shift admits a unique labelled graph representation that is minimal and right-resolving (every vertex emits only edges with distinct labels). This is not the case for general sofic shifts. We now refer to this as the *Fischer graph* or the *Fischer cover.* The Fischer graph of the even shift is depicted in Example 2.4.

Krieger [**Kr84**] later associated to *any* sofic shift a cover that is canonical in the sense that any conjugacy at the level of sofic systems lifts to a unique conjugacy on the covers. The Fischer graph is also canonical and the two graphs need not coincide.

DEFINITION 2.7. Let (σ_X, X) be a one-sided subshift over the symbol set $\{1,\ldots,N\}$. The *past set* of $x \in X$ is the set $P(x) = \{\nu \in L(X) : \nu x \in X\}$ The *(left) Krieger graph* of (σ_X, X) is a labelled graph whose vertices is the collection of past sets, and there is an edge from P to P' with label $a \in \{1,\ldots,N\}$ if there exists $x \in X$ such that $P = P(ax)$ and $P' = P(x)$.

EXAMPLE 2.8. The labelled graph

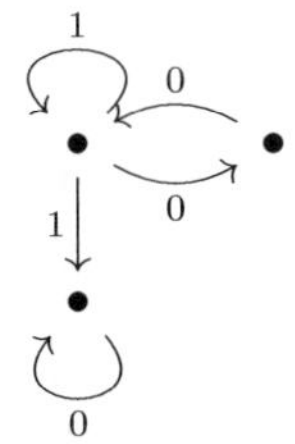

is the Krieger cover of the even shift. There are three distinct past sets e.g. represented by 0^∞, 10^∞, and 010^∞:

$$\begin{aligned} P(10^\infty) &= \{0,1,00,01,11,\ldots\} \\ P(010^\infty) &= \{0,00,10,\ldots\} \\ P(0^\infty) &= \{0,1,00,01,10,11,\ldots\}. \end{aligned}$$

The reader may verify that the above graph is indeed the Krieger graph of the even shift.

Observe that if μ is a word in the even shift that contains 1 and if ν and ω are words such that $\nu\mu$ and $\mu\omega$ are allowed, then $\nu\mu\omega$ is also allowed. However, if $\mu = 000$, then we can choose $\nu = 01$ and $\omega = 10$ and observe that both $\nu\mu$ and $\mu\omega$ are allowed but $\nu\mu\omega = 0100010$ is not allowed. A similar argument works for any word consisting only of 0s.

Every word occuring in 0^∞ is thus not synchronising, and if we remove this part of the graph we obtain the Fischer graph of the even shift in Example 2.4. This is no coincidence.

A word μ in a subshift (σ_x, X) is *intrinsically synchronising* if whenever ν and ω are words in X and $\nu\mu$ and $\mu\omega$ then $\nu\mu\omega$ is also allowed. In the even shift, a word is intrinsically synchronising if and only if it contains a 1. A right-infinite sequence $x \in X$ is synchronising if it contains an intrinsically synchronising word. Krieger [**Kr84**] showed that the Fischer graph arises as the subgraph of the Krieger graph whose vertices are the past sets of synchronising elements.

Below we mention a class of subshifts that is very different from the sofic shifts.

EXAMPLE 2.9. Let R_α be the rigid rotation on $\mathbb{R}/\mathbb{Z}$ by an irrational parameter $\alpha \in (0,1) \setminus \mathbb{Q}$. Define a coding map $\nu\colon \mathbb{R}/\mathbb{Z} \to \{0,1\}$ by $\nu(t) = 1$ if $t \in [0,\alpha)$, and $\nu(t) = 0$ if $t \in (\alpha, 1]$. The two-sided Sturmian shift is then the closure of $\{\nu(R_\alpha^i(t))_{i\in\mathbb{Z}} : t \in \mathbb{R}/\mathbb{Z}\}$ in $\{0,1\}^{\mathbb{Z}}$. This is an example of a minimal homeomorphism on the Cantor space, cf. Section 8.1, it has no periodic points and zero entropy, cf. [**LM95**, Section 13.7]. A one-sided Sturmian shift is defined analogously.

2.2. C*-algebras. Throughout the paper, we let $\mathbb{K}$ denote the compact operators on separable Hilbert space and c_0 the commutative subalgebra of diagonal operators.

Let A be an irreducible and nonpermutation $\{0,1\}$-matrix of dimension n. Cuntz and Krieger [**CK80**] define a C*-algebra $\mathcal{O}_A$ (today known as the *Cuntz–Krieger algebra*) as *the* C*-algebra generated by n partial isometries $s_1, \dots, s_n$ satisfying the relation

$$1 = \sum_{i=1}^{n} s_i s_i^*, \qquad \text{and} \qquad s_j^* s_j = \sum_{i=1}^{n} A(i,j) s_i s_i^* \tag{2.1}$$

for all $j = 1, \dots, n$. Really they assume A satisfy a technical condition (I) but irreducible and nonpermutation matrices are examples of this. They show that $\mathcal{O}_A$ is simple (it contains no nontrivial closed two-sided ideals), and that it is defined independently of the specific Hilbert space on which the partial isometries act (up to canonical *-isomorphism), so the above *the* is justified. They provide a large class of examples of *purely infinite* C*-algebras, cf. [**Cu81a**]. When the dimension of A is $n \geq 2$ and every entry of A is 1 then the partial isometries are honest isometries, and $\mathcal{O}_A$ is canonically isomorphic to Cuntz' algebra $\mathcal{O}_n$, cf. [**Cu77**]. A diagonal subalgebra $\mathcal{D}_A$ of $\mathcal{O}_A$ is generated by projections of the form $s_\mu s_\mu^*$ where $\mu = \mu_1 \cdots \mu_m$ is a finite word in X_A and $s_\mu = s_{\mu_1} \cdots s_{\mu_m}$; $\mathcal{O}_A$ also admits a *gauge action* γ^A of $\mathbb{T}$ determined by $\gamma_z^A(s_i) = z s_i$ for all $z \in \mathbb{T}$ and generators s_i.

When A is not irreducible the C*-algebras generated by similar relations is not simple and will generally depend on the Hilbert space on which the generators act. It is however possible to define $\mathcal{O}_A$ as a universal C*-algebra subject to *Cuntz–Krieger families* [**aHR97**, Section2]: a Cuntz–Krieger family for A is a collection of partial isometries $\{S_i\}_i$ satisfying (2.1), and the C*-algebra $\mathcal{O}_A$ is universal if it is generated by a Cuntz–Krieger family $\{s_i\}_i$, and for any other family $\{S_i\}_i$ on a Hilbert space H there is a *-representation $\mathcal{O}_A \to B(H)$ sending $s_i \mapsto S_i$, for all i.

The construction of C*-algebras was extended to graphs in [**EW80**, **KPRR97**] where the latter used topological groupoids whose unit space is the right-infinite paths on the graphs. This path space admits a shift operation which is a local homeomorphism. We shall not dig deeper into graph C*-algebras here but simply refer the interested reader to Raeburn's monograph [**Ra05**] which constructs universal C*-algebras from generators and relations.

How can these constructions of C*-algebras be generalised to general subshifts e.g. sofic shifts?

Matsumoto was the first to associate C*-algebras to general subshifts in [**Ma97**] and subsequent papers using a Fock space construction. Unfortunately, there was a mistake in a technical lemma which caused some confusion (this is however nicely clarified in [**CM04**]).

For sofic shifts (under some technical conditions), Carlsen [**Ca03**] and Samuel [**Sa98**] independently observed that Matsumoto's C*-algebra is isomorphic to the Cuntz–Krieger algebras of their Krieger graphs. In fact, we may now *define* the C*-algebra of a sofic shift to be the Cuntz–Krieger algebra of its Krieger cover. Similarly, we can define a C*-algebra of an irreducible sofic shift to be the Cuntz–Krieger algebra of its Fischer cover; this will in general be a quotient of the former C*-algebra.

The approach in [**BC20b**] is to construct a canonical cover $(\sigma_{\tilde{X}}, \tilde{X})$ to any subshift (σ_X, X) that generalises the path space of the Krieger graph. This is a local homeomorphism from which we can construct an étale groupoid and therefore a C*-algebra $\mathcal{O}_X$. The cover is useful because the shift on a general subshift is usually not a local homeomorphism (only if it is of finite type), so the associated groupoid will usually not be étale. Alternatively, Matsumoto [**Ma**] has constructed C*-algebras of so-called normal subshifts that generalise the Fischer cover.

The covers, groupoids, and C*-algebras of Sturmian shifts were studied in [**Br21**].

REMARK 2.10. We mention here briefly the advantage of étale groupoids. Conceptually, groupoids can encode the orbit structure of a dynamical system and they should be thought of as the non-commutative orbit space. The interested reader should consult [**Si20**] for a friendly introduction to the topic (see also [**Re80**] for the source of everything groupoid C*-algebras).

A local homeomorphism $T\colon X \to X$ on a locally compact Hausdorff space X determines a groupoid

$$G_T = \bigcup_{m,n\in\mathbb{N}} \{(x, m-n, y) \in X \times \mathbb{Z} \times X : T^m x = T^n y\}$$

with unit space $G^{(0)} = \{(x,0,x) : x \in X\}$ canonically identified with X, and range and source maps $r(x,p,y) = x$ and $s(x,p,y) = y$ for all $(x,p,y) \in G_\sigma$. A pair of elements (x,p,y) and (y',q,z) are composable precisely if $y = y'$ in which case $(x,p,y)(y,q,z) = (x,p+q,z)$, and inversion is given as $(x,p,y)^{-1} = (y,-p,x)$. Under a suitable topology inherited from X, G_T is a locally compact Hausdorff, étale and amenable groupoid. As such it defines a groupoid C*-algebra $C^*(T) = C^*(G_T)$ which contains the functions on the unit space $C_0(X)$ (the *diagonal*), and the groupoid homomorphism $c_\sigma\colon G_T \to \mathbb{Z}$ given by $c_\sigma(x,p,y) = p$ for all $(x,p,y) \in G_\sigma$, induces a *gauge action* γ^T of $\mathbb{T}$ on $C^*(G_T)$. When $T = \sigma_A$ is a shift of finite type, then $C^*(G_{\sigma_A})$ is canonically isomorphic to $\mathcal{O}_A$ in a way that is diagonal-preserving and gauge-equivariant.

A wonderful theorem of Renault [**Re08**] allows for a reconstruction of a topological groupoid (up to isomorphism) from its C*-algebra with diagonal subalgebra, see also [**Re80, Ku86**]. Below we state a generalised version of the result from [**CRST21**, Section 3]. The technical conditions are satisfied for all examples relevant to this paper and we shall not elaborate on them here.

THEOREM 2.11. *Let G_1 and G_2 be a second-countable, Hausdorff, étale groupoids whose interior of isotropy is abelian and torsion-free. If there exists a *-isomorphism $\varphi\colon C_r^*(G_1) \to C_r^*(G_2)$ satisfying $\varphi(C_0(G_1^{(0)})) = C_0(G_2^{(0)})$, then G_1 and G_2 are isomorphic as topological groupoids.*

This is an immensely useful tool in questions related to dynamical systems and their C*-algebras. Many variations on Renault's reconstruction theory are now available in the literature but Renault's paper is still worth studying today.

3. Two-sided conjugacy

When are two dynamical systems the same? We can illustrate the immense complexity of this question by asking when two-sided shifts of finite type are conjugate. This fundamental problem of symbolic dynamics has still not been satisfactorily solved, and the question of whether there exists a decision algorithm to determine conjugacy is still open.

We cannot address this problem without mentioning the efforts of Williams in the seminal work [**Wi73**]. In it, he addressed a question of Bowen (are shifts of finite type determined by their zeta function?) and provided an algebraic classification of two-sided shifts of finite type in terms of $\mathbb{N}$-matrices; this is the notion of strong shift equivalence.

3.1. Strong shift equivalence. Let us start with an example.

EXAMPLE 3.1 (In-splitting). Consider the graph below with adjacency matrix A

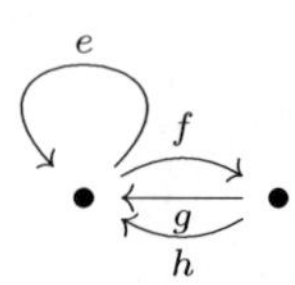

$$A = \begin{pmatrix} 1 & 1 \\ 2 & 0 \end{pmatrix}.$$

The left-most vertex v has three edges going *in* (e, g and h), and we split v into two vertices (aligned on top of each other below) and distribute the edges e and h to the top copy of v and g to the bottom copy; the outgoing edges e and f are copied so there is one emitted from each of the two copies of v. We then get the graph below with adjacency matrix B

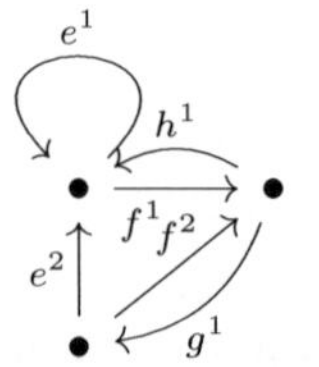

$$B = \begin{pmatrix} 1 & 0 & 1 \\ 1 & 0 & 1 \\ 1 & 1 & 0 \end{pmatrix}.$$

Note that the loop e is both incoming and outgoing for v which is why we also have two copies. The two-sided shifts are conjugate, and it is possible to write down an explicit conjugacy.

Alternatively, the in-split can be encoded into the two matrices

$$R = \begin{pmatrix} 1 & 0 \\ 1 & 0 \\ 1 & 1 \end{pmatrix}, \qquad \text{and} \qquad S = \begin{pmatrix} 1 & 0 & 1 \\ 0 & 1 & 0 \end{pmatrix},$$

for which we have $B = RS$ and $SR = A$. This is an example of an (elementary) strong shift equivalent between A and B.

Based on such observations, Williams introduced strong shift equivalence.

DEFINITION 3.2. A pair of square $\mathbb{N}$-matrices A and B are *elementary strong shift equivalent* if there are rectangular $\mathbb{N}$-matrices R and S such that $A = RS$ and $SR = B$, and *strong shift equivalent* if there are $\ell \in \mathbb{N}$ and matrices $A = C_0, \ldots, C_\ell = B$ such that C_i and C_{i-1} are elementary strong shift equivalent for all $i = 1, \ldots, \ell$.

The classification of two-sided shifts of finite type is from [**Wi73**, Theorem A].

THEOREM 3.3. *A pair of two-sided shifts of finite type $(\sigma_A, \underline{X}_A)$ and $(\sigma_B, \underline{X}_B)$ are conjugate if and only if the matrices A and B are strong shift equivalent.*

For the general definition and notation of in-splits, we refer the reader to [**LM95**, Section 2.4] (or [**BP04**, Section 4] for general graphs). The proof of the classification theorem (see e.g. [**LM95**, Proof of Theorem 7.2.7]) is really a matter of decomposing an arbitrary conjugacy into a composition of elementary conjugacies coming from splittings such as the in-split above and the dual notion of out-splits (see Example 4.1 below) which also produces a strong shift equivalence. As such, the theorem says that conjugacies can only arise by a sequence of splittings.

EXAMPLE 3.4. Let $k \in \mathbb{N}$ and consider the matrices

$$A = \begin{pmatrix} 1 & k \\ k-1 & 1 \end{pmatrix}, \quad \text{and} \quad B = \begin{pmatrix} 1 & k(k-1) \\ 1 & 1 \end{pmatrix}.$$

For $k = 3$ the matrices are strong shift equivalent (in seven steps, cf. [**LM95**, Example 7.3.12]), but for $k \geq 4$ this is still an open problem. This example shows that the complexity of strong shift equivalence is not just in the dimensions of the adjacency matrices but in the number of elementary strong shift equivalences connecting the matrices as well as the sizes of these connecting matrices.

REMARK 3.5. Nasu [**Na86**] (see also [**HN88**]) has generalised the notion of strong shift equivalent to cover so-called representation matrices of sofic shifts. The paper shows that strong shift equivalence for these representation matrices coincides with conjugacy of the two-sided sofic systems.

3.2. Shift equivalence. Seeing that strong shift equivalence is hard to determine, Williams also introduced shift equivalence as an a priori weaker algebraic relation that seems easier to compute.

DEFINITION 3.6. A pair of square $\mathbb{N}$-matrices A and B are *shift equivalent* if there are $\ell \in \mathbb{N}_+$ and rectangular $\mathbb{N}$-matrices R and S such that

$$A^\ell = RS, \quad B^\ell = SR, \quad AR = RB, \quad BS = SA. \tag{3.1}$$

It is not hard to verify that strong shift equivalence implies shift equivalence (which seems to justify the terminology).

If we consider the matrices as linear maps $A\colon Z^{|A|} \to Z^{|A|}$ and $B\colon Z^{|B|} \to Z^{|B|}$ then the shift equivalence relations are concisely expressed (in the case $\ell = 2$) in the following diagram

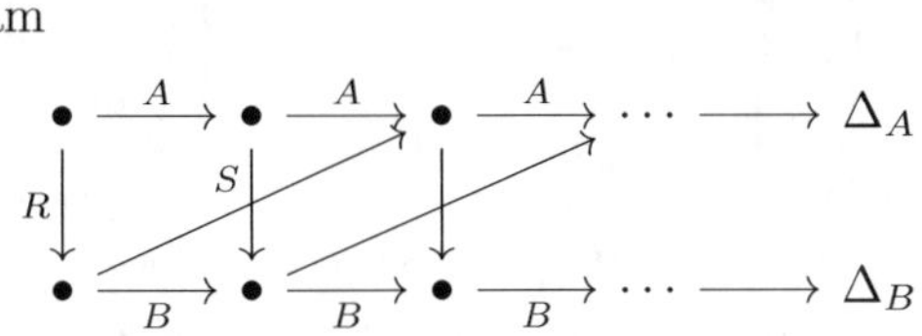

where all the downwards vertical maps are R and the upwards maps are S. Here, Δ_A is the inductive limit which inherits the order structure Δ_A^+ from the positive order on $\mathbb{Z}^{|A|}$, and the matrix A induces an automorphism δ_A on Δ_A which is essentially multiplication by A. The matrices R and S induce group isomorphisms (that we also denote by R and S) between Δ_A and Δ_B, and these isomorphisms preserve the order structure and intertwine the automorphisms δ_A and δ_B. This means that the module structure $(\Delta_A, \Delta_A^+, \delta_A)$ is an invariant of shift equivalence.

In fact, the dimension module is a *complete* invariant for shift equivalence. This was proved by Krieger in [**Kr80a, Kr80b**] (although in the latter paper, the result is only stated for primitive matrices, see also [**LM95**, Theorem 7.5.8]).

THEOREM 3.7. *Let A and B be square $\mathbb{N}$-matrices. Then $(\sigma_A, \underline{X}_A)$ and $(\sigma_B, \underline{X}_B)$ are shift equivalent if and only if the dimension triples $(\Delta_A, \Delta_A^+, \delta_A)$ and $(\Delta_B, \Delta_B^+, \delta_B)$ are isomorphic.*

EXAMPLE 3.8. The matrices in Example 3.4 are shift equivalent for all $k \geq 3$. They are in fact similar (over $\mathbb{Z}$) and for primitive matrices this implies shift equivalent; explicitly for $k = 3$ the matrices

$$R = \begin{pmatrix} 8 & 3 \\ 1 & 16 \end{pmatrix}, \qquad \text{and} \qquad S = \begin{pmatrix} 2 & 3 \\ 1 & 1 \end{pmatrix}$$

define a shift equivalence between A and B.

Williams' motivation for introducing shift equivalence was in part because it was *more computable* than strong shift equivalence. This intuition was formalised by Kim and Roush in [**KR88**] (see also [**KR79**]) where they showed that shift equivalence is decidable. In [**Wi73**, Theorem F], it is claimed that shift equivalence coincides with strong shift equivalence and therefore the classification problem for shifts of finite type should seemingly be resolved. Alas. Parry found a mistake in the proof and the question of whether shift equivalence implies strong shift equivalence came to be known as the *Williams problem*, the *Shift equivalence problem*, or the *Williams conjecture*.

Fast forward some twenty years, Kim and Roush exhibited a two-component counterexample to the Williams problem in [**KR92**], and all hope to salvage the equality of shift equivalence and strong shift equivalence was extinguished a few years later when they constructed an irreducible example.

EXAMPLE 3.9. The (in)famous primitive counterexample to the Williams problem by Kim and Roush [**KR99**, Section 7] is given by the matrices

$$A = \begin{pmatrix} 0 & 0 & 1 & 1 & 3 & 0 & 0 \\ 1 & 0 & 0 & 0 & 3 & 0 & 0 \\ 0 & 1 & 0 & 0 & 3 & 0 & 0 \\ 0 & 0 & 1 & 0 & 3 & 0 & 0 \\ 0 & 0 & 0 & 0 & 0 & 0 & 1 \\ 1 & 1 & 1 & 1 & 10 & 0 & 0 \\ 1 & 1 & 1 & 1 & 0 & 1 & 0 \end{pmatrix} \qquad \text{and} \qquad B = \begin{pmatrix} 0 & 0 & 1 & 1 & 3 & 0 & 0 \\ 1 & 0 & 0 & 0 & 0 & 0 & 0 \\ 0 & 1 & 0 & 0 & 0 & 0 & 0 \\ 0 & 0 & 1 & 0 & 0 & 0 & 0 \\ 0 & 0 & 0 & 0 & 0 & 0 & 1 \\ 4 & 5 & 6 & 3 & 10 & 0 & 0 \\ 4 & 5 & 6 & 3 & 0 & 1 & 0 \end{pmatrix}.$$

In the paper, an explicit shift equivalence is given with *integer* matrices, and for primitive matrices this implies shift equivalence with $\mathbb{N}$-matrices. Kim and Roush provide conditions that shift equivalent matrices must satisfy (based on the complicated *sign-gyration* invariant) to not be strong shift equivalent (and thereby be counterexamples to the Williams problem). The bulk of the problem is then to find concrete examples satisfying those conditions; indeed they remark that it *was difficult for us.*

The paper of Kim and Roush [**KR99**] provide abstract conditions that counterexamples to the Williams problem must satisfy, and the concrete examples of Kim and Roush show that the conditions are meaningful. However, the precise relationship between shift equivalence and strong shift equivalence, in particular how one relation *refines* the other, seems to still be a mystery.

Since the classification of two-sided shifts of finite type is very complicated it is useful to have computable invariants to distinguish certain systems apart. Bowen and Franks [**BF77**] studied the two groups

$$\ker(\mathrm{Id}-A), \qquad \text{and} \qquad \mathrm{BF}(A)=\mathbb{Z}^{|A|}/(\mathrm{Id}-A)\mathbb{Z}^{|A|}, \tag{3.2}$$

where $\mathrm{Id}-A$ is viewed as a map on $\mathbb{Z}^{|A|}$, and showed that they are invariants of conjugacy. Today, we usually only refer to the cokernel as the *Bowen–Franks group.* The groups actually arose in their study of flow equivalence, cf. Section 7. For example, the matrices in Example 3.1 have Bowen–Franks group $\mathbb{Z}/2\mathbb{Z}$, and the Kim–Roush example Example 3.9 have Bowen–Franks group isomorphic to $\mathbb{Z}/99\mathbb{Z}$.

3.3. C*-algebras. Cuntz and Krieger introduced their C*-algebras from irreducible shifts of finite type (actually satisfying a technical condition (I)), and they were already aware [**CK80**, Theorem 4.1] that the stabilised C*-algebra (together with the diagonal subalgebra) is an invariant of two-sided conjugacy. In fact, they showed that the pair is invariant of flow equivalence, cf. Section 7.

Using the Pimsner–Voiculescu sequence, Cuntz [**Cu81b**, Proposition 3.1] identified the K-theory of the Cuntz–Krieger algebra $\mathcal{O}_A$ with the invariants of Bowen and Franks (of the *transposed* matrix A^t):

$$K_0(\mathcal{O}_A)\cong \mathrm{BF}(A^t) \qquad \text{and} \qquad K_1(\mathcal{O}_A)\cong \ker(\mathrm{Id}-A^t).$$

See also [**Cu81a**] where $K_0(\mathcal{O}_A)$ is almost identified. Strictly speaking, this was done under the hypothesis of Cuntz and Krieger's condition (I), but the argument has later been generalised to e.g. graphs, cf. [**Ra05**, Chapter 7]. This provides strong evidence that the C*-algebras actually remember a lot of the underlying dynamics.

Rørdam [**Rø95**] later classified the simple Cuntz–Krieger algebras (those $\mathcal{O}_A$ for which A is irreducible and nonpermutation) up to stable isomorphism by their K_0, i.e. by the Bowen–Franks group. This deep result involved KK-theory and Huang's work on flow equivalence [**Hu94**]. Cuntz then observed that simple Cuntz–Krieger algebras are classified up to *-isomorphism by the Bowen–Franks group together with the class of the unit.

Further evidence is provided by considering the fixed-point algebra of the gauge action γ^A on the Cuntz–Krieger algebra $\mathcal{O}_A$. This is an approximately finite-dimensional (AF) C*-algebra, the K_0-group of which is identified with the dimension group as an ordered group (again of the transposed matrix A^t). In fact the equivariant K-theory of the gauge action coincides with Krieger's dimension triple (the underlying group is the K_0-group of the fixed-point algebras).

The theorem below is from Bratteli and Kishimoto [**BK00**, Corollary 1.5 and Theorem 4.3], and its proof relies on classification of actions on C*-algebras from

K-theoretic data. The importance of the matrices being primitive is to ensure that the fixed-point algebra is simple with one-dimensional lattices of traces.

THEOREM 3.10. *Let A and B be primitive matrices. They are shift equivalent if and only if there is a *-isomorphism $\varphi\colon \mathcal{O}_A \otimes \mathbb{K} \to \mathcal{O}_B \otimes \mathbb{K}$ satisfying $\varphi \circ (\gamma^A \otimes \mathrm{Id}) = (\gamma^B \otimes \mathrm{Id}) \circ \varphi$.*

REMARK 3.11. It is not known to which extend such a characterisation holds. The issue is really going *from* shift equivalent matrices *to* equivariantly *-isomorphic C*-algebras; the other direction follows from the identification of the K-theory with the dimension triple. Recently, Eilers and Szabó have announced that they can generalise the result to cover irreducible and nonpermutation matrices.

Quite surprisingly, it is possible to characterise strong shift equivalence (or, equivalently, two-sided conjugacy of shifts of finite type) using C*-algebras. Matsumoto [**Mat17a**, Theorem 2.18] made progress on this for irreducible and nonpermutation matrices and characterised it in terms of Morita equivalence of the triple $(\mathcal{O}_A, C(X_A), \gamma^A)$, the Cuntz–Krieger algebra, its diagonal subalgebra, and the canonical gauge action. Below we record a slightly different characterisation from [**CR17**, Theorem 5.1]; this result has no irreducibility conditions.

THEOREM 3.12. *Let A and B be square nonnegative $\mathbb{N}$-matrices with no zero rows or columns. The two-sided shifts $(\sigma_A, \underline{X}_A)$ and $(\sigma_B, \underline{X}_B)$ are conjugate if and only if there exists a *-isomorphism $\varphi\colon \mathcal{O}_A \otimes \mathbb{K} \to \mathcal{O}_B \otimes \mathbb{K}$ satisfying $\varphi(C(X_A) \otimes c_0) = C(X_B) \otimes c_0$ and $\varphi \circ (\gamma^A_z \otimes \mathrm{Id}) = (\gamma^B_z \otimes \mathrm{Id}) \circ \varphi$ for all $z \in \mathbb{T}$.*

A similar theorem holds for general subshifts (not necessarily of finite type) by [**BC20b**, Theorem 7.5].

REMARK 3.13. It is curious that shift equivalence seems to be harder to encode into C*-algebras than strong shift equivalence. Perhaps this follows from the fact that the former relation is not an orbit equivalence and so groupoid techniques do not seem fit for the task.

4. One-sided conjugacy

A conjugacy of one-sided shifts lifts to a conjugacy of their natural extensions (the correponding two-sided shifts) but the converse need not be the case. In fact, the conjugacy question for noninvertible systems turns out to be starkly different from that of the invertible systems. In his influencial paper [**Wi73**], Williams devises an algorithm to solve the conjugacy problem for one-sided shifts of finite type and observes *almost as a by-product we obtain a classification of 1-sided shifts, up to topological conjugacy.*

4.1. Williams' amalgamation algorithm. The in-spliting process of Example 3.1 has a dual notion called *out-splitting.*

EXAMPLE 4.1 (Out-splitting). Consider the graph below with adjacency matrix A

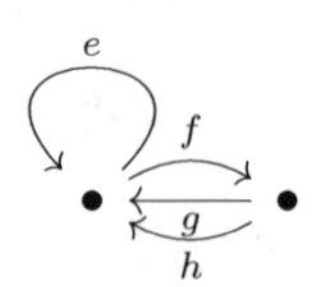

$$A = \begin{pmatrix} 1 & 1 \\ 2 & 0 \end{pmatrix}.$$

The left-most vertex v has two edges going *out* (e and f), and we split v into two vertices (aligned on top of each other below) and distribute the two edges to each copy of v; the incoming edges g and h are then copied so there is one entering each of the two copies of v. We then get the graph below with adjacency matrix B

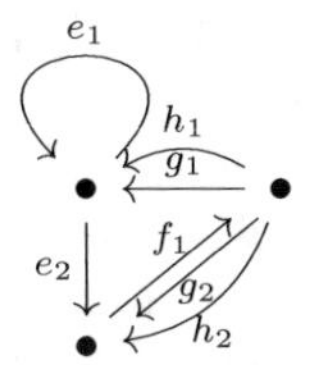

$$C = \begin{pmatrix} 1 & 1 & 0 \\ 0 & 0 & 1 \\ 2 & 2 & 0 \end{pmatrix}.$$

Note that the loop e is both incoming and outgoing for v which is why we also have two copies. As in Example 3.1 it is easy to write down matrices R and S that verify that A and C are strong shift equivalent.

The reader may verify that an explicit one-sided conjugacy $h\colon X_A \to X_C$ is given as follows: any occurrence of ee is mapped to e^1, ef is mapped to e^2, f is mapped to f^1, ge is mapped to g^1, gf is mapped to g^2, he is mapped to h^1, and hf is mapped to h^2. For example,

$$h(eefgefhf\cdots) = e^1e^2f^1g^1e^2f^1h^2f^1\cdots$$

For the general definition and notation of out-splits, we refer the reader to [**LM95**, Section 2.4] (or [**BP04**, Section 3] for general graphs). Any out-split produces conjugate graphs.

The inverse operation to an out-split is called an *out-amalgamation* (here we will just say amalgamation) and naturally this operation also preserves the one-sided conjugacy class of the system. The surprising fact is that these two operations exhaust all conjugacies of one-sided shifts of finite type.

At the level of matrices the amalgamation may be described as follows: an *amalgamation* of A is obtained by collapsing identical columns of A and adding up the corresponding rows entrywise. An example is the matrix C in Example 4.1 in which the first two columns are identical. By keeping only one of the columns and adding the first are second rows, we arrive at the matrix A. The dimension of the amalgamated matrix is then less than the dimension of A. The *total amalgamation* of a square $\mathbb{N}$-matrix A is obtained by iteratively collapsing identical columns and adding corresponding rows until the resulting matrix contains no identical columns. Williams showed that this always exists (it is independent of the order of amalagamations) and that it is unique (up to permutation). The classification for one-sided shifts of finite type is from [**Wi73**, Theorem G].

THEOREM 4.2. *Any square $\mathbb{N}$-matrix admits a unique total amalgamation (up to permutation), and if A and B are square $\mathbb{N}$-matrices, then their one-sided shifts are conjugate if and only if the total amalgamations of A and B agree (up to permutation).*

Observe that in Example 4.1 above, A has no identical columns, so it is its own total amalgamation. However, the first two columns of C are identical and the total amalgamation of C coincides with A. Conversely, both of the graphs in the in-split example Example 3.1 contain no identical columns, and they are distinct, so this shows that the one-sided shifts X_A and X_B are not one-sided conjugate (though they are two-sided conjugate).

REMARK 4.3. For graphs with finitely many vertices (but potentially infinitely many edges) a generalisation of Williams' algorithm works to show that conjugacy for such graphs is decidable. This result will appear in forthcoming work [**ABCEW**].

4.2. General C*-description for local homeomorphisms. We saw in Section 3.3 that Cuntz and Krieger showed that $\mathcal{O}_A$ is an invariant of the two-sided shift of finite type $(\sigma_A, \underline{X}_A)$ *up to stable isomorphism.* Above we also saw that one-sided conjugacy is completely classified using out-splits and amalgamations. In [**BP04**], Bates and Pask described how to generalise Williams' splitting operations to possibly infinite graphs, and they observed that in-splits produce graph C*-algebras that are stably isomorphic while out-splits produce graph C*-algebras that are honestly *-isomorphic. From Theorem 4.2, we then know that $\mathcal{O}_A$ is really an invariant of *one-sided conjugacy.* This was in fact already observed by Cuntz and Krieger, cf. [**CK80**, Proposition 2.17], and the point is also made explicit by Katsura [**Ka09**].

The main result in [**BC20a**] is a converse to [**CK80**, Proposition 2.17] in which one-sided conjugacy of shifts of finite type is characterised in terms of diagonal-preserving *-isomorphism of Cuntz–Krieger algebras that intertwine a certain completely positive map. Working with irreducible shifts of finite type in [**Mat22**], Matsumoto dispensed with the completely positive map and instead required that the *-isomorphism intertwine a whole family a gauge actions. The following theorem from [**ABCE22**, Theorem 3.1] is the appropriate generalisation of Matsumoto's idea to general local homeomorphisms. It illustrates an example of a connection between general dynamical systems in the form of local homeomorphisms and their C*-algebras that is directly inspired by shifts of finite type.

THEOREM 4.4. *Let $T\colon X \to X$ and $S\colon Y \to Y$ be local homeomorphisms. Then T and S are conjugate if and only there is a *-isomorphism $\varphi\colon C^*(T) \to C^*(S)$ satisfying $\varphi(C_0(X)) = C_0(Y)$ (implemented by a homeomorphism $h\colon X \to Y$), and*

$$\varphi \circ \gamma_t^{X,f} = \gamma_t^{Y,f\circ h^{-1}} \circ \varphi,$$

for all $t \in \mathbb{R}$ and $f \in C_0(X, \mathbb{R})$.

When the space is totally disconnected, we may replace $\mathbb{R}$ by $\mathbb{Z}$ (this applies e.g. to (infinite) directed graphs).

Below we specify it to general subshifts. This has not appeared elsewhere and it complements Matsumoto's work on conjugacy of normal subshifts [**Mat21**, Theorem 1.5] and [**BC20b**, Theorem 4.4]. We emphasise that the result below applies to all general subshifts. This class of dynamical systems is not immediately covered by the above theorem because the action is not by local homeomorphisms. We provide only a sketch of proof; the details are an adaptation of the proof of the above theorem to general subshifts as in [**BC20b**].

COROLLARY 4.5. *Let (σ_X, X) and (σ_Y, Y) be general one-sided shift spaces. The systems are conjugate if and only if there is a *-isomorphism $\varphi\colon \mathcal{O}_X \to \mathcal{O}_Y$ satisfying $\varphi(C(X)) = C(Y)$ (implemented by a homeomorphism $h\colon X \to Y$) and*

$$\varphi \circ \gamma_z^{X,f\circ\pi_X} = \gamma_z^{Y,(f\circ h^{-1})\circ\pi_Y} \circ \varphi, \tag{4.1}$$

for all $f \in C(X, \mathbb{Z})$ and $z \in \mathbb{T}$. Here, $\pi_X\colon \tilde{X} \to X$ and $\pi_Y\colon \tilde{Y} \to Y$ are the canonical factor maps associated to the systems.

SKETCH OF PROOF. Assume there exist a homeomorphism $h\colon X \to Y$ and a *-isomorphism $\varphi\colon \mathcal{O}_X \to \mathcal{O}_Y$ satisfying $\varphi(C(X)) = C(Y)$ with $\varphi(f) = f \circ h^{-1}$ for all $f \in C(X)$, and (4.1). By [**BC20b**, Theorem 3.3], we have $\varphi(D_X) = D_Y$, and this means there is a homeomorphism $\tilde{h}\colon \tilde{X} \to \tilde{Y}$ of the covers satisfying $h \circ \pi_X = \pi_Y \circ \tilde{h}$. Applying (4.1) for $f = 1$, it follows that h and $\tilde{h}$ are one-sided eventual conjugacies (cf. Definition 5.1 below). Let $k \in \mathbb{N}$ be the constant associated to h and $\tilde{h}$ according to the definition.

Now, fix $x \in X$ and let $\tilde{x} \in \pi_X^{-1}(x) \subset \tilde{X}$. The hypothesis (4.1) implies (this requires some work) that for any $f \in C(X, \mathbb{Z})$

$$f(x) = \sum_{i=0}^{k} f \circ h^{-1} \circ \pi_Y(\sigma_{\tilde{Y}}^i(\tilde{h}(\tilde{x}))) - \sum_{j=0}^{k} f \circ h^{-1} \circ \pi_Y(\sigma_{\tilde{Y}}^j(\tilde{h}(\sigma_{\tilde{X}}(\tilde{x})))).$$

If we let $u = \sigma_Y(h(x))$ and $v = h(\sigma_X(x))$, then this simplifies to

$$\sum_{i=0}^{k} f \circ h^{-1}(\sigma_Y^i(u)) = \sum_{j=0}^{k} f \circ h^{-1}(\sigma_Y^j(v))$$

for all $f \in C(X, \mathbb{Z})$. Now [**ABCE22**, Lemma 3.7] (this lemma is stated for systems of local homeomorphisms but this is not needed) allows us to conclude that $u = v$ and so h is a conjugacy. □

5. Matsumoto's one-sided eventual conjugacy

The conjugacy problem for one-sided shifts of finite type is understood via out-splits and amalgamations, and there is a user-friendly decision procedure to determine when two such systems are conjugate. In this section, we discuss Matsumoto's notion of eventual conjugacy [**Mat17a**], its relation to C*-algebras, and how it relates to strong shift equivalence. We shall be careful to call this eventual conjugacy *in Matsumoto's sense* because there is an arguably more natural notion of eventual conjugacy (which we shall discuss in Section 6), and the two do not coincide. Matsumoto's eventual conjugacy was first defined for irreducible shifts of finite type, but here we state it for local homeomorphisms.

DEFINITION 5.1. Let $T\colon X \to X$ and $S\colon Y \to Y$ be local homeomorphism. A homeomorphism $h\colon X \to Y$ is a *Matsumoto eventual conjugacy* if there are continuous maps $k\colon X \to \mathbb{N}$ and $k'\colon Y \to \mathbb{N}$ satisfying

$$S^{k(x)+1} \circ h(x) = S^{k(x)} \circ h \circ T(x), \tag{5.1}$$

$$T^{k'(y)+1} \circ h^{-1}(y) = T^{k'(y)} \circ h^{-1} \circ S(y), \tag{5.2}$$

for all $x \in X$ and $y \in Y$. When X and Y are compact, we may equivalently choose k and k' to be fixed nonnegative integers.

Note that h is a conjugacy if we can choose $k = 0$.

EXAMPLE 5.2. The example below from [**BC20a**] shows that Matsumoto's eventual conjugacy is nontrivial in the sense that it does not coincide with one-sided conjugacy. Let X and Y be the one-sided shift spaces determined by the graphs

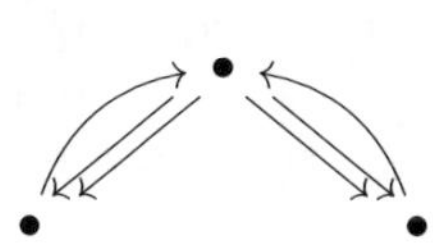

and

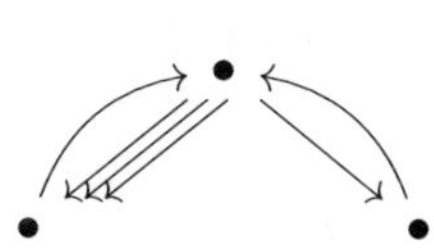

respectively. A concrete eventual conjugacy is given as follows: label the edges pointing downwards in the first graph from left to right by a, b, c, d and edges pointing upwards e, f; similarly, label the edges of the second graph by a', b', c', d', e', f'. Define $h\colon X \to Y$ by sending an infinite path $x = x_0x_1\cdots$ to $y = y_0y_1\cdots$ where $y_i = e'$ if $x_{i-1} = c$, and $y_i = x_i'$ otherwise. The reader may verify that this is a Matsumoto eventual conjugacy. On the other hand, the two graphs have distinct total amalgamations which can easily be checked so the one-sided systems are not conjugate.

The following C*-algebraic characterisation has its origin in [**Mat17a**, Theorem 1.2] for infinite irreducible shifts of finite type. It was later developed for graphs in [**CR17**, Theorem 4.1], and found its final form for general local homeomorphisms in [**CRST21**, Theorem 8.10]. The C*-algebras that appear are the groupoid C*-algebras built from the local homeomorphisms as in Section 2; for shifts of finite type they are just the Cuntz–Krieger algebras.

THEOREM 5.3. *Let $T\colon X \to X$ and $S\colon Y \to Y$ be local homeomorphisms. Then T and S are Matsumoto eventually conjugate if and only if there is a *-isomorphism $\varphi\colon C^*(T) \to C^*(S)$ satisfying $\varphi(C_0(X)) = C_0(Y)$ and $\varphi\circ\gamma_z^T = \gamma_z^S\circ\varphi$, for all $z \in \mathbb{T}$.*

The proof encodes an eventual conjugacy into groupoids and then uses the reconstruction theory for topological groupoids with its canonical cocycle; this is a variation on Theorem 2.11. A similar result for general shift spaces can be found in [**BC20b**, Theorem 5.3].

5.1. Balanced strong shift equivalence. Now we discuss *balanced strong shift equivalence* as introduced in [**Br22**, Section 5]. This is an equivalence relation on square $\mathbb{N}$-matrices which should be thought of as the one-sided analogue of Williams' strong shift equivalence.

The pair of graphs in Example 5.2 inspired a new move on graphs introduced in [**ER**], the *balanced in-split*: one graph G produces two graphs E and F by performing two in-splits at G in which the vertices are split into *the same* number of new vertices, but the incoming edges may be distributed *differently*. The pair of graphs E and F are then said to be a balanced in-split of G.

EXAMPLE 5.4. Consider the graph

The bottom vertex has four incoming edges, and the reader can verify that the two graphs in Example 5.2 arise as in-splits of this graph in two different ways:

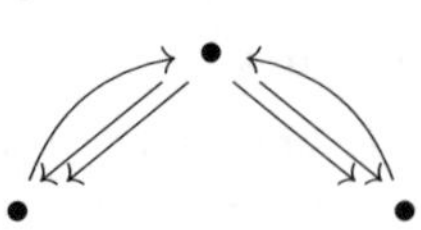

and

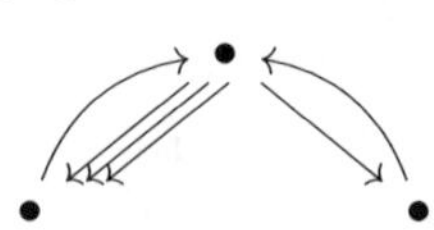

In both cases, the bottom vertex is split into two vertices but the four edges are distributed differently: in the left-most graph the two vertices are given two edges each, while in the right-most graph the edges are distributed with three edges to one vertex and a single edge to the other vertex. As is always the case for an in-split, the outgoing edge is copied.

It is shown in [**ER**, Corollary 3.7] that the balanced in-split induces a *-isomorphism of graph C*-algebras that is both diagonal-preserving and gauge-equivariant. Via Theorem 5.3, this means that this move induces an eventual conjugacy in Matsumoto's sense on the one-sided shift spaces. The paper [**Br22**] gives a direct proof of this fact and also shows the converse for finite graphs with no sinks: any eventual conjugacy is the result of balanced in-splits up to one-sided conjugacy. Rephrasing the balanced in-split into the language of matrices we arrive at the definition of balanced strong shift equivalence.

DEFINITION 5.5. Let A and B be square $\mathbb{N}$-matrices. They are *elementary balanced strong shift equivalent* if there are rectangular $\mathbb{N}$-matrices R_1, R_2, and S such that

$$A = SR_1, \qquad B = SR_2, \qquad R_1 S = R_2 S.$$

The matrices are *balanced strong shift equivalent* if there are conjugate copies A' and B' of A and B, respectively, and $n \in \mathbb{N}_+$ and matrices $A' = A_0, \dots, A_n = B'$ such that A_{i-1} and A_i are elementary balanced strong shift equivalent for all $i = 1, \dots, n$.

The main result of [**Br22**] is then the following characterisation.

THEOREM 5.6. *Let A and B be square $\mathbb{N}$-matrices with no zero rows (the graphs have no sinks). The one-sided shifts (σ_A, X_A) and (σ_B, X_B) are Matsumoto eventually conjugate if and only if A and B are balanced strong shift equivalent.*

Analogous to Williams' classification of two-sided conjugacies, this result says that (σ_A, X_A) and (σ_B, X_B) are Matsumoto eventually conjugate if and only the underlying graphs can be connected by out-splits and balanced in-splits. The assumption that the graphs admit no sinks is merely a technical one; it should be possible to prove a similar results for a large class of graphs.

Noting the connection to two-sided conjugacy and strong shift equivalence, we ask the obvious decidability question which was also mentioned in [**Br22**].

QUESTION 5.7. Is balanced strong shift equivalence decidable?

6. A discrepancy in one-sided eventual conjugacy

Williams introduced shift equivalence because it is tamer than strong shift equivalence and easier to determine. This is an *algebraic* relation on matrices, and the analogous *dynamical* description was only later discovered to be two-sided eventual conjugacy: all higher powers are two-sided conjugate. One-sided eventual conjugacy (in the sense of having one-sided conjugate higher powers) was studied in depth in [**BFF97**], and the authors show that this relation is decidable by a very user-friendly algorithm. We observe in this section the unfortunate fact that this does not coincide with Matsumoto's eventual conjugacy from Section 5.

We can in fact also make sense of a shift equivalence for one-sided shifts (we call this *unital shift equivalence*), and it turns out that this does not coincide with either of the two version of one-sided eventual conjugacy. This will be elaborated on in future work [**Br**].

6.1. Conjugate higher powers. We first define the notion of having conjugate higher powers. This is usually called eventual conjugacy.

Definition 6.1. Let $\sigma_X\colon X \to X$ and $\sigma_Y\colon Y \to Y$ be dynamical systems (e.g. σ_X and σ_Y are both (local) homeomorphisms). The systems are *eventually conjugate* (or have *conjugate higher powers*) if the higher powers σ_X^i and σ_Y^i are conjugate for all but finitely many $i \in \mathbb{N}_+$.

For shifts of finite type (one-sided or two-sided), the higher power (σ_A^i, X_A) is canonically conjugate to (σ_{A^i}, X_{A^i}).

Example 6.2. It is important that all sufficiently high powers are conjugate. The matrices

$$A = \begin{pmatrix} 0 & 2 \\ 2 & 0 \end{pmatrix} \quad \text{and} \quad B = \begin{pmatrix} 2 & 0 \\ 0 & 2 \end{pmatrix}$$

satisfy $A^{2n} = B^{2n}$ for all $n \in \mathbb{N}$ but one is irreducible while the other is reducible.

In the case of two-sided shifts of finite type, having conjugate higher powers coincides with shift equivalence, cf. Section 3.2. One direction was observed by Williams [**Wi73**], and the converse was proved by Kim and Roush [**KR79**] (see also [**LM95**, Theorem 7.5.15]).

Theorem 6.3. *Let $(\sigma_A, \underline{X}_A)$ and $(\sigma_B, \underline{X}_B)$ be two-sided shifts of finite type. The matrices A and B are shift equivalent if and only if the systems have conjugate higher powers.*

This result provides the dynamical analogue of the algebraic notion of shift equivalence. A similar correspondence does not occur in the one-sided setting; this will be elaborated on elsewhere.

Remark 6.4. A similar result applies to sofic shifts. Boyle and Krieger [**BK88**] introduce and study shift equivalence for two-sided sofic systems and show that it coincides with having conjugate higher powers.

In [**BFF97**], Boyle, D. Fiebig, and U. Fiebig study one-sided shifts of finite type that have conjugate higher powers (they refer to this as one-sided eventual conjugacy). We present here the main result related to this notion.

Let $N(A)$ be the least positive integer satisfying $\operatorname{rank}(A^n) = \operatorname{rank}(A^{n+1})$ which is at most the dimension of A. The surprisingly nice characterisation below is from [**BFF97**, Section 8].

Theorem 6.5. *Let A and B be square $\mathbb{N}$-matrices and let $n \geq \max\{N(A), N(B)\}$. The following are equivalent.*

(1) *A and B determine one-sided shifts of finite type that have conjugate higher powers;*
(2) *the total amalgamations of A^m and B^m agree (up to the same permutation) for $m = n, n+1$; and*
(3) *the total amalgamations of A^m and B^m agree (up to the same permutation) for all $m \geq n$.*

An immediate consequence of this theorem is that having conjugate higher powers for one-sided shifts of finite type is decidable. Simply construct the total amalgamations of A^m and B^m and check whether they are the same (up to permutation). For very large systems, the task of determining graph isomorphism of the total amalgamations can of course be formidable, but often times in practice this is very user-friendly.

Boyle–Fiebig–Fiebig also provide an example to show that having conjugate higher powers does not coincide with one-sided conjugacy. Below we generalise this example slightly and show that the one-sided systems determined from the matrices are Matsumoto eventually conjugate.

EXAMPLE 6.6. For each $k \in \mathbb{N}_+$, we consider the matrix

$$A_k = \begin{pmatrix} 2k & 0 & 4k \\ k & 2k & 0 \\ k & 2k & 0 \end{pmatrix}.$$

For $k = 1$, this is [**BFF97**, Example 5.6]. The total amalgamation of A_k^2 is the matrix $(4k)^2$ while the total amalgamation of A_k^3 is the matrix $(4k)^3$. Therefore, it follows from Theorem 6.5 that A_k and $(4k)$ have conjugate higher powers. On the other hand, A_k is already in the form of a total amalgamation. This means that A_k and $(4k)$ are not one-sided conjugate for any k.

Next we observe that the matrices A_k and $(4k)$ are in fact balanced strong shift equivalent via the matrix

$$C = \begin{pmatrix} 2k & 4k \\ k & 2k \end{pmatrix}.$$

Explicitly, we consider the matrices

$$S = \begin{pmatrix} 1 & 0 \\ 0 & 1 \\ 0 & 1 \end{pmatrix}, \quad R_1 = \begin{pmatrix} 2k & 0 & 4k \\ k & 2k & 0 \end{pmatrix}, \quad \text{and} \quad R_2 = \begin{pmatrix} 2k & 2k & 2k \\ k & k & k \end{pmatrix}.$$

Direct computations show that

$$A_k = SR_1, \qquad SR_2 = \begin{pmatrix} 2k & 2k & 2k \\ k & k & k \\ k & k & k \end{pmatrix}, \quad \text{and} \quad R_1 S = C = R_2 S.$$

Since the total amalgamation of SR_2 is the matrix $(4k)$, the matrices S, R_1, R_2 implement a concrete balanced strong shift equivalence between A_k and $(4k)$ for each $k \in \mathbb{N}$, so the one-sided shifts are Matsumoto eventually conjugate.

The above observations lead us to ask whether one-sided eventual conjugacy in the sense of having conjugate higher powers and in Matsumoto's sense simply coincide. This is not the case.

EXAMPLE 6.7. The adjacency matrices of the graphs in Example 5.2 are

$$A = \begin{pmatrix} 0 & 2 & 2 \\ 1 & 0 & 0 \\ 1 & 0 & 0 \end{pmatrix} \quad \text{and} \quad B = \begin{pmatrix} 0 & 3 & 1 \\ 1 & 0 & 0 \\ 1 & 0 & 0 \end{pmatrix}.$$

The total amalgamation of any power of A is always a 2×2-matrix, while the total amalgamation of any power of B is a 3×3-matrix. Therefore, the one-sided systems do not have conjugate higher powers.

The examples so far have shown us that both having conjugate higher powers and being balanced strong shift equivalent are nontrivial in the sense that they do not coincide with one-sided conjugacy. We have also seen that the two notions do not agree, in particular balanced strong shift equivalence does not imply having conjugate higher powers. This is why we do not simply refer to the relations as *one-sided eventual conjugacy.* We leave the following problem open.

QUESTION 6.8. Does having conjugate higher powers imply balanced strong shift equivalence?

EXAMPLE 6.9. Consider Ashley's graph

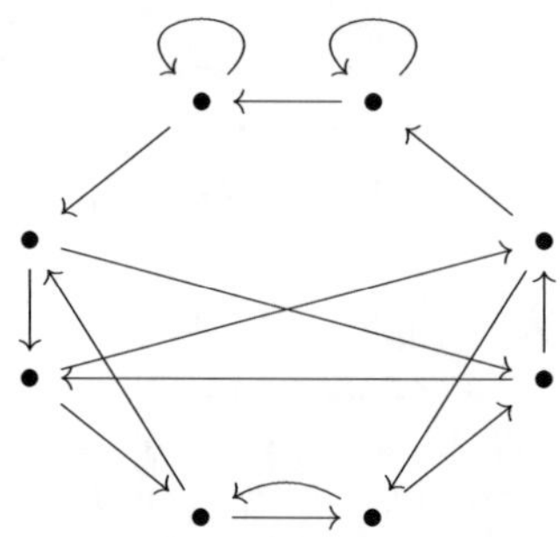

The characteristic polynomial of the adjacency matrix of this graph is $x^7(x-2)$ and a result of Williams (see e.g. [**Ki98**, Lemma 2.2.6]) implies that the two-sided edge shift is shift equivalent to the matrix (2), the full 2-shift. According to [**Ki98**, Example 2.2.7 and the notes to that section] it has been an open problem since 1989 when the graph was discovered to determine whether they are two-sided conjugate.

A computation of the higher powers of the adjacency matrix of Ashley's graph shows that all the higher powers are one-sided conjugate to the higher powers of the full 2-shift (using Theorem 6.5), so they have conjugate higher powers. This makes the above question of whether this implies (balanced) strong shift equivalence even more pressing.

EXAMPLE 6.10. Consider Rourke's example of shift equivalent primitive matrices

$$A = \begin{pmatrix} 1 & 2 & 1 \\ 1 & 1 & 0 \\ 1 & 0 & 1 \end{pmatrix} \qquad \text{and} \qquad B = \begin{pmatrix} 1 & 0 & 1 & 0 & 1 \\ 0 & 1 & 1 & 1 & 0 \\ 1 & 1 & 1 & 0 & 0 \\ 1 & 0 & 0 & 0 & 1 \\ 0 & 1 & 0 & 1 & 0 \end{pmatrix}$$

from the errata of [**Wi73**]. Williams remarks that it is not known if they are strong shift equivalent. A computation of the higher powers shows that the total amalgamation of B^n is A^n for $n \geq 2$, so the matrices have conjugate higher powers. I do not know if the question of strong shift equivalence has been resolved, or if they are balanced strong shift equivalent.

7. Flow equivalence

In this section we return to two-sided dynamics. Let us first define the notion of flow equivalence as in e.g. [**LM95**, Section 13.6].

Let X be a compact metric space equipped with a homeomorphism $\sigma\colon X \to X$. The product $X \times \mathbb{R}$ carries an obvious $\mathbb{R}$-action (a flow) and the *suspension space* is the quotient of $X \times \mathbb{R}$ by the identification $(\sigma(x), t) \sim (x, t+1)$ for all $x \in X$ and $t \in \mathbb{R}$. Let $[x,t] \in X \times \mathbb{R}/\sim$ be the class of $(x,t) \in X \times \mathbb{R}$. The quotient is a compact metric space and it carries an induced flow (the suspension flow) $s.[x,t] = [x, t+s]$ for all $x \in X$ and $t, s \in \mathbb{R}$. A pair of homeomorphisms $\sigma_X : X \to X$ and $\sigma_Y : Y \to Y$ are *flow equivalent* if their suspension flows are topologically equivalent, i.e. there is a homeomorphism $h\colon X\times\mathbb{R}/\sim\to X\times\mathbb{R}/\sim$ that takes orbits (flow lines) to orbits in an orientation-preserving way. Note that if the space X is totally disconnected, then the flow lines are exactly the connected components of $X\times\mathbb{R}/\sim$, so flow equivalence is a matter of 'oriented homeomorphism' on the suspension spaces.

The system $\sigma_X : X\to X$ appears as a cross section of the flow on the suspension space. Equivalently, two homeomorphisms are flow equivalent if they are cross sections of a common flow. As Franks mentions in [**Fr84**], symbolic dynamics is useful as a tool for studying the qualitative behaviour of certain systems of smooth dynamics. By restricting to a transversal with its first return map (a cross section), often times the resulting homeomorphism is an irreducible shift of finite type. Restricting to a different transversal will result in a different system but they are flow equivalent.

In the beginning of the 1970s about the time when Williams was working on conjugacy, Bowen [**Bo72**] studied the flows coming from shifts of finite type. The hugely influential (and notoriously short) paper by Parry and Sullivan [**PS75**] followed a few years later in which the authors prove that flow equivalence of shifts of finite type is generated by conjugacy and *symbol expansion* (the reader *with a less direct pipeline to topological truth* may want to consult e.g. [**BCE17**]). We illustrate symbol expansion with the following example that we saw in Example 2.2.

EXAMPLE 7.1. Consider the graphs E and F below

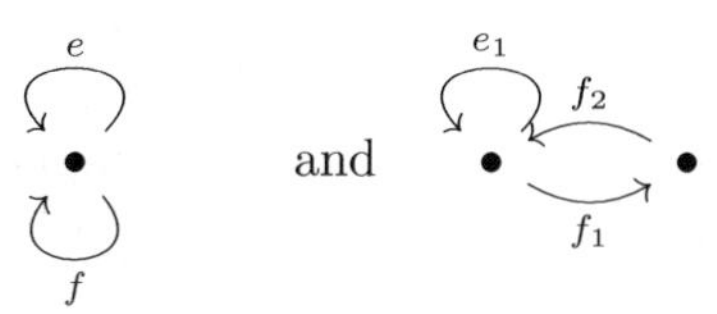

We construct F from E by performing a symbol expansion on the symbol f in E. Choose a new symbol, in this case f_2, and demand that it always follows f. In this way, we have slowed down the time it takes to walk along f. By identifying f in E with f_1 in F we obtain the graph F. The reader may chack that the association $e\mapsto e_1$ and $f\mapsto f_1f_2$ induces a flow equivalence between E and F.

Note that the edge shift of E has no obstructions on its allowed paths (the full 2-shift), so its entropy is $\log 2$, while it can be shown that the entropy of F is $(1+\sqrt{5})/2$ (the golden mean), cf. [**LM95**, Example 4.1.4]. In particular, the systems are not conjugate and entropy is not invariant under flow equivalence.

Parry and Sullivan also showed that the integer $\det(\mathrm{Id}-A)$ is an invariant of the flow equivalence class of $(\sigma_A,\underline{X}_A)$ (in Example 7.1 this value is -1 for both graphs). Shortly after, Bowen and Franks [**BF77**] showed that the kernel and cokernel of $\mathrm{Id}-A$ viewed as a linear map on $\mathbb{Z}^{|A|}$ (cf. (3.2)) are also invariant under flow equivalence

The cokernel $\mathrm{BF}(A)=\mathbb{Z}^{|A|}/(\mathrm{Id}-A)\mathbb{Z}^{|A|}$ (which today is usually referred to as the Bowen–Franks group) almost subsumes the information of the determinant since

$$|\det(\mathrm{Id}-A)|=\begin{cases}|\,\mathrm{BF}(A)| & \text{if } \mathrm{BF}(A) \text{ is finite,}\\ 0 & \text{if } \mathrm{BF}(A) \text{ is infinite,}\end{cases}$$

so only the *sign* of the determinant contains new information not retrievable from the group, cf. Example 7.3. We may therefore say that the *signed Bowen–Franks group*

$$(\mathrm{BF}(A), \operatorname{sign}\det(\mathrm{Id} - A))$$

is an invariant of flow equivalent. It is a surprising fact that for irreducible and nonpermutation matrices this is a *complete* invariant. This is Franks' classification result [**Fr84**].

THEOREM 7.2. *Let A and B be irreducible and nonpermutation $\mathbb{N}$-matrices. The two-sided shifts $(\sigma_A, \underline{X}_A)$ and $(\sigma_B, \underline{X}_B)$ are flow equivalent if and only if* $\mathrm{BF}(A) \cong \mathrm{BF}(B)$ *and* $\operatorname{sign}\det(\mathrm{Id} - A)) = \operatorname{sign}\det(\mathrm{Id} - A)$.

Using a different approach, Boyle and Handelman [**BH96**] studied the *ordered cohomology* of a shift of finite type $(\sigma_A, \underline{X}_A)$ given as

$$C(\underline{X}_A, \mathbb{Z})/\{f - f \circ \sigma_A : f \in C(\underline{X}_A, \mathbb{Z})\}$$

with the order $[f] \geq 0$ if there exists a nonnegative representative $f \geq 0$. For Cantor minimal systems, this is the group that Giordano, Putnam, and Skau used to classify Cantor minimal systems up to orbit equivalence, cf. Section 8.1. Boyle and Handelman initiated a more general study of this ordered group, and proved that it classifies irreducible shifts of finite type up to flow equivalence [**BH96**, Theorem 1.12]. It might be the case that the ordered cohomology with an order unit is relevant for further classification of shifts of finite type, cf. [**Bo08**, Problem 23.4].

A complete invariant of flow equivalence for general shifts of finite type exists, but it is *much* more complicated and involves the K-web. The work of Boyle [**Bo02**] and Boyle–Huang [**BH03**] develops this in a way that emphasises certain positivity and algebraic aspects separately. For two-component shifts of finite type, Huang [**Hu94**] has a satisfying classification of flow equivalence using Frank's classification on the irreducible components as well as a distinguished element in a tensor product of Bowen–Franks groups, the latter of which he refers to as the *Cuntz group* since Cuntz found this to be an invariant of flow equivalence in [**Cu81b**, Section 4].

We mention also here that flow equivalence was proven to be decidable by Boyle and Steinberg [**BS**, Corollary 2.4] although their proof only provides the existence of a decision algorithm.

Since the inception of the Cuntz–Krieger algebras, flow equivalence has played an important role. Cuntz and Krieger showed [**CK80**, Theorem 4.1] that the stabilised C*-algebra $\mathcal{O}_A \otimes \mathbb{K}$ together with its diagonal subalgebra $C(X_A) \otimes c_0$ is an invariant of the flow class of $(\sigma_A, \underline{X}_A)$. They point out that using Parry and Sullivan's characterisation (the symbol expansion), this is an *instant computational proof.* The following example should be seen in this light.

EXAMPLE 7.3. Consider the graphs

and note that the Bowen–Franks groups are trivial for both graphs. The two-sided shifts are however not flow equivalent as the reader may check by computing the sign of the determinant.

The C*-algebra of the graph with a single vertex and two loops is Cuntz' algebra $\mathcal{O}_2$, and the C*-algebra of the right-most graph is usually known as $\mathcal{O}_{2-}$. For a long time it was not known whether the two were stably isomorphic, and it was only with Rørdam's clasification of simple Cuntz–Krieger algebras it was finally concluded that $\mathcal{O}_2 \cong \mathcal{O}_{2-}$.

For the cognoscenti, this is an example of a *Cuntz splice*; a move designed to flip the sign of the determinant and thereby change the flow class of the graph.

The fact that two systems that are dynamically far apart (they are not flow equivalent) can have *-isomorphic C*-algebras was a mystery. How much is actually remembered (or lost) in the transition from dynamical systems or graphs to C*-algebras?

The mystery was resolved only much later by Matsumoto and Matui. Their C*-algebraic characterisation of flow equivalence for infinite irreducible shifts of finite type is from [**MM14**, Corollary 3.8]. The result was later generalised to include the reducible shifts of finite type in [**CEOR19**].

THEOREM 7.4. *Let A and B be square $\mathbb{N}$-matrices. The two-sided shifts $(\sigma_A, \underline{X}_A)$ and $(\sigma_B, \underline{X}_B)$ are flow equivalent if and only there exists a stable *-isomorphism $\varphi\colon \mathcal{O}_A \otimes \mathbb{K} \to \mathcal{O}_B \otimes \mathbb{K}$ satisfying $\varphi(C(X_A) \otimes c_0) = C(X_B) \otimes c_0$.*

This means that although $\mathcal{O}_2 \cong \mathcal{O}_{2-}$, cf. Theorem 7.3, this *-isomorphism cannot be diagonal-preserving.

As mentioned above, one direction was already observed by Cuntz and Krieger. The proof of the other direction (assuming stable diagonal *-isomorphism and showing flow equivalence) is remarkable in that it used Matusmoto's notion of continuous orbit equivalence (see Section 8), its connection to C*-algebras, and the fact that it implies flow equivalence, étale topological groupoids and their cohomology which the authors identify with the ordered cohomology of Boyle and Handelman, Huang's theorem [**Hu94**, Theorem 2.15] that any automorphism of the Bowen–Franks group is induced by a flow equivalence, and finally Franks' classification in terms of the signed Bowen–Franks invariant.

This proof strategy does not quite work for general shifts of finite type, and the interesting aspect of the generalisation to the reducible case is the construction of an explicit flow equivalence [**CEOR19**, Section 3.1]. It remains an open problem if a similar result holds for general subshifts although some progress was made in [**BC20b**, Section 8].

QUESTION 7.5. Let $(\sigma_X, \underline{X})$ and $(\sigma_Y, \underline{Y})$ be general subshifts and suppose there exists a *-isomorphism $\varphi\colon \mathcal{O}_X \otimes \mathbb{K} \to \mathcal{O}_Y \otimes \mathbb{K}$ satisfying $\varphi(C(X) \otimes c_0) = C(Y) \otimes c_0$. Does it follow that $(\sigma_X, \underline{X})$ and $(\sigma_Y, \underline{Y})$ are flow equivalent?

Eilers, Restorff, and Ruiz [**ERR09**, Example 4.4] exhibit an example of two substitution shifts that are not flow equivalent but whose C*-algebras are stably isomorphic.

7.1. Back to shift equivalence. It is easy to verify that conjugacy implies flow equivalence (from the definitions). Let us briefly here mention a connection to shift equivalence, cf. Section 3.2. The Bowen–Franks group is invariant under shift equivalence, and so is the spectrum away from zero, so $\det(\mathrm{Id} - A)$ is also preserved. By Franks' classification we therefore have the following observation.

COROLLARY 7.6. *For irreducible and nonpermutation $\mathbb{N}$-matrices, shift equivalence implies flow equivalence.*

Interestingly, Huang [**Hu94**, Corollary 3.4] uses the classification of flow equivalence of two-component shifts of finite type to show that shift equivalence implies flow equivalence also in this case; Huang also shows [**Hu95**, Corollary 3.12] that this happens whenever the Bowen–Franks group is finite (I thank Efren Ruiz for alerting me to this result). It is still not known whether this happens for all reducible shifts of finite type.

A somewhat surprising way of rephrasing Corollary 7.6 is by combining Bratteli and Kishimoto's characterisation of shift equivalence Theorem 3.10 with Matsumoto and Matui's result on flow equivalence Theorem 7.4.

COROLLARY 7.7. *Let A and B be primitive $\mathbb{N}$-matrices. If there is a *-isomorphism $\varphi\colon \mathcal{O}_A \otimes \mathbb{K} \to \mathcal{O}_B \otimes \mathbb{K}$ satisfying $\varphi \circ (\gamma^A \otimes \mathrm{Id}) = (\gamma^B \otimes \mathrm{Id}) \circ \varphi$, then there is a (possibly different) *-isomorphism $\varphi'\colon \mathcal{O}_A \otimes \mathbb{K} \to \mathcal{O}_B \otimes \mathbb{K}$ satisfying $\varphi'(C(X_A) \otimes c_0) = C(X_B) \otimes c_0$.*

This means that the *-isomorphism in Example 7.3 cannot be equivariant, since this would imply flow equivalence.

It is important to note that we cannot expect φ and φ' to be identical; that is, we cannot expect an equivariant *-isomorphism to automatically also be diagonal-preserving. If this were the case, then we would have a two-sided conjugacy, cf. Theorem 3.12, but we know from the counterexample of Kim and Roush Example 3.9 that this is not necessarily the case.

8. Continuous orbit equivalence

Finally, we address the notion of one-sided continuous orbit equivalence. This should be thought of as the one-sided analogue of flow equivalence which we just discussed. We should emphasise that the one-sided case is quite distinct from the two-sided case.

8.1. Cantor minimal systems. For a minimal homeomorphism $\phi\colon X \to X$ on the Cantor space X (short: *Cantor minimal systems*), Giordano, Putnam, and Skau [**GPS95**] managed to completely classify ϕ up to (variations of) orbit equivalence. Important examples of Cantor minimal systems include odometers and Sturmian shifts. The crossed product C*-algebra $C(X) \rtimes_\phi \mathbb{Z}$ is a simple A$\mathbb{T}$-algebra (inductive limit of circle algebras) and its K_0-group is a simple dimension group that classifies these crossed products up to *-isomorphism when considered as an ordered group with a distinguished order unit (the K_1-group is always isomorphic to the integers). The K_0-group is order isomorphic to (and may hence be described concretely as) the coinvariants

$$K^0(X,\phi) = C(X,\mathbb{Z})/\{f - f \circ \phi^{-1} : f \in C(X,\mathbb{Z})\} \tag{8.1}$$

with a distinguished order unit given by the class of the function 1. An element $a \in K^0(X,\phi)$ is infinitesimal if $-\varepsilon 1 \leq a \leq \varepsilon 1$ for all rational $0 < \varepsilon$, and the infinitesimals form a subgroup $\mathrm{Inf}(K^0(X,\phi))$ of $K^0(X,\phi)$.

An orbit equivalence is a homeomorphism $h\colon X_1 \to X_2$ that sends the orbit of any $x \in X_1$ onto the orbit of $h(x) \in X_2$. Equivalently, there are cocycle functions $n\colon X_1 \to \mathbb{Z}$ and $m\colon X_2 \to \mathbb{Z}$ such that $h \circ \phi_1(x) = \phi_2^{n(x)} \circ h(x)$ for all $x \in X_1$ and $h^{-1} \circ \phi_2(y) = \phi_1^{m(y)} \circ h^{-1}(y)$ for all $y \in X_2$. Strong orbit equivalence assumes that the cocycle functions have at most one point of discontinuity, and continuous orbit equivalence simply assumes that the cocycle functions are continuous.

The *full group* $[\phi]$ of (X, ϕ) is the group consisting of all self-orbit equivalences, i.e. a homeomorphism ψ is in $[\phi]$ if there is a function $n\colon X \to \mathbb{Z}$ such that $\psi(x) = \phi^{n(x)}(x)$ for all $x \in X$. The *topological full group* $[[\phi]]$ is the subgroup of orbit equivalences for which the cocycle function n can be taken to be continuous. Giordano, Putnam, and Skau proved that Cantor minimal systems (X_1, ϕ_1) and (X_2, ϕ_2) are

- topologically orbit equivalent precisely when their dimension groups with distinguished unit modulo the infinitesimals are isomorphic [**GPS95**, Theorem 2.2], and if and only the full groups are isomorphic [**GPS99**, Corollary 4.6];
- strong orbit equivalent precisely when their dimension groups with units are isomorphic, and that this happens if and only if the crossed products are *-isomorphic [**GPS95**, Theorem 2.2]; and
- continuously orbit equivalent precisely when their crossed products are *-isomorphic in a diagonal-preserving way [**GPS95**, Theorem 2.4], and if and only if their topological full groups are isomorphic [**GPS99**, Corollary 4.4].

In the third bullet above, we emphasise the definition of continuous orbit equivalence because it provides a nice analogue to the one-sided case that we shall discuss shortly. However, Boyle [**Bo83**] showed in his thesis that this coincides with flip conjugacy (i.e. that ϕ_1 is conjugate to either ϕ_2 or ϕ_2^{-1}), and the facts that continuous orbit equivalence implies flip conjugacy and that this coincides with diagonal-preserving *-isomorphism of crossed products was later generalised to the setting of topologically transitive and topologically free homeomorphisms of compact Hausdorff spaces by Boyle and Tomiyama [**BT98**].

The topological full groups from Cantor minimal systems admit interesting group theoretic properties. Indeed, Jushenko and Monod [**JM13**, Theorem A] showed that these groups are always amenable, and by restricting to those Cantor minimal systems that are conjugate to minimal shifts, they exhibited the first examples of (actually uncountably many) infinite finitely generated simple and amenable groups, cf. [**Ma06**].

If we consider homeomorphisms on compact Hausdorff spaces as $\mathbb{Z}$-actions then flip conjugacy is really the correct notion of conjugacy of such actions, so the Boyle–Tomiyama result says that topologically transitive and topologically free $\mathbb{Z}$-actions on compact Hausdorff spaces are continuously orbit equivalence rigid. Continuous orbit equivalence for general group actions and its connection to diagonal-preserving *-isomorphism of crossed products was thoroughly studied by Li in [**Li18**] in which examples that are not continuously orbit equivalence rigid also appear.

8.2. One-sided continuous orbit equivalence. For irreducible shifts of finite type, the notion of one-sided continuous orbit equivalence was introduced by Matsumoto [**Ma10**]. The general definition can be found e.g. in [**CRST21**, Definition 8.1].

DEFINITION 8.1. Let $T\colon X \to X$ and $S\colon Y \to Y$ be local homeomorphisms on locally compact Hausdorff spaces X and Y, respectively. Then T and S are *continuously orbit equivalent* if there exist a homeomorphism $h\colon X \to Y$ and continuous functions $k_X, l_X : X \to \mathbb{N}$ and $k_Y, l_Y : Y \to \mathbb{N}$ such that

$$S^{l_X(x)} \circ h(x) = S^{k_X(x)} \circ h \circ T(x), \tag{8.2}$$

$$T^{l_Y(y)} \circ h^{-1}(y) = T^{k_Y(y)} \circ h^{-1} \circ S(y), \tag{8.3}$$

for all $x \in X$ and $y \in Y$.

REMARK 8.2. Note that if in the above definition, the cocycles for h can be chosen to be $l_X = k_X + 1$ and $l_Y = k_Y + 1$, then h is a Matsumoto eventual conjugacy, cf. Definition 5.1.

EXAMPLE 8.3. The graphs in Example 7.1 are continuously orbit equivalent; an explicit map can be constructed using the same symbol expansion as in that example, see [**Ma10**, p. 203].

Matsumoto had a clear motivation from C*-algebras, and [**Ma10**, Theorem 1.1] shows that continuous orbit equivalence is the same as diagonal-preserving *-isomorphism of Cuntz–Krieger algebras. Later in [**MM14**, Theorem 3.6], Matsumoto and Matui characterised continuous orbit equivalence in terms of the *unital* signed Bowen–Franks group. The *unital* Bowen–Franks group is the cokernel $\mathrm{BF}(A) = \mathbb{Z}^{|A|}/(\mathrm{Id} - A)\mathbb{Z}^{|A|}$ together with the class of the vector $(1, \ldots, 1) \in \mathbb{Z}^{|A|}$ in $\mathrm{BF}(A)$, we denote the class as u_A. We summarise these results.

THEOREM 8.4. *Let A and B be irreducible and nonpermutation $\mathbb{N}$-matrices. The one-sided shifts of finite type (σ_A, X_A) and (σ_B, X_B) are continuously orbit equivalent if and only if* $(\mathrm{BF}(A), u_A) \cong (\mathrm{BF}(B), u_B)$ *and* $\det(\mathrm{Id} - A) = \det(\mathrm{Id} - B)$, *and if and only if there is a *-isomorphism $\varphi\colon \mathcal{O}_A \to \mathcal{O}_B$ satisfying $\varphi(C(X_A)) = C(X_B)$.*

Part of the proof involves showing that continuous orbit equivalence implies flow equivalence (see Section 7). The connection between continuous orbit equivalence and diagonal-preserving *-isomorphism of Cuntz–Krieger algebras and the fact that continuous orbit equivalence implies flow equivalence was again generalised to reducible shifts of finite type in [**CEOR19**].

EXAMPLE 8.5. Ashley's graphs Example 6.9 are continuously orbit equivalent: since the two systems are primitive and flow equivalent, and the Bowen–Franks group are trivial (so the condition on the class of the unit is trivially satisfied), this follows from the above theorem.

EXAMPLE 8.6. Kim and Roush's primitive counterexample to the Williams problem Theorem 3.9 are continuously orbit equivalent, cf. [**ER**, Example 4.9].

EXAMPLE 8.7. The matrices

$$A = \begin{pmatrix} 1 & 1 & 1 \\ 1 & 1 & 1 \\ 1 & 0 & 0 \end{pmatrix}, \qquad \text{and} \qquad B = \begin{pmatrix} 1 & 1 & 1 \\ 1 & 1 & 0 \\ 1 & 1 & 0 \end{pmatrix}.$$

are strong shift equivalent: A is an out-split and B is an in-split of $\begin{pmatrix} 1 & 2 \\ 1 & 1 \end{pmatrix}$. However, $\mathcal{O}_A \cong \mathcal{O}_3$ and $\mathcal{O}_B \cong \mathcal{O}_3 \otimes M_2$, so they are not continuously orbit equivalent, cf. [**Ma17b**, Section 3].

An algebraic characterisation of continuous orbit equivalence for reducible shifts of finite type that generalises the unital signed Bowen–Franks invariant seems to still be missing. Therefore, it is still not clear whether this is decidable.

For local homeomorphisms, [**CRST21**, Theorem 8.2] shows that continuous orbit equivalence is characterised by diagonal-preserving *-isomorphism. The proof of this relies on groupoid reconstruction, cf. Theorem 2.11.

THEOREM 8.8. *Let $T\colon X \to X$ and $S\colon Y \to Y$ be local homeomorphisms. Then T and S are (stabiliser-preserving) continuously orbit equivalent if and only if there is a *-isomorphism $\varphi\colon C^*(T) \to C^*(S)$ satisfying $\varphi(C_0(X)) = C_0(Y)$.*

This is used to show in [**BC20b**, Theorem 6.10] (see also [**Ma20**, Theorem 1.2]) that (under certain technical conditions) shift spaces (σ_X, X) and (σ_Y, Y) are continuously orbit equivalent if and only if there is a *-isomorphism $\varphi\colon \mathcal{O}_X \to \mathcal{O}_Y$ satisfying $\varphi(C(X)) = C(Y)$. Note that this is not immediate from Theorem 8.8: the shifts σ_X and σ_Y are generally *not* local homeomorphisms (only when they are of finite type) but the actions on the covers $(\sigma_{\tilde{X}}, \tilde{X})$ and $(\sigma_{\tilde{Y}}, \tilde{Y})$ are, so the theorem can only be applied to the covers; the difficulty is then to understand how a continuous orbit equivalence among shift spaces passes to the covers (and back again). At the moment, it is not clear if the technical conditions can be removed.

We also state a question related to flow equivalence.

QUESTION 8.9. Does continuous orbit equivalence imply flow equivalence?

EXAMPLE 8.10. The labelled graphs

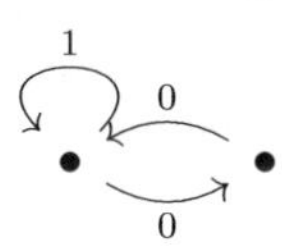

and

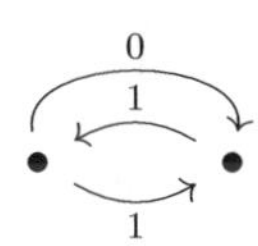

represent the even and odd shifts, respectively. We encountered the left-most graph as representing a sofic shift in Example 2.4. In the right-most labelled path space, only an odd number of 0s are allowed between any two 1s, so this is again a sofic shifts. A continuous orbit equivalence between the even and odd shifts is given by sending any occurrence of 1 to 10, cf. [**BC20b**, Example 6.15] (a similar argument shows that they are flow equivalent).

Following the ideas of Giordano, Putnam, and Skau, cf. Section 8.1, it is also possible to characterise the one-sided shifts of finite type up to continuous orbit equivalence using a notion of topological full group. Let $[[\sigma_A]]$ be the subgroup of homeomorphisms h of X_A for which there exist continuous cocycles $k_X, l_X\colon X_A \to \mathbb{N}$ such that relations similar to those in Definition 8.1 are satisfied. Infinite and irreducible shifts of finite type X_A and X_B are then continuously orbit equivalent precisely if there exists a homeomorphism $h\colon X_A \to X_B$ such that $h \circ [[\sigma_A]] \circ h^{-1} = [[\sigma_B]]$, cf. [**Ma10**, Theorem 1.1].

For the full 2-shift, the topological full group $[[\sigma_{(2)}]]$ is isomorphic to Thompson's group V, and $[[\sigma_{(n)}]]$ is isomorphic to the Higman–Thompson group V_n, cf. [**Ma15**, Remark 6.3]. Matui comments that from this perspective, the topological full groups of shifts of finite type may be viewed as generalisations of Higman–Thompson groups.

In fact, Matui vastly generalised the notion of topological full group to the setting of étale groupoids [**Ma12**]. Suppose G is a locally compact Hausdorff and étale groupoid whose unit space $G^{(0)}$ is the Cantor space. The compact open bisections (the subsets of G on which the range and source maps are injective) form an inverse semigroup, and the topological full group $[[G]]$ of G is the subgroup consisting of the global bisections (those for which the range and source is the whole unit space).

Matui proves that the (abstract) isomorphism class of the topological full groups characterise the isomorphism class of the groupoids, cf. [**Ma15**, Theorem 3.10]. Again we refer the reader to [**Si20**] for the basics of étale groupoids. The groupoid G_A of a one-sided shift of finite type X_A is minimal and effective precisely when A is an irreducible and nonpermutation matrix.

THEOREM 8.11. *Let G_1 and G_2 be second-countable, minimal, effective, and étale groupoids whose unit spaces are Cantor spaces. Then G_1 and G_2 are isomorphic as topological groupoids if and only if the topological full groups $[[G_1]]$ and $[[G_2]]$ are abstractly isomorphic as groups.*

Comparing this to Renault's reconstruction theorem Theorem 2.11 we see how this vastly generalises Matsumoto's observations. The groups arising from continuous orbit equivalences of dynamical systems thus exhibit interesting group theoretic properties as well as interesting connections to dynamics, groupoids, and C*-algebras.

References

[ABCE22] B. Armstrong, K. A. Brix, T. M. Carlsen, and S. Eilers, *Conjugacy of local homeomorphisms via groupoids and* C*-*algebras*, Ergodic Theory Dynam. Systems **43** (2023), no. 8, 2516–2537, DOI 10.1017/etds.2022.50. MR4611740

[ABCEW] B. Armstrong, K.A. Brix, T.M. Carlsen, S. Eilers, and J. Winkel, in preparation.

[BP04] T. Bates and D. Pask, *Flow equivalence of graph algebras*, Ergodic Theory Dynam. Systems **24** (2004), no. 2, 367–382, DOI 10.1017/S0143385703000348. MR2054048

[Bo72] R. Bowen, *One-dimensional hyperbolic sets for flows*, J. Differential Equations **12** (1972), 173–179, DOI 10.1016/0022-0396(72)90012-5. MR336762

[Bo78] R. Bowen, *On Axiom A diffeomorphisms*, Regional Conference Series in Mathematics, No. 35, American Mathematical Society, Providence, RI, 1978. MR482842

[BF77] R. Bowen and J. Franks, *Homology for zero-dimensional nonwandering sets*, Ann. of Math. (2) **106** (1977), no. 1, 73–92, DOI 10.2307/1971159. MR458492

[Bo83] M. Boyle, *Topological Orbit Equivalence and Factor Maps in Symbolic Dynamics*, ProQuest LLC, Ann Arbor, MI, 1983. Thesis (Ph.D.)–University of Washington. MR2632783

[Bo02] M. Boyle, *Flow equivalence of shifts of finite type via positive factorizations*, Pacific J. Math. **204** (2002), no. 2, 273–317, DOI 10.2140/pjm.2002.204.273. MR1907894

[Bo08] M. Boyle, *Open problems in symbolic dynamics*, Geometric and probabilistic structures in dynamics, Contemp. Math., vol. 469, Amer. Math. Soc., Providence, RI, 2008, pp. 69–118, DOI 10.1090/conm/469/09161. MR2478466

[BCE17] M. Boyle, T. M. Carlsen, and S. Eilers, *Flow equivalence and isotopy for subshifts*, Dyn. Syst. **32** (2017), no. 3, 305–325, DOI 10.1080/14689367.2016.1207753. MR3669803

[BFF97] M. Boyle, D. Fiebig, and U.-R. Fiebig, *A dimension group for local homeomorphisms and endomorphisms of onesided shifts of finite type*, J. reine angew. Math. **487** (1997), 27–59. MR1454258

[BH96] M. Boyle and D. Handelman, *Orbit equivalence, flow equivalence and ordered cohomology*, Israel J. Math. **95** (1996), 169–210, DOI 10.1007/BF02761039. MR1418293

[BH03] M. Boyle and D. Huang, *Poset block equivalence of integral matrices*, Trans. Amer. Math. Soc. **355** (2003), no. 10, 3861–3886, DOI 10.1090/S0002-9947-03-02947-7. MR1990568

[BK88] M. Boyle and W. Krieger, *Almost Markov and shift equivalent sofic systems*, Dynamical systems (College Park, MD, 1986), Lecture Notes in Math., vol. 1342, Springer, Berlin, 1988, pp. 33–93, DOI 10.1007/BFb0082823. MR970547

[BS] M. Boyle and B. Steinberg, *Decidability of flow equivalence and isomorphism problems for graph C^*-algebras and quiver representations*, Proc. Amer. Math. Soc. **151** (2023), 5151–5161. DOI: https://doi.org/10.1090/proc/15148

[BT98] M. Boyle and J. Tomiyama, *Bounded topological orbit equivalence and C^*-algebras*, J. Math. Soc. Japan **50** (1998), no. 2, 317–329, DOI 10.2969/jmsj/05020317. MR1613140

[BK00] O. Bratteli and A. Kishimoto, *Trace scaling automorphisms of certain stable AF algebras. II*, Q. J. Math. **51** (2000), no. 2, 131–154, DOI 10.1093/qjmath/51.2.131. MR1765786

[Br22] K. A. Brix, *Balanced strong shift equivalence, balanced in-splits, and eventual conjugacy*, Ergodic Theory Dynam. Systems **42** (2022), no. 1, 19–39, DOI 10.1017/etds.2020.126. MR4348408

[Br21] K.A. Brix, *Sturmian subshifts and their C^*-algebras*, J. Operator Theory **90**, no. 2, Autumn 2023, pp. 523–544.

[Br] K.A. Brix, *Unital shift equivalence*, in preparation.

[BC20a] K. A. Brix and T. M. Carlsen, *Cuntz-Krieger algebras and one-sided conjugacy of shifts of finite type and their groupoids*, J. Aust. Math. Soc. **109** (2020), no. 3, 289–298, DOI 10.1017/S1446788719000168. MR4190081

[BC20b] K. A. Brix and T. M. Carlsen, C^**-algebras, groupoids and covers of shift spaces*, Trans. Amer. Math. Soc. Ser. B **7** (2020), 134–185, DOI 10.1090/btran/53. MR4168660

[Ca03] T. M. Carlsen, *On C^*-algebras associated with sofic shifts*, J. Operator Theory **49** (2003), no. 1, 203–212. MR1978330

[CEOR19] T. M. Carlsen, S. Eilers, E. Ortega, and G. Restorff, *Flow equivalence and orbit equivalence for shifts of finite type and isomorphism of their groupoids*, J. Math. Anal. Appl. **469** (2019), no. 2, 1088–1110, DOI 10.1016/j.jmaa.2018.09.056. MR3860463

[CM04] T. M. Carlsen and K. Matsumoto, *Some remarks on the C^*-algebras associated with subshifts*, Math. Scand. **95** (2004), no. 1, 145–160, DOI 10.7146/math.scand.a-14453. MR2091486

[CR17] T. M. Carlsen and J. Rout, *Diagonal-preserving gauge-invariant isomorphisms of graph C^*-algebras*, J. Funct. Anal. **273** (2017), no. 9, 2981–2993, DOI 10.1016/j.jfa.2017.06.018. MR3692327

[CRST21] T. M. Carlsen, E. Ruiz, A. Sims, and M. Tomforde, *Reconstruction of groupoids and C^*-rigidity of dynamical systems*, Adv. Math. **390** (2021), Paper No. 107923, 55 pp., DOI 10.1016/j.aim.2021.107923. MR4292963

[Cu77] J. Cuntz, *Simple C^*-algebras generated by isometries*, Comm. Math. Phys. **57** (1977), no. 2, 173–185. MR467330

[Cu81a] J. Cuntz, *K-theory for certain C^*-algebras*, Ann. of Math. (2) **113** (1981), no. 1, 181–197, DOI 10.2307/1971137. MR604046

[Cu81b] J. Cuntz, *A class of C^*-algebras and topological Markov chains. II. Reducible chains and the Ext-functor for C^*-algebras*, Invent. Math. **63** (1981), no. 1, 25–40, DOI 10.1007/BF01389192. MR608527

[CK80] J. Cuntz and W. Krieger, *A class of C^*-algebras and topological Markov chains*, Invent. Math. **56** (1980), no. 3, 251–268, DOI 10.1007/BF01390048. MR561974

[ERR09] S. Eilers, G. Restorff, and E. Ruiz, *Classification of extensions of classifiable C^*-algebras*, Adv. Math. **222** (2009), no. 6, 2153–2172, DOI 10.1016/j.aim.2009.07.014. MR2562779

[ERRS16] S. Eilers, G. Restorff, E. Ruiz, and A. P. W. Sørensen, *The complete classification of unital graph C^*-algebras: geometric and strong*, Duke Math. J. **170** (2021), no. 11, 2421–2517, DOI 10.1215/00127094-2021-0060. MR4302548

[ER] S. Eilers and E. Ruiz, *Refined moves for structure-preserving isomorphism of graph C^*-algebras*, arXiv:1908.03714v1 [math.OA], 46 pages.

[EW80] M. Enomoto and Y. Watatani, *A graph theory for C^*-algebras*, Math. Japon. **25** (1980), no. 4, 435–442. MR594544

[Fi75] R. Fischer, *Sofic systems and graphs*, Monatsh. Math. **80** (1975), no. 3, 179–186, DOI 10.1007/BF01319913. MR407235

[Fr84] J. Franks, *Flow equivalence of subshifts of finite type*, Ergodic Theory Dynam. Systems **4** (1984), no. 1, 53–66, DOI 10.1017/S0143385700002261. MR758893

[GPS95] T. Giordano, I. F. Putnam, and C. F. Skau, *Topological orbit equivalence and C^*-crossed products*, J. Reine Angew. Math. **469** (1995), 51–111. MR1363826

[GPS99] T. Giordano, I. F. Putnam, and C. F. Skau, *Full groups of Cantor minimal systems*, Israel J. Math. **111** (1999), 285–320, DOI 10.1007/BF02810689. MR1710743

[HN88] T. Hamachi and M. Nasu, *Topological conjugacy for 1-block factor maps of subshifts and sofic covers*, Dynamical systems (College Park, MD, 1986), Lecture Notes in Math., vol. 1342, Springer, Berlin, 1988, pp. 251–260, DOI 10.1007/BFb0082835. MR970559

[Hu94] D. Huang, *Flow equivalence of reducible shifts of finite type*, Ergodic Theory Dynam. Systems **14** (1994), no. 4, 695–720, DOI 10.1017/S0143385700008129. MR1304139

[Hu95] D. Huang, *Flow equivalence of reducible shifts of finite type and Cuntz-Krieger algebras*, J. Reine Angew. Math. **462** (1995), 185–217, DOI 10.1515/crll.1995.462.185. MR1329907

[aHR97] A. an Huef and I. Raeburn, *The ideal structure of Cuntz-Krieger algebras*, Ergodic Theory Dynam. Systems **17** (1997), no. 3, 611–624, DOI 10.1017/S0143385797079200. MR1452183

[JM13] K. Juschenko and N. Monod, *Cantor systems, piecewise translations and simple amenable groups*, Ann. of Math. (2) **178** (2013), no. 2, 775–787, DOI 10.4007/annals.2013.178.2.7. MR3071509

[Ka09] T. Katsura, *Cuntz-Krieger algebras and C^*-algebras of topological graphs*, Acta Appl. Math. **108** (2009), no. 3, 617–624, DOI 10.1007/s10440-009-9535-0. MR2563503

[KR79] K. H. Kim and F. W. Roush, *Some results on decidability of shift equivalence*, J. Combin. Inform. System Sci. **4** (1979), no. 2, 123–146. MR564188

[KR88] K. H. Kim and F. W. Roush, *Decidability of shift equivalence*, Dynamical systems (College Park, MD, 1986), Lecture Notes in Math., vol. 1342, Springer, Berlin, 1988, pp. 374–424, DOI 10.1007/BFb0082843. MR970567

[KR92] K. H. Kim and F. W. Roush, *Williams's conjecture is false for reducible subshifts*, J. Amer. Math. Soc. **5** (1992), no. 1, 213–215, DOI 10.2307/2152756. MR1130528

[KR99] K. H. Kim and F. W. Roush, *The Williams conjecture is false for irreducible subshifts*, Ann. of Math. (2) **149** (1999), no. 2, 545–558, DOI 10.2307/120975. MR1689340

[Ki98] B. P. Kitchens, *Symbolic dynamics: One-sided, two-sided and countable state Markov shifts*, Universitext, Springer-Verlag, Berlin, 1998, DOI 10.1007/978-3-642-58822-8. MR1484730

[Ku86] A. Kumjian, *On C^*-diagonals*, Canad. J. Math. **38** (1986), no. 4, 969–1008, DOI 10.4153/CJM-1986-048-0. MR854149

[Kr80a] W. Krieger, *On a dimension for a class of homeomorphism groups*, Math. Ann. **252** (1979/80), no. 2, 87–95, DOI 10.1007/BF01420115. MR593623

[Kr80b] W. Krieger, *On dimension functions and topological Markov chains*, Invent. Math. **56** (1980), no. 3, 239–250, DOI 10.1007/BF01390047. MR561973

[Kr84] W. Krieger, *On sofic systems. I*, Israel J. Math. **48** (1984), no. 4, 305–330, DOI 10.1007/BF02760631. MR776312

[KPRR97] A. Kumjian, D. Pask, I. Raeburn, and J. Renault, *Graphs, groupoids, and Cuntz-Krieger algebras*, J. Funct. Anal. **144** (1997), no. 2, 505–541, DOI 10.1006/jfan.1996.3001. MR1432596

[Li18] X. Li, *Continuous orbit equivalence rigidity*, Ergodic Theory Dynam. Systems **38** (2018), no. 4, 1543–1563, DOI 10.1017/etds.2016.98. MR3789176

[LM95] D. Lind and B. Marcus, *An introduction to symbolic dynamics and coding*, Cambridge Mathematical Library, Cambridge University Press, Cambridge, 2021. Second edition [of 1369092], DOI 10.1017/9781108899727. MR4412543

[Ma97] K. Matsumoto, *On C^*-algebras associated with subshifts*, Internat. J. Math. **8** (1997), no. 3, 357–374, DOI 10.1142/S0129167X97000172. MR1454478

[Ma10] K. Matsumoto, *Orbit equivalence of topological Markov shifts and Cuntz-Krieger algebras*, Pacific J. Math. **246** (2010), no. 1, 199–225, DOI 10.2140/pjm.2010.246.199. MR2645883

[Mat17a] K. Matsumoto, *Continuous orbit equivalence, flow equivalence of Markov shifts and circle actions on Cuntz-Krieger algebras*, Math. Z. **285** (2017), no. 1-2, 121–141, DOI 10.1007/s00209-016-1700-3. MR3598806

[Ma17b] K. Matsumoto, *Uniformly continuous orbit equivalence of Markov shifts and gauge actions on Cuntz-Krieger algebras*, Proc. Amer. Math. Soc. **145** (2017), no. 3, 1131–1140, DOI 10.1090/proc/13387. MR3589313

[Ma20] K. Matsumoto, *A groupoid approach to C^*-algebras associated with λ-graph systems and continuous orbit equivalence of subshifts*, Dyn. Syst. **35** (2020), no. 3, 398–429, DOI 10.1080/14689367.2020.1716952. MR4149088

[Ma] K. Matsumoto, *Subshifts, λ-graph bisystems and C^*-algebras*, J. Math. Anal. Appl. **485** (2020), no. 2, 123843, 55, DOI 10.1016/j.jmaa.2020.123843. MR4052571

[Mat21] K. Matsumoto, *One-sided topological conjugacy of normal subshifts and gauge actions on the associated C^*-algebras*, Dyn. Syst. **36** (2021), no. 4, 586–607, DOI 10.1080/14689367.2021.1970115. MR4338820

[Mat22] K. Matsumoto, *On one-sided topological conjugacy of topological Markov shifts and gauge actions on Cuntz-Krieger algebras*, Ergodic Theory Dynam. Systems **42** (2022), no. 8, 2575–2582, DOI 10.1017/etds.2021.53. MR4448399

[MM14] K. Matsumoto and H. Matui, *Continuous orbit equivalence of topological Markov shifts and Cuntz-Krieger algebras*, Kyoto J. Math. **54** (2014), no. 4, 863 877, DOI 10.1215/21562261-2801849. MR3276420

[Ma06] H. Matui, *Some remarks on topological full groups of Cantor minimal systems*, Internat. J. Math. **17** (2006), no. 2, 231–251, DOI 10.1142/S0129167X06003448. MR2205435

[Ma12] H. Matui, *Homology and topological full groups of étale groupoids on totally disconnected spaces*, Proc. Lond. Math. Soc. (3) **104** (2012), no. 1, 27–56, DOI 10.1112/plms/pdr029. MR2876963

[Ma15] H. Matui, *Topological full groups of one-sided shifts of finite type*, J. Reine Angew. Math. **705** (2015), 35–84, DOI 10.1515/crelle-2013-0041. MR3377390

[MvN43] F. J. Murray and J. von Neumann, *On rings of operators. IV*, Ann. of Math. (2) **44** (1943), 716–808, DOI 10.2307/1969107. MR9096

[Na86] M. Nasu, *Topological conjugacy for sofic systems*, Ergodic Theory Dynam. Systems **6** (1986), no. 2, 265–280, DOI 10.1017/S0143385700003448. MR857201

[vN31] J. von Neumann, *Proof of the quasi-ergodic hypothesis*, Proc. N.A.S. **18** (1) (1931), 70–82.

[Pa66] W. Parry, *Symbolic dynamics and transformations of the unit interval*, Trans. Amer. Math. Soc. **122** (1966), 368–378, DOI 10.2307/1994554. MR197683

[PS75] B. Parry and D. Sullivan, *A topological invariant of flows on 1-dimensional spaces*, Topology **14** (1975), no. 4, 297–299, DOI 10.1016/0040-9383(75)90012-9. MR405385

[Ra05] I. Raeburn, *Graph algebras*, CBMS Regional Conference Series in Mathematics, vol. 103, Published for the Conference Board of the Mathematical Sciences, Washington, DC; by the American Mathematical Society, Providence, RI, 2005, DOI 10.1090/cbms/103. MR2135030

[Re80] J. Renault, *A groupoid approach to C^*-algebras*, Lecture Notes in Mathematics, vol. 793, Springer, Berlin, 1980. MR584266

[Re08] J. Renault, *Cartan subalgebras in C^*-algebras*, Irish Math. Soc. Bull. **61** (2008), 29–63. MR2460017

[Rø95] M. Rørdam, *Classification of Cuntz-Krieger algebras*, K-Theory **9** (1995), no. 1, 31–58, DOI 10.1007/BF00965458. MR1340839

[Sa98] J. N. Samuel, *C^*-algebras of sofic shifts*, ProQuest LLC, Ann Arbor, MI, 1999. Thesis (Ph.D.)–University of Victoria (Canada). MR2699259

[Si20] A. Sims, G. Szabó, and D. Williams, *Operator algebras and dynamics: groupoids, crossed products, and Rokhlin dimension*, Advanced Courses in Mathematics. CRM Barcelona, Birkhäuser/Springer, Cham, [2020]. Lecture notes from the Advanced Course held at Centre de Recerca Matemàtica (CRM) Barcelona, March 13–17, 2017; Edited by Francesc Perera, DOI 10.1007/978-3-030-39713-5. MR4321941

[SW16] A. Sims and D. P. Williams, *The primitive ideals of some étale groupoid C^*-algebras*, Algebr. Represent. Theory **19** (2016), no. 2, 255–276, DOI 10.1007/s10468-015-9573-4. MR3489096

[We73] B. Weiss, *Subshifts of finite type and sofic systems*, Monatsh. Math. **77** (1973), 462–474, DOI 10.1007/BF01295322. MR340556

[Wi73] R. F. Williams, *Classification of subshifts of finite type*, Ann. of Math. (2) **98** (1973), 120–153; errata, ibid. (2) 99 (1974), 380–381, DOI 10.2307/1970908. MR331436

School of Mathematics and Statistics, University of Glasgow, University Place, Glasgow G12 8QQ, United Kingdom

Email address: kabrix.math@fastmail.com

Contemporary Mathematics
Volume **798**, 2024
https://doi.org/10.1090/conm/798/15981

$\mathcal{PT}$-symmetric couplings of dual pairs

Volodymyr Derkach and Carsten Trunk

Abstract. We apply the boundary triple technique to construct a coupling (A, B) of two dual pairs (A_+, B_+) and (A_-, B_-) relative to some boundary triples. The notion of a real dual pair with respect to the time reversal operator $\mathcal{T}$ is introduced and it is shown that the coupling of two real dual pairs corresponding to real boundary triples is also real. If the operator $\mathcal{PT}$ intertwines the dual pairs (A_+, B_+) and (A_-, B_-) for some parity operator $\mathcal{P}$, then it is shown that there exists a coupling (A, B) of two dual pairs (A_+, B_+) and (A_-, B_-) such that the operator A is $\mathcal{PT}$-symmetric and $\mathcal{P}$-symmetric in the Krein space $(\mathfrak{H}, \langle\cdot,\cdot\rangle)$ with the fundamental symmetry $\mathcal{P}$. As the main result we describe proper extensions of A which are $\mathcal{PT}$-symmetric and $\mathcal{P}$-selfadjoint.

We apply this result to interpret the $\mathcal{PT}$-symmetric Hamiltonian considered in Bender & Boettcher (1998) as a member of a family of $\mathcal{PT}$-symmetric and $\mathcal{P}$-selfadjoint extensions of the corresponding minimal operator.

Keywords: dual pair; boundary triple; coupling; $\mathcal{PT}$-symmetric operator; non-Hermitian Hamiltonian

1. Introduction

In the seminal paper by Bender and Boettcher [**6**] a new view at Quantum Mechanics was proposed which adopts all its axioms except the one that restricts the Hamiltonian to be Hermitian, relaxing it to the assumption that the Hamiltonian is $\mathcal{PT}$-symmetric. Here, $\mathcal{P}$ is parity and $\mathcal{T}$ is time reversal. Since 1998, $\mathcal{PT}$-symmetric Hamiltonians have been analyzed intensively by many authors. In [**24**] $\mathcal{PT}$-symmetry was embedded into a more general mathematical framework: pseudo-Hermiticity or, what is the same, self-adjoint operators in Krein spaces, ([**20**], [**4**], [**17**], [**21**]) For a general introduction into $\mathcal{PT}$-symmetric Quantum Mechanics we refer to the overview paper of Mostafazadeh [**26**] and to the books of Moiseyev [**23**] and Bender [**5**].

A prominent class consists of the $\mathcal{PT}$-symmetric Hamiltonians

$$H := \frac{1}{2}p^2 - (iz)^{N+2},$$

where N is a positive integer [**8**]. The associated eigenvalue problem is defined on a contour Γ in the complex plane which is contained in a specific area in the complex plane, the so-called Stokes wedges, see [**6**],

$$-y''(z) - (iz)^{N+2}y(z) = \lambda y(z), \quad z \in \Gamma, \tag{1.1}$$

2020 *Mathematics Subject Classification.* Primary 47B28, Secondary 47B50, 81Q12.

where $\lambda \in \mathbb{C}$ is the eigenvalue parameter. Recall ([**7**]) that a *Stokes wedge* S_k, $k = 0, \ldots, N+3$, is an open sector in the plane with vertex zero,

$$S_k := \left\{ z \in \mathbb{C} \,\middle|\, -\frac{N+2}{2N+8}\pi + \frac{2k-2}{4+N}\pi < \arg(z) < -\frac{N+2}{2N+8}\pi + \frac{2k}{4+N}\pi \right\}.$$

The boundary of S_k is called *Stokes lines* and consists of two rays through the origin. $\mathcal{PT}$-symmetry forces Γ to lie in two Stokes wedges, which are symmetric with respect to the imaginary axis.

In [**25**] the problem was mapped back to the real axis using a real parametrization. In [**7**] and in [**18**] this approach was extended to different parameterizations and contours. For simplicity, we choose here Γ to be a wedge-shaped contour,

$$\Gamma := \{ x e^{i\phi \operatorname{sgn} x} \,|\, x \in \mathbb{R} \}, \tag{1.2}$$

for some angle $\phi \in (-\pi/2, \pi/2)$. Let $z : \mathbb{R} \to \mathbb{C}$ parameterize Γ via $z(x) := xe^{i\phi \operatorname{sgn} x}$. Then y solves (1.1) for $z \neq 0$ if and only if the pair of functions u_+, u_- given by $u_\pm(x) := y(z(x))$, $x \in \mathbb{R}_\pm$, solve

$$\mathfrak{a}_-[u_-] = \lambda u_-, \quad x \in \mathbb{R}_-, \qquad \mathfrak{a}_+[u_+] = \lambda u_+, \quad x \in \mathbb{R}_+, \tag{1.3}$$

where the differential expressions $\mathfrak{a}_\pm$ are given by

$$\mathfrak{a}_\pm[u_\pm] = -e^{\mp 2i\phi} u_\pm'' - (ix)^{N+2} e^{\pm i(N+2)\phi} u_\pm. \tag{1.4}$$

In what follows we assume that Γ lies in Stokes wedges and then by Leben & Trunk [**21**] the differential expressions $\mathfrak{a}_\pm$ are in the limit-point cases at $\pm\infty$ according to the classification in Brown et al. [**11**] (which is a refinement of the classification in Sims [**27**]). We mention, that the limit-circle case can be treated in a similar way as in [**3**],[**4**].

The theory of $\mathcal{PT}$-symmetry claims that the main object, the Hamiltonian, commutes under the joint action of the parity $\mathcal{P}$ and the time reversal $\mathcal{T}$,

$$(\mathcal{P}f)(x) := f(-x), \qquad (\mathcal{T}f)(x) := \overline{f(x)}. \tag{1.5}$$

The time reversal $\mathcal{T}$ applied to $\mathfrak{a}_\pm$ gives rise to new differential expressions $\mathfrak{b}_\pm = \mathcal{T}\mathfrak{a}_\pm\mathcal{T}$ defined on $\mathbb{R}_\pm$ of the form

$$\mathfrak{b}_\pm[v_\pm] = -e^{\pm 2i\phi} v_\pm'' - (-ix)^{N+2} e^{\mp i(N+2)\phi} v_\pm. \tag{1.6}$$

In Section 7 we introduce the minimal operators $A_\pm$ and $B_\pm$ associated with $\mathfrak{a}_\pm$ and $\mathfrak{b}_\pm$ in $L^2(\mathbb{R}_\pm)$. Due to their (in general) non-real coefficients, the operators $A_\pm$ and $B_\pm$ are neither selfadjoint nor symmetric. But they satisfy

$$\langle A_\pm f, g \rangle_\pm = \langle f, B_\pm g \rangle_\pm \tag{1.7}$$

for all $f \in \operatorname{dom} A_\pm$ and $g \in \operatorname{dom} B_\pm$. Here $\langle \cdot, \cdot \rangle_\pm$ stands for the usual inner products in the Hilbert spaces $L^2(\mathbb{R}_\pm)$. Condition (1.7) shows that the pairs (A_+, B_+) and (A_-, B_-) form *dual pairs* (see Section 2 for details). An extension theory for dual pairs based on the boundary triples technique was developed by Malamud and Mogilevskii in [**22**]. This is a generalization of the boundary triple approach to the extension theory of symmetric operators which was elaborated by Kochubei [**19**], M. & V. Gorbachuk [**16**], Derkach & Malamud [**14**] and others.

Following this approach we construct in Theorem 7.1 boundary triples for dual pairs (A_+, B_+) and (A_-, B_-). As our interest is focused on the Hamiltonian in $L^2(\mathbb{R})$ and not on the differential expressions $\mathfrak{a}_\pm$, $\mathfrak{b}_\pm$, which are defined on the semi-axes, we extend the coupling method for symmetric operators from [**13**] to the case of dual pairs and create a new dual pair (A, B) of operators defined on $\mathbb{R}$. This dual pair (A, B) is called the *coupling of the dual pairs* (A_+, B_+) and (A_-, B_-) (see Theorem 3.1 below).

We show that the operator $\mathcal{PT}$ intertwines the dual pairs (A_+, B_+), and (A_-, B_-), i.e.,

$$\mathcal{PT}A_+ = A_-\mathcal{PT}, \qquad \mathcal{PT}B_+ = B_-\mathcal{PT}.$$

Due to our construction of the coupling these relations imply that the operator A is $\mathcal{PT}$*-symmetric*, i.e.,

$$\mathcal{PT}A = A\mathcal{PT}.$$

Moreover, the operator A turns out to be $\mathcal{P}$-symmetric in the Krein space $(\mathfrak{H}, [\cdot,\cdot])$ with the fundamental symmetry $\mathcal{P}$ in $\mathfrak{H} = L^2(\mathbb{R})$. In Trunk & Leben [**21**] it was shown that the extension H_0 of A defined as a restriction of the adjoint A^+ to the domain

$$\text{dom } H_0 = \left\{u_+ \oplus u_- \in \text{dom } A^+ \,|\, u_+(0) - u_-(0) = e^{-2i\phi}u_+'(0) - e^{2i\phi}u_-'(0) = 0\right\}$$

is a $\mathcal{PT}$-symmetric and $\mathcal{P}$-selfadjoint operator in the Krein space $(\mathfrak{H}, [\cdot,\cdot])$. Here A^+ stands for the adjoint with respect to the Krein space inner product $[\cdot,\cdot]$. In Theorem 7.2 we find a one-parameter family $\{H_\alpha\}_{\alpha\in\mathbb{R}}$ of $\mathcal{PT}$-symmetric and $\mathcal{P}$-selfadjoint extensions of A in the Krein space $(\mathfrak{H}, [\cdot,\cdot])$. with domain

$$\text{dom } H_\alpha = \left\{u_+ \oplus u_- \in \text{dom } A^+ \,|\, u_+(0) - u_-(0) = 0,\, e^{-2i\phi}u_+'(0) - e^{2i\phi}u_-'(0) = \alpha u_+(0)\right\}.$$

The result of Theorem 7.2 is based on the abstract construction of the coupling (A, B) of two dual pairs (A_+, B_+) and (A_-, B_-) in Theorem 3.1 and the description of all $\mathcal{PT}$-symmetric and $\mathcal{P}$-selfadjoint extensions of A given in Theorem 5.5.

Summing up, the results presented here promote the use of boundary triple techniques for dual pairs and techniques from Sturm–Liouville theory for complex potentials in the study of $\mathcal{PT}$-symmetric Quantum Mechanics. This is in a line with [**21**]. The here presented approach via dual pairs is the tool to deal with the various symmetries $\mathcal{P}$, $\mathcal{T}$ and $\mathcal{PT}$. It is the aim of this paper to recall those techniques and, hence, provide a mathematically sound setting of the (nowadays) classical Bender–Boettcher-theory.

Notation. By $\mathbb{R}_+$ (resp. $\mathbb{R}_-$) we denote the set of all positive (resp. negative) reals. For $z \in \mathbb{C}$, $\overline{z}$ denotes the complex conjugate of z and for a subset $U \subset \mathbb{C}$ we set $\overline{U} := \{\overline{z} |\, z \in U\}$.

All operators in this paper are densely defined linear operators in some Hilbert spaces. For such an operator T, we use the common notation dom T, ran T and $\ker T$ for the domain, the range and the null-space, respectively, of T. Moreover, as usual, $\rho(T)$, $\sigma(T)$ and $\sigma_p(T)$ stands for the resolvent set, the spectrum and the point spectrum, respectively, of T. The inner product in a Hilbert space is usually denoted by $\langle\cdot,\cdot\rangle$ and the adjoint of the operator T by T^*. The set of all bounded and everywhere defined operators in a Hilbert space $\mathfrak{H}$ is denoted by $\mathcal{L}(\mathfrak{H})$.

2. Preliminaries: Dual pairs of linear operators and Weyl functions

In this section we remind known facts about dual pairs of linear operators, their boundary triples and the corresponding Weyl functions from [**22**].

DEFINITION 2.1. *A pair (A, B) of densely defined closed linear operators A and B in a Hilbert space $\mathfrak{H}$ with inner product $\langle\cdot,\cdot\rangle$ is called a* dual pair, *if*

$$\langle Af, g\rangle - \langle f, Bg\rangle = 0 \quad \text{for all} \quad f \in \operatorname{dom} A, \quad g \in \operatorname{dom} B. \tag{2.1}$$

The equality (2.1) means that

$$A \subseteq B^*, \quad B \subseteq A^*. \tag{2.2}$$

Clearly, if (A, B) is a dual pair, then (B, A) is also a dual pair.

DEFINITION 2.2. *Let $\mathcal{H}_1$, $\mathcal{H}_2$ be auxiliary Hilbert spaces with inner products $\langle\cdot,\cdot\rangle_{\mathcal{H}_1}$ and $\langle\cdot,\cdot\rangle_{\mathcal{H}_2}$, respectively. Let*

$$\Gamma^B = \begin{pmatrix}\Gamma_1^B\\ \Gamma_2^B\end{pmatrix} : \operatorname{dom} B^* \to \begin{pmatrix}\mathcal{H}_1\\ \mathcal{H}_2\end{pmatrix}, \quad \Gamma^A = \begin{pmatrix}\Gamma_1^A\\ \Gamma_2^A\end{pmatrix} : \operatorname{dom} A^* \to \begin{pmatrix}\mathcal{H}_1\\ \mathcal{H}_2\end{pmatrix} \tag{2.3}$$

be linear operators. The triple $(\mathcal{H}_1 \times \mathcal{H}_2, \Gamma^A, \Gamma^B)$ is called a boundary triple for the dual pair *(A, B), if:*

(1) *the mappings Γ^B and Γ^A in (2.3) are surjective;*
(2) *the following identity holds for every $f \in \operatorname{dom} B^*$, $g \in \operatorname{dom} A^*$*

$$\langle B^* f, g\rangle - \langle f, A^* g\rangle = \langle\Gamma_1^B f, \Gamma_1^A g\rangle_{\mathcal{H}_1} - \langle\Gamma_2^B f, \Gamma_2^A g\rangle_{\mathcal{H}_2}, \tag{2.4}$$

It is easily seen that if a triple $(\mathcal{H}_1 \times \mathcal{H}_2, \Gamma^A, \Gamma^B)$ is a boundary triple for a dual pair (A, B), then the following identity holds

$$\langle A^* g, f\rangle - \langle g, B^* f\rangle = \langle\Gamma_2^A g, \Gamma_2^B f\rangle_{\mathcal{H}_2} - \langle\Gamma_1^A g, \Gamma_1^B f\rangle_{\mathcal{H}_1}, \quad f \in \operatorname{dom} B^*,\ g \in \operatorname{dom} A^*$$

and hence the triple

$$(\mathcal{H}_2 \times \mathcal{H}_1, (\Gamma^B)^T, (\Gamma^A)^T) := \left(\mathcal{H}_2 \times \mathcal{H}_1, \begin{pmatrix}\Gamma_2^B\\ \Gamma_1^B\end{pmatrix} \begin{pmatrix}\Gamma_2^A\\ \Gamma_1^A\end{pmatrix}\right) \tag{2.5}$$

is a boundary triple for the dual pair (B, A). The boundary triple (2.5) is called *transposed* with respect to the boundary triple $(\mathcal{H}_1 \times \mathcal{H}_2, \Gamma^A, \Gamma^B)$. Moreover, it follows from (2.4), $A^{**} = A$ and $B^{**} = B$ that

$$\operatorname{dom} A = \{f \in \operatorname{dom} B^* \,|\, \Gamma^B f = 0\} \quad \text{and} \quad \operatorname{dom} B = \{f \in \operatorname{dom} A^* \,|\, \Gamma^A f = 0\}.$$

A linear operator $\widetilde{A}$ is called a proper extension of A and is written $\widetilde{A} \in \operatorname{Ext}(A, B)$, if

$$A \subset \widetilde{A} \subset B^* \quad \text{and} \quad A \neq \widetilde{A}. \tag{2.6}$$

Define the proper extensions $A_1, A_2 \in \operatorname{Ext}(A, B)$ of A as restrictions of B^* to the sets

$$\operatorname{dom} A_1 = \{f \in \operatorname{dom} B^* \,|\, \Gamma_1^B f = 0\}, \quad \text{and} \quad \operatorname{dom} A_2 = \{f \in \operatorname{dom} B^* \,|\, \Gamma_2^B f = 0\}. \tag{2.7}$$

Similarly, the proper extensions $B_1, B_2 \in \operatorname{Ext}(B, A)$ of B are defined as restrictions of A^*,

$$\operatorname{dom} B_1 = \{g \in \operatorname{dom} A^* \,|\, \Gamma_1^A g = 0\} \quad \text{and} \quad \operatorname{dom} B_2 = \{g \in \operatorname{dom} A^* \,|\, \Gamma_2^A g = 0\}. \tag{2.8}$$

In the following lemma we collect some statements from **[22]** and provide short proofs in the present notations for the convenience of the reader.

LEMMA 2.3. *Let us consider* dom A^* *and* dom B^* *as Hilbert spaces with the graph norms*

$$(\|g\|^2 + \|A^*g\|^2)^{1/2}, \quad (\|f\|^2 + \|B^*f\|^2)^{1/2}, \quad g \in \text{dom } A^*, \quad f \in \text{dom } B^*$$

and let $(\mathcal{H}_1 \times \mathcal{H}_2, \Gamma^A, \Gamma^B)$ *be a boundary triple for the dual pair* (A, B). *Then*

(i) *The operators* $\Gamma^A : \text{dom } A^* \to \mathcal{H}_1 \times \mathcal{H}_2$ *and* $\Gamma^B : \text{dom } B^* \to \mathcal{H}_1 \times \mathcal{H}_2$ *are bounded.*

(ii) $B_1 = A_2^*$. *Hence* $\rho(B_1) = \overline{\rho(A_2)}$ *and* (A_2, B_1) *is a dual pair.*

(iii) $B_2 = A_1^*$. *Hence* $\rho(B_2) = \overline{\rho(A_1)}$ *and* (A_1, B_2) *is a dual pair.*

(iv) *For every* $z \in \rho(A_j)$, $j = 1, 2$, *the following direct decomposition holds*

$$\text{dom } B^* = \text{dom } A_j \dotplus \mathfrak{N}_z(B^*), \quad \textit{where} \quad \mathfrak{N}_z(B^*) = \ker(B^* - zI), \tag{2.9}$$

and the mapping $\Gamma_j^B|_{\mathfrak{N}_z(B^*)} : \mathfrak{N}_z(B^*) \to \mathcal{H}_j$, $j = 1, 2$, *is boundedly invertible.*

(v) *For every* $z \in \rho(B_j)$, $j = 1, 2$, *the following direct decomposition holds*

$$\text{dom } A^* = \text{dom } B_j \dotplus \mathfrak{N}_z(A^*), \quad \textit{where} \quad \mathfrak{N}_z(A^*) = \ker(A^* - zI), \tag{2.10}$$

and the mapping $\Gamma_j^A|_{\mathfrak{N}_z(A^*)} : \mathfrak{N}_z(A^*) \to \mathcal{H}_j$, $j = 1, 2$, *is boundedly invertible.*

PROOF. (i) Notice first that the operator $\Gamma^A : \text{dom } A^* \to \mathcal{H}_1 \times \mathcal{H}^2$ is closable. Indeed, assume that

$$g_n \in \text{dom } A^*, \quad g_n \to 0, \quad A^*g_n \to 0, \quad \Gamma^A g_n \to \begin{pmatrix} h_1 \\ h_2 \end{pmatrix}.$$

Since $\text{ran } \Gamma^B = \mathcal{H}_1 \times \mathcal{H}_2$ there exists $f \in \text{dom } B^*$ such that $\Gamma^B f = \begin{pmatrix} h_1 \\ -h_2 \end{pmatrix}$. Considering (2.4) with $g := g_n$ and taking the limit as $n \to \infty$ we obtain

$$0 = \|h_1\|^2 + \|h_2\|^2 \Rightarrow h_1 = h_2 = 0.$$

Hence the operator $\Gamma^A : \text{dom } A^* \to \mathcal{H}_1 \times \mathcal{H}^2$ is closable and by the Closed Graph Theorem it is bounded.

(ii) The inclusion $B_1 \subseteq A_2^*$ follows from (2.4). Let $g \in \text{dom } A_2^*$. Then for every $f \in \text{dom } A_2$ one obtains from (2.4)

$$0 = \langle A_2^* g, f \rangle - \langle g, A_2 f \rangle = \langle \Gamma_1^A g, \Gamma_1^B f \rangle$$

Since $\Gamma_1^B(\text{dom } A_2) = \mathcal{H}_1$ this implies $\Gamma_1^A g = 0$ and hence $g \in \text{dom } B_1$. This proves (ii).

The proof of (iii) is similar.

(iv) Let $z \in \rho(A_j)$, $j = 1, 2$. Then for every $f \in \text{dom } B^*$ there exists $f_j \in \text{dom } A_j$ such that

$$(B^* - zI)f = (A_j - zI)f_j \Rightarrow f_z := f - f_j \in \mathfrak{N}_z(B^*)$$

and f admits the decomposition $f = f_j + f_z$. Clearly, the decomposition (2.9) is direct, since $\text{dom } A_j \cap \mathfrak{N}_z(B^*) = \{0\}$ for $z \in \rho(A_j)$. It follows from (2.9) and Definition 2.2 that for $z \in \rho(A_j)$ the mapping $\Gamma_j^B|_{\mathfrak{N}_z(B^*)} : \mathfrak{N}_z(B^*) \to \mathcal{H}_j$ is surjective and hence boundedly invertible.

The proof of (v) is similar. □

DEFINITION 2.4. *The operator functions*

$$\gamma(z) = (\Gamma_2^B|_{\mathfrak{N}_z(B^*)})^{-1} \text{ and } M(z) := \Gamma_1^B(\Gamma_2^B|_{\mathfrak{N}_z(B^*)})^{-1}, \quad z \in \rho(A_2) \tag{2.11}$$

are called the γ-field *and* the Weyl function, *respectively, of the dual pair* (A, B), *corresponding to the boundary triple* $(\mathcal{H}_1 \times \mathcal{H}_2, \Gamma^A, \Gamma^B)$.

Clearly, the operator functions

$$\gamma^T(z) = (\Gamma_1^A|_{\mathfrak{N}_z(A^*)})^{-1} \text{ and } M^T(z) := \Gamma_2^A(\Gamma_1^A|_{\mathfrak{N}_z(A^*)})^{-1}, \quad z \in \rho(B_1) \tag{2.12}$$

are the γ–field and the Weyl function, respectively, of the dual pair (B, A), corresponding to the transposed boundary triple $(\mathcal{H}_2 \times \mathcal{H}_1, (\Gamma^B)^T, (\Gamma^A)^T)$.

LEMMA 2.5. *Let* $(\mathcal{H}_1 \times \mathcal{H}_2, \Gamma^A, \Gamma^B)$ *be a boundary triple for the dual pair* (A, B) *and let* M, M^T, γ *and* γ^T *be defined by* (2.11) *and* (2.12). *Then the following statements hold.*

(i) *The Weyl functions* M *and* M^T *satisfy the relations*

$$M^T(\bar{z}) = M(z)^*, \quad z \in \rho(A_2) = \overline{\rho(B_1)}. \tag{2.13}$$

$$M(z) - M^T(\zeta)^* = (z - \bar{\zeta})\gamma^T(\zeta)^*\gamma(z), \quad z \in \rho(A_2), \zeta \in \rho(B_1). \tag{2.14}$$

(ii) *The following identities hold*

$$\Gamma_1^B(A_2 - zI)^{-1} = \gamma^T(\bar{z})^*, \quad z \in \rho(A_2); \tag{2.15}$$

$$\Gamma_2^A(B_1 - zI)^{-1} = \gamma(\bar{z})^*, \quad z \in \rho(B_1). \tag{2.16}$$

PROOF. (i) It follows from (2.4) with $f = \gamma(z)u \in \mathfrak{N}_z(B^*)$, $z \in \rho(A_2)$, $u \in \mathcal{H}_2$ and $g = \gamma^T(\zeta)v \in \mathfrak{N}_\zeta(A^*)$, $\zeta \in \rho(B_1)$, $v \in \mathcal{H}_1$ that

$$(z-\bar{\zeta}) \langle \gamma(z)u, \gamma^T(\zeta)v\rangle = \langle M(z)u, v\rangle_{\mathcal{H}_1} - \langle u, M^T(\zeta)v\rangle_{\mathcal{H}_2} = \langle (M(z) - M^T(\zeta)^*)u, v\rangle_{\mathcal{H}_1}.$$

This proves (2.14) and (2.13) when setting $\zeta = \bar{z}$.

(ii) Let us set in (2.4)

$$f = (A_2 - zI)^{-1}h, \quad g = \gamma^T(\bar{z})v \in \mathfrak{N}_{\bar{z}}(A^*), \quad h \in \mathfrak{H}, \quad v \in \mathcal{H}_1, \quad z \in \rho(A_2).$$

Since $\Gamma_2^B f = 0$ one obtains from (2.4)

$$\langle A_2(A_2 - zI)^{-1}h, \gamma^T(\bar{z})v\rangle - \langle (A_2 - zI)^{-1}h, \bar{z}\gamma^T(\bar{z})v\rangle = \langle \Gamma_1^B(A_2 - zI)^{-1}h, v\rangle_{\mathcal{H}_1}$$

and hence

$$\langle \Gamma_1^B(A_2 - zI)^{-1}h, v\rangle_{\mathcal{H}_1} = \langle \gamma^T(\bar{z})^*h, v\rangle_{\mathcal{H}_1}.$$

This proves (2.15). The proof of (2.16) is similar. □

In order to describe all proper extension $\widetilde{A} \in \operatorname{Ext}(A, B)$ of a dual pair (A, B), we are using the notion of a linear relation. Recall [**1**] that a linear relation Θ from $\mathcal{H}_1$ to $\mathcal{H}_2$ is understood as a linear subspace $\Theta \subset \mathcal{H}_1 \times \mathcal{H}_2$. For a linear relation Θ the symbols dom Θ, $\ker\Theta$, ran Θ and $\rho(\Theta)$ stand for the domain, kernel, range and the resolvent set, respectively, [**1**]. The adjoint Θ^* is the closed linear relation from $\mathfrak{H}_2$ to $\mathfrak{H}_1$ defined by

$$\Theta^* = \left\{ \begin{pmatrix} v_2 \\ v_1 \end{pmatrix} \in \mathfrak{H}_2 \times \mathfrak{H}_1 \,|\, (v_1, u_1)_{\mathfrak{H}_1} = (v_2, u_2)_{\mathfrak{H}_2}, \begin{pmatrix} u_1 \\ u_2 \end{pmatrix} \in \Theta \right\}. \tag{2.17}$$

If $\mathcal{H}_1 = \mathcal{H}_2$ then the linear relation Θ is called selfadjoint if $\Theta = \Theta^*$.

In what follows we suppose that $\mathcal{H}_1 = \mathcal{H}_2 = \mathcal{H} := \mathbb{C}^d$, $d \in \mathbb{N}$. Every linear relation in $\mathcal{H}$ of rank d can be defined by the equality (see [**13**])

$$\Theta = \ker \begin{pmatrix} C & D \end{pmatrix} = \left\{ \begin{pmatrix} u_1 \\ u_2 \end{pmatrix} \,|\, Cu_1 + Du_2 = 0 \right\}. \tag{2.18}$$

where $C, D \in \mathbb{C}^{d \times d}$ are two $d \times d$-matrices, such that

$$\det(CC^* + DD^*) \neq 0. \tag{2.19}$$

The adjoint linear relation Θ^* takes the form

$$\Theta^* = \operatorname{ran} \begin{pmatrix} D^* \\ -C^* \end{pmatrix} = \left\{ \begin{pmatrix} D^* v \\ -C^* v \end{pmatrix} \,|\, v \in \mathbb{C}^d \right\} \tag{2.20}$$

and therefore the linear relation Θ is self-adjoint if and only if

$$CD^* = DC^*. \tag{2.21}$$

For a linear relation Θ in $\mathcal{H}$ denote by A_Θ the proper extension of A defined on

$$\operatorname{dom} A_\Theta = \{ f \in \operatorname{dom} B^* \,|\Gamma^B f \in \Theta \} \tag{2.22}$$

as the restriction of B^*: $A_\Theta = B^* \restriction_{\operatorname{dom} A_\Theta} \in \operatorname{Ext}(A, B)$. The following statement describes spectral properties of A_Θ, cf. [**22**, Proposition 5.2 and Theorem 5.5].

LEMMA 2.6. *Let (A, B) be a dual pair in a Hilbert space $\mathfrak{H}$, let $(\mathcal{H}^2, \Gamma^A, \Gamma^B)$ be a boundary triple for the dual pair (A, B), $\mathcal{H} = \mathbb{C}^d$, let M be the corresponding Weyl function, let Θ be a linear relation in $\mathcal{H}$ defined by* (2.18), (2.19) *and $z \in \rho(A_2)$. Then the following statements hold.*

(i) *$A_\Theta^* \in \operatorname{Ext}(B, A)$ is the restriction of A^* to*

$$\operatorname{dom} A_\Theta^* = \{ g \in \operatorname{dom} A^* \,|\, (\Gamma^A)^T g \in \Theta^* \}. \tag{2.23}$$

(ii) *$z \in \sigma_p(A_\Theta) \iff 0 \in \sigma_p(CM(z) + D)$ and for such z*

$$\ker(A_\Theta - zI) = \gamma(z) \ker(CM(z) + D).$$

(iii) *$z \in \rho(A_\Theta) \iff 0 \in \rho(CM(z)+D)$ and for such z the resolvent $(A_\Theta - zI)^{-1}$ takes the form*

$$(A_\Theta - zI)^{-1} = (A_2 - zI)^{-1} - \gamma(z)(CM(z) + D)^{-1} C \gamma^T(\bar{z})^*, \quad z \in \rho(A_\Theta) \cap \rho(A_2). \tag{2.24}$$

The formula (2.24) is obtained by comparing [**22**, (5.14),(5.20)] with (2.18).

3. Coupling of dual pairs

Here we recall and extend some results for the coupling of two dual pairs and operators. Items (i)–(iii) in the following theorem are from [**15**]. For the sake of completeness, we also give a proof here.

THEOREM 3.1. *Let (A_+, B_+) and (A_-, B_-) be dual pairs in Hilbert spaces $\mathfrak{H}_+$ and $\mathfrak{H}_-$, respectively, let $(\mathcal{H}^2, \Gamma^{A_\pm}, \Gamma^{B_\pm})$ be a boundary triple for the dual pair $(A_\pm, B_\pm)$, $\mathcal{H} = \mathbb{C}^d$ and let $M_\pm$ be the corresponding Weyl function. Denote by A^* and B^* the restrictions of the operators $A_+^* \oplus A_-^*$ and $B_+^* \oplus B_-^*$, respectively, to the domains*

$$\operatorname{dom} A^* = \{ g_+ \oplus g_- | g_\pm \in \operatorname{dom} A_\pm^*,\, \Gamma_1^{A_+} g_+ = \Gamma_1^{A_-} g_- \}; \tag{3.1}$$

(3.2) $$\operatorname{dom} B^* = \{f_+ \oplus f_- | f_\pm \in \operatorname{dom} B_\pm^*,\ \Gamma_2^{B_+} f_+ = \Gamma_2^{B_-} f_-\}.$$

Then the following statements hold.

(i) *The operators $A := (A^*)^*$ and $B := (B^*)^*$ are restrictions of the operators B^* and A^*, respectively, to the domains*

(3.3)
$$\operatorname{dom} A = \{f_+ \oplus f_- | f_\pm \in \operatorname{dom} B_\pm^*,\ \Gamma_2^{B_+} f_+ = \Gamma_2^{B_-} f_- = \Gamma_1^{B_+} f_+ + \Gamma_1^{B_-} f_- = 0\};$$

(3.4) $$\operatorname{dom} B = \{g_+ \oplus g_- | g_\pm \in \operatorname{dom} A_\pm^*,\ \Gamma_1^{A_+} g_+ = \Gamma_1^{A_-} g_- = \Gamma_2^{A_+} g_+ + \Gamma_2^{A_-} g_- = 0\},$$

and (A, B) is a dual pair in $\mathfrak{H}_+ \oplus \mathfrak{H}_-$.

(ii) *The triple $(\mathcal{H}^2, \Gamma^A, \Gamma^B)$ with Γ^A and Γ^B defined on $g \in \operatorname{dom} A^*$ and $f \in \operatorname{dom} B^*$ by the equalities*

(3.5)
$$\Gamma^A g = \begin{pmatrix} \Gamma_1^A g \\ \Gamma_2^A g \end{pmatrix} := \begin{pmatrix} \Gamma_1^{A_+} g_+ \\ \Gamma_2^{A_+} g_+ + \Gamma_2^{A_-} g_- \end{pmatrix}, \quad \Gamma^B f = \begin{pmatrix} \Gamma_1^B f \\ \Gamma_2^B f \end{pmatrix} := \begin{pmatrix} \Gamma_1^{B_+} f_+ + \Gamma_1^{B_-} f_- \\ \Gamma_2^{B_+} f_+ \end{pmatrix},$$

is a boundary triple for (A, B).

(iii) *The operators A_2, $A_{+,2}$, $A_{-,2}$ and B_1, $B_{+,1}$ and $B_{-,1}$ defined by (2.7) and (2.8), respectively, are related by*

(3.6) $$A_2 = A_{+,2} \oplus A_{-,2}, \quad \text{and} \quad B_1 = B_{+,1} \oplus B_{-,1}.$$

The Weyl function M and the γ-field corresponding to the boundary triple $(\mathcal{H}^2, \Gamma^A, \Gamma^B)$ are given by

(3.7) $$M(z) = M_+(z) + M_-(z), \quad \gamma(z) = \begin{pmatrix} \gamma_+(z) \\ \gamma_-(z) \end{pmatrix}, \quad z \in \rho(A_2).$$

(iv) *The restrictions A_1 of B^* and B_2 of A^* defined in (2.7) and (2.8) have the domains*

(3.8) $$\operatorname{dom} A_1 = \{f_+ \oplus f_- | f_\pm \in \operatorname{dom} B_\pm^*,\ \Gamma_2^{B_+} f_+ - \Gamma_2^{B_-} f_- = \Gamma_1^{B_+} f_+ + \Gamma_1^{B_-} f_- = 0\}$$

(3.9) $$\operatorname{dom} B_2 = \{f_+ \oplus f_- | f_\pm \in \operatorname{dom} A_\pm^*,\ \Gamma_1^{A_+} f_+ - \Gamma_1^{A_-} f_- = \Gamma_2^{A_+} f_+ + \Gamma_2^{A_-} f_- = 0\}.$$

They are proper extensions of A and B, respectively, i.e., $A_1 \in \operatorname{Ext}(A, B)$, $B_2 \in \operatorname{Ext}(B, A)$, and fulfill $A_1 = B_2^*$.

(v) *If $z \in \rho(A_2)$, then*

(3.10) $$z \in \rho(A_1) \Leftrightarrow 0 \in \rho(M_+(z) + M_-(z))$$

and

(3.11)
$$(A_1 - zI)^{-1} = (A_2 - zI)^{-1} - \gamma(z)(M_+(z) + M_-(z))^{-1}\gamma^T(\bar{z})^*, \quad z \in \rho(A_1) \cap \rho(A_2).$$

Proof. (i)&(ii) Let $f = f_+ \oplus f_- \in \operatorname{dom}(B_+^* \oplus B_-^*)$, $g = g_+ \oplus g_- \in \operatorname{dom}(A_+^* \oplus A_-^*)$. Then it follows from the equalities

$$\langle B_+^* f_+, g_+ \rangle - \langle f_+, A_+^* g_+ \rangle = \langle \Gamma_1^{B_+} f_+, \Gamma_1^{A_+} g_+ \rangle_{\mathcal{H}_1} - \langle \Gamma_2^{B_+} f_+, \Gamma_2^{A_+} g_+ \rangle_{\mathcal{H}_2},$$
$$\langle B_-^* f_-, g_- \rangle - \langle f_-, A_-^* g_- \rangle = \langle \Gamma_1^{B_-} f_-, \Gamma_1^{A_-} g_- \rangle_{\mathcal{H}_1} - \langle \Gamma_2^{B_-} f_-, \Gamma_2^{A_-} g_- \rangle_{\mathcal{H}_2}.$$

that

(3.12)

$$\langle (B_+^* \oplus B_-^*)f, g\rangle - \langle f, (A_+^* \oplus A_-^*)g\rangle = \langle \Gamma_1^{B_+} f_+, \Gamma_1^{A_+} g_+\rangle_{\mathcal{H}} - \langle \Gamma_2^{B_+} f_+, \Gamma_2^{A_+} g_+\rangle_{\mathcal{H}} + \langle \Gamma_1^{B_-} f_-, \Gamma_1^{A_-} g_-\rangle_{\mathcal{H}} - \langle \Gamma_2^{B_-} f_-, \Gamma_2^{A_-} g_-\rangle_{\mathcal{H}}.$$

The equality (3.3) follows from (3.12) since the mappings $\Gamma^{A_\pm} : \operatorname{dom} A_\pm^* \to \mathcal{H}^2$ are surjective. Similarly, (3.4) follows from (3.12) since the mappings $\Gamma^{B_\pm} : \operatorname{dom} B_\pm^* \to \mathcal{H}^2$ are surjective. Next, for $f \in \operatorname{dom} B^*$, $g \in \operatorname{dom} A^*$ the equation (3.12) takes the form

(3.13)

$$\langle B^* f, g\rangle - \langle f, A^* g\rangle = \langle \Gamma_1^{B_+} f_+ + \Gamma_1^{B_-} f_-, \Gamma_1^{A_+} g_+\rangle_{\mathcal{H}} - \langle \Gamma_2^{B_+} f_+, \Gamma_2^{A_+} g_+ + \Gamma_2^{A_-} g_-\rangle_{\mathcal{H}}.$$

This proves that (A, B) is a dual pair in $\mathfrak{H}_+ \oplus \mathfrak{H}_-$ and that (ii) holds.

(iii) From (3.5) we obtain $\operatorname{dom} A_2 = \ker \Gamma_2^B = \ker \Gamma_2^{B_+}$ and, as $A_2 \subset B^*$ by (3.2), formula (3.6) follows. Moreover, (3.2) and (3.5) imply that the γ–field of (A, B) corresponding to the boundary triple $(\mathcal{H}^2, \Gamma^A, \Gamma^B)$ takes the form

$$\gamma(z) = \begin{pmatrix} \gamma_+(z) \\ \gamma_-(z) \end{pmatrix} \tag{3.14}$$

where $\gamma_\pm(z)$ are the γ–fields of $(A_\pm, B_\pm)$ corresponding to the boundary triples $(\mathcal{H}^2, \Gamma^{A_\pm}, \Gamma^{B_\pm})$. Now formula (3.7) follows from the definition of the Weyl function, see Definition 2.4.

(iv)&(v) The extension A_1 of A given by (3.8) coincides with the extension A_Θ of A defined by (2.22) with $\Theta = \ker\begin{pmatrix} C & D\end{pmatrix} = \ker\begin{pmatrix} I_{\mathcal{H}} & 0\end{pmatrix}$. Then the equivalence (3.10) follows from Lemma 2.6 (iii) and formula (3.11) follows from (2.24). □

DEFINITION 3.2. (1) *The dual pair (A, B) given by (3.3), (3.4) is called the* coupling *of the dual pairs (A_+, B_+) and (A_-, B_-) relative to the triples $(\mathcal{H}^2, \Gamma^{A_+}, \Gamma^{B_+})$ and $(\mathcal{H}^2, \Gamma^{A_-}, \Gamma^{B_-})$.*

(2) *The operator A_1 from (3.8) is called the* coupling *of the operators A_+ and A_- relative to the triples $(\mathcal{H}^2, \Gamma^{A_+}, \Gamma^{B_+})$, $(\mathcal{H}^2, \Gamma^{A_-}, \Gamma^{B_-})$ and the operator $A_2 = A_{+,2} \oplus A_{-,2}$ from (3.6) is called the* decoupled operator.

According to this definition the operator B_2 from (3.9) is the coupling of the operators B_+ and B_- relative to the triples

$$(\mathcal{H}^2, (\Gamma^{B_+})^T, (\Gamma^{A_+})^T), (\mathcal{H}^2, (\Gamma^{B_-})^T, (\Gamma^{A_-})^T),$$

see also (2.5). Since the operators A_1 and B_2 are connected by the formula $A_1 = B_2^*$, in what follows we will be interested only in the coupling A_1.

4. Real dual pairs and real boundary triples

Let $\mathcal{T}$ be a conjugation (time reversal) operator in $\mathfrak{H}$, i.e., $\mathcal{T}$ is antilinear, $\mathcal{T}^2 = I_{\mathfrak{H}}$ and

$$(\mathcal{T}f, \mathcal{T}g)_{\mathfrak{H}} = (g, f)_{\mathfrak{H}} \quad \text{for all} \quad f, g \in \mathfrak{H}. \tag{4.1}$$

In what follows, besides the conjugation $\mathcal{T}$ in $\mathfrak{H}$, we will also consider a conjugation in $\mathcal{H}$, which will be denoted by $j_{\mathcal{H}}$.

DEFINITION 4.1. *A dual pair* (A,B) *in* $\mathfrak{H}$ *is called* $\mathcal{T}$-real, *if*

$$\mathcal{T}\,\mathrm{dom}\, A = \mathrm{dom}\, B \quad \textit{and} \quad \mathcal{T}A = B\mathcal{T}. \tag{4.2}$$

A boundary triple $(\mathcal{H}^2, \Gamma^A, \Gamma^B)$ *for a* $\mathcal{T}$*-real pair* (A,B) *is called* $(j_{\mathcal{H}}, \mathcal{T})$-real, *if*

$$j_{\mathcal{H}}\Gamma_1^B = \Gamma_2^A\mathcal{T}, \quad j_{\mathcal{H}}\Gamma_2^B = \Gamma_1^A\mathcal{T}. \tag{4.3}$$

Observe that the conditions (4.2) are clearly equivalent to

$$\mathcal{T}\,\mathrm{dom}\, A^* = \mathrm{dom}\, B^* \quad \text{and} \quad \mathcal{T}A^* = B^*\mathcal{T}. \tag{4.4}$$

LEMMA 4.2. *Let* (A,B) *be a* $\mathcal{T}$*-real dual pair and let* $(\mathcal{H}^2, \Gamma^A, \Gamma^B)$ *be a* $(j_{\mathcal{H}}, \mathcal{T})$*-real boundary triple for* (A,B)*. Then the corresponding Weyl function* $M(z)$ *satisfies the condition*

$$M(z) = j_{\mathcal{H}} M(z)^* j_{\mathcal{H}}, \quad z \in \rho(A_2). \tag{4.5}$$

PROOF. Notice that for $f_{\bar z} \in \mathfrak{N}_{\bar z}(A^*)$ one has

$$\mathcal{T} f_{\bar z} \in \mathfrak{N}_z(B^*).$$

Indeed, for every $h \in \mathrm{dom}\, B$ one gets, by (4.2), $\mathcal{T}h \in \mathrm{dom}\, A$ and hence

$$\langle \mathcal{T} f_{\bar z}, (B-\bar z)h\rangle = \langle \mathcal{T}(B-\bar z)h, f_{\bar z}\rangle = \langle (A-z)\mathcal{T}h, f_{\bar z}\rangle = 0.$$

Moreover, by a similar argument, we have also $\mathcal{T}\mathfrak{N}_z(B^*) \subseteq \mathfrak{N}_{\bar z}(A^*)$ and therefore

$$\mathcal{T}\mathfrak{N}_{\bar z}(A^*) = \mathfrak{N}_z(B^*).$$

Next by (4.3), (2.11) and by (2.12), one obtains for $z \in \rho(A_2)$

$$\begin{aligned} M(z)\Gamma_2^B\mathcal{T} f_{\bar z} &= \Gamma_1^B \mathcal{T} f_{\bar z} = j_{\mathcal{H}}\Gamma_2^A f_{\bar z} \\ &= j_{\mathcal{H}} M^T(\bar z)\Gamma_1^A f_{\bar z} = j_{\mathcal{H}} M^T(\bar z) j_{\mathcal{H}} \Gamma_2^B \mathcal{T} f_{\bar z} \end{aligned} \tag{4.6}$$

Since $\Gamma_2^B\mathcal{T}\mathfrak{N}_{\bar z}(A^*) = \mathcal{H}$ for $z \in \rho(A_2) = \overline{\rho(B_1)}$, this implies

$$M(z) = j_{\mathcal{H}} M^T(\bar z) j_{\mathcal{H}}.$$

By (2.13), one obtains (4.5). □

In what follows we consider a Hilbert space $\mathfrak{H}$ decomposed into an orthogonal sum

$$\mathfrak{H} = \mathfrak{H}_+ \oplus \mathfrak{H}_- \tag{4.7}$$

of two subspaces $\mathfrak{H}_\pm$ with conjugations $\mathcal{T}_\pm \in \mathcal{L}(\mathfrak{H}_\pm)$. Then the orthogonal sum

$$\mathcal{T} = \mathcal{T}_+ \oplus \mathcal{T}_- \tag{4.8}$$

is a conjugation in $\mathfrak{H}$.

THEOREM 4.3. *Let* $\mathfrak{H}$ *be a Hilbert space with a conjugation* $\mathcal{T}$ *such that* (4.7) *and* (4.8) *hold. Moreover, let* $(A_\pm, B_\pm)$ *be* $\mathcal{T}_\pm$*-real dual pairs in Hilbert spaces* $\mathfrak{H}_\pm$*. Finally, with* $j_{\mathcal{H}}$ *a conjugation in* $\mathcal{H}$*, let* $(\mathcal{H}^2, \Gamma^{A_\pm}, \Gamma^{B_\pm})$ *be* $(j_{\mathcal{H}}, \mathcal{T}_\pm)$*-real boundary triples for* $(A_\pm, B_\pm)$*, and let*

$$A_0 := A_+ \oplus A_-, \quad B_0 := B_+ \oplus B_-. \tag{4.9}$$

Then the following statements hold.

(i) *The dual pair* (A_0, B_0) *is* $\mathcal{T}$*-real and the boundary triple* $((\mathcal{H}\oplus\mathcal{H})^2, \Gamma^{A_0}, \Gamma^{B_0})$ *with*

$$\Gamma^{A_0} := \begin{pmatrix} \Gamma_1^{A_0} \\ \Gamma_2^{A_0} \end{pmatrix} := \begin{pmatrix} \Gamma_1^{A_+} \oplus \Gamma_1^{A_-} \\ \Gamma_2^{A_+} \oplus \Gamma_2^{A_-} \end{pmatrix}, \quad \Gamma^{B_0} := \begin{pmatrix} \Gamma_1^{B_0} \\ \Gamma_2^{B_0} \end{pmatrix} := \begin{pmatrix} \Gamma_1^{B_+} \oplus \Gamma_1^{B_-} \\ \Gamma_2^{B_+} \oplus \Gamma_2^{B_-} \end{pmatrix} \tag{4.10}$$

is $(j_{\mathcal{H}\oplus\mathcal{H}}, \mathcal{T})$*-real, where* $j_{\mathcal{H}\oplus\mathcal{H}} := j_{\mathcal{H}} \oplus j_{\mathcal{H}}$.

(ii) *The coupling* (A, B) *of the dual pairs* (A_+, B_+) *and* (A_-, B_-) *constructed in* (3.3), (3.4) *is* $\mathcal{T}$*-real.*

(iii) *The boundary triple* $(\mathcal{H}^2, \Gamma^A, \Gamma^B)$ *from Theorem* 3.1 *is* $(j_{\mathcal{H}}, \mathcal{T})$*-real.*

(iv) *The dual pairs* (A_1, B_2) *and* (A_2, B_1) *are* $\mathcal{T}$*-real.*

PROOF. Since the dual pairs $(A_\pm, B_\pm)$ are $\mathcal{T}_\pm$-real, one has

$$\mathcal{T}_+ A_+ = B_+ \mathcal{T}_+, \quad \mathcal{T}_- A_- = B_- \mathcal{T}_-. \tag{4.11}$$

Since the boundary triples $(\mathcal{H} \oplus \mathcal{H}, \Gamma^{B_\pm}, \Gamma^{A_\pm})$ are $(j_{\mathcal{H}}, \mathcal{T})$-real, one has

$$\begin{aligned} j_{\mathcal{H}}\Gamma_1^{B_+} f_+ &= \Gamma_2^{A_+} \mathcal{T}_+ f_+, & j_{\mathcal{H}}\Gamma_1^{B_-} f_- &= \Gamma_2^{A_-} \mathcal{T}_- f_-, \\ j_{\mathcal{H}}\Gamma_2^{B_+} f_+ &= \Gamma_1^{A_+} \mathcal{T}_+ f_+, & j_{\mathcal{H}}\Gamma_2^{B_-} f_- &= \Gamma_1^{A_-} \mathcal{T}_- f_-, \end{aligned} \qquad f_\pm \in \operatorname{dom} B_\pm^*. \tag{4.12}$$

(i) From (4.12) we see $j_{\mathcal{H}\oplus\mathcal{H}}\Gamma_1^{B_0} = \Gamma_2^{A_0}\mathcal{T}$ and $j_{\mathcal{H}\oplus\mathcal{H}}\Gamma_2^{B_0} = \Gamma_1^{A_0}\mathcal{T}$. This shows the $\mathcal{T}$-realness of the dual pair (A_0, B_0) and the $(j_{\mathcal{H}\oplus\mathcal{H}}, \mathcal{T})$-realness of the boundary triple $((\mathcal{H} \oplus \mathcal{H})^2, \Gamma^{A_0}, \Gamma^{B_0})$ follows from (4.11) and (4.12).

(ii) It follows from (4.11) and (4.4) that

$$\mathcal{T}_\pm \operatorname{dom} B_\pm^* = \operatorname{dom} A_\pm^* \quad \text{and} \quad A_+^* \mathcal{T}_+ = \mathcal{T}_+ B_+^*, \quad A_-^* \mathcal{T}_- = \mathcal{T}_- B_-^* \tag{4.13}$$

Hence

$$(A_+^* \oplus A_-^*)\mathcal{T} = \mathcal{T}(B_+^* \oplus B_-^*).$$

Let $f = f_+ \oplus f_- \in \operatorname{dom} B^*$, $f_\pm \in \operatorname{dom} B_\pm^*$. Then, by (3.2),

$$\Gamma_2^{B_+} f_+ = \Gamma_2^{B_-} f_-. \tag{4.14}$$

In view of (4.12) the condition (4.14) takes the form

$$j_{\mathcal{H}}\Gamma_1^{A_+}\mathcal{T}_+ f_+ = \Gamma_2^{B_+} f_+ = \Gamma_2^{B_-} f_- = j_{\mathcal{H}}\Gamma_1^{A_-}\mathcal{T}_- f_-$$

and, by (3.1), $\mathcal{T}f = \mathcal{T}_+ f_+ \oplus \mathcal{T}_- f_- \in \operatorname{dom} A^*$. Therefore,

$$A^*\mathcal{T}f = (A_+^*\mathcal{T}_+ f_+) \oplus (A_-^*\mathcal{T}_- f_-) = \mathcal{T}(B_+^* f_+ \oplus B_-^* f_-) = \mathcal{T}B^* f$$

and thus (A, B) is a $\mathcal{T}$-real dual pair in $\mathfrak{H}$.

(iii) Because of (ii) for $f \in \operatorname{dom} B^*$ we have $\mathcal{T}f \in \operatorname{dom} A^*$. It follows from (3.5) and (4.12) that

$$\Gamma_1^A \mathcal{T} f = \Gamma_1^{A_+}\mathcal{T}_+ f_+ = j_{\mathcal{H}}\Gamma_2^{B_+} f_+ = j_{\mathcal{H}}\Gamma_2^B f,$$

and

$$\Gamma_2^A \mathcal{T} f = \Gamma_2^{A_+}\mathcal{T}_+ f_+ + \Gamma_2^{A_-}\mathcal{T}_- f_- = j_{\mathcal{H}}(\Gamma_1^{B_+} f_+ + \Gamma_1^{B_-} f_-) = j_{\mathcal{H}}\Gamma_1^B f.$$

Thus the boundary triple $(\mathcal{H}^2, \Gamma^A, \Gamma^B)$ is real.

(iv) Assume that $f \in \operatorname{dom} A_1 = \{f \in \operatorname{dom} B^* | \Gamma_1^B f = 0\}$. Then, due to (4.3), $\Gamma_2^A \mathcal{T} f = j_{\mathcal{H}}\Gamma_1^B f = 0$ and hence $\mathcal{T}f \in \operatorname{dom} B_2$. By (4.4),

$$\mathcal{T} \operatorname{dom} A_1 = \operatorname{dom} B_2 \quad \text{and} \quad \mathcal{T}B_2 = A_1\mathcal{T}. \tag{4.15}$$

Similarly, it is shown that

$$\mathcal{T} \operatorname{dom} A_2 = \operatorname{dom} B_1 \quad \text{and} \quad \mathcal{T}B_1 = A_2\mathcal{T}. \tag{4.16}$$

□

5. Parity and $\mathcal{PT}$-symmetric operators

DEFINITION 5.1. *Let $\mathfrak{H}_\pm$ be Hilbert spaces and $\mathfrak{H} = \mathfrak{H}_+ \oplus \mathfrak{H}_-$. A linear operator $\mathcal{P} \in \mathcal{L}(\mathfrak{H})$ will be called an* abstract parity operator, *if*

$$\mathcal{P} = \mathcal{P}^*, \quad \mathcal{P}^2 = I_{\mathfrak{H}} \quad \text{and} \quad \mathcal{P}\mathfrak{H}_\pm = \mathfrak{H}_\mp. \tag{5.1}$$

Now consider a Hilbert space $\mathfrak{H} = \mathfrak{H}_+ \oplus \mathfrak{H}_-$ with a parity operator $\mathcal{P}$ and a conjugation $\mathcal{T} \in \mathcal{L}(\mathfrak{H})$, such that

$$\mathcal{TP} = \mathcal{PT} \quad \text{and} \quad \mathcal{T}\mathfrak{H}_\pm = \mathfrak{H}_\pm. \tag{5.2}$$

The conditions (5.2) mean that the operator $\mathcal{T}$ admits the representation as an orthogonal sum $\mathcal{T} = \mathcal{T}_+ \oplus \mathcal{T}_-$ of two conjugations $\mathcal{T}_+$ and $\mathcal{T}_-$ in the Hilbert spaces $\mathfrak{H}_+$ and $\mathfrak{H}_-$, respectively. In what follows we identify the spaces $\mathfrak{H}_+$ and $\mathfrak{H}_-$ with the subspaces $\mathfrak{H}_+ \oplus \{0\}$ and $\{0\} \oplus \mathfrak{H}_-$ of $\mathfrak{H}$, and hence we will set

$$\mathcal{T}f_+ := \mathcal{T}\begin{pmatrix} f_+ \\ 0 \end{pmatrix} = \begin{pmatrix} \mathcal{T}_+ f_+ \\ 0 \end{pmatrix}, \quad \mathcal{T}f_- := \mathcal{T}\begin{pmatrix} 0 \\ f_- \end{pmatrix} = \begin{pmatrix} 0 \\ \mathcal{T}_- f_- \end{pmatrix} \quad \text{for all} \quad f_\pm \in \mathfrak{H}_\pm.$$

Similarly, we set

$$\mathcal{P}f_+ := \mathcal{P}\begin{pmatrix} f_+ \\ 0 \end{pmatrix}, \quad \mathcal{P}f_- := \mathcal{P}\begin{pmatrix} 0 \\ f_- \end{pmatrix} \quad \text{for all} \quad f_\pm \in \mathfrak{H}_\pm. \tag{5.3}$$

The operator $\mathcal{P}$ maps the subspace $\mathfrak{H}_+$ onto $\mathfrak{H}_-$, and vice versa. It acts by the formula

$$\mathcal{P}\begin{pmatrix} f_+ \\ f_- \end{pmatrix} = \begin{pmatrix} \mathcal{P}f_- \\ \mathcal{P}f_+ \end{pmatrix}, \quad f_\pm \in \mathfrak{H}_\pm. \tag{5.4}$$

We will say that the parity $\mathcal{P}$ *intertwines* two given dual pairs $(A_\pm, B_\pm)$ in two Hilbert spaces $\mathfrak{H}_\pm$, if

$$\mathcal{P}A_+ = B_-\mathcal{P}, \quad \mathcal{P}B_+ = A_-\mathcal{P}. \tag{5.5}$$

LEMMA 5.2. *Let $\mathcal{P}$ be a parity operator in $\mathfrak{H} = \mathfrak{H}_+ \oplus \mathfrak{H}_-$ and let $\mathcal{T}$ be a conjugation in $\mathfrak{H}$, such that (5.2) holds. Let $(A_\pm, B_\pm)$ be $\mathcal{T}_\pm$-real dual pairs in Hilbert spaces $\mathfrak{H}_\pm$ intertwined by the parity $\mathcal{P}$. With $\jmath_{\mathcal{H}}$ a conjugation in $\mathcal{H}$ let $(\mathcal{H}^2, \Gamma^{A_\pm}, \Gamma^{B_\pm})$ be $(\jmath_{\mathcal{H}}, \mathcal{T})$-real boundary triples for $(A_\pm, B_\pm)$, such that*

$$\begin{pmatrix} \Gamma_1^{B_+} \\ \Gamma_2^{B_+} \end{pmatrix} f_+ = \begin{pmatrix} \Gamma_2^{A_-} \\ \Gamma_1^{A_-} \end{pmatrix} \mathcal{P}f_+, \quad \begin{pmatrix} \Gamma_1^{B_-} \\ \Gamma_2^{B_-} \end{pmatrix} f_- = \begin{pmatrix} \Gamma_2^{A_+} \\ \Gamma_1^{A_+} \end{pmatrix} \mathcal{P}f_-, \quad f_\pm \in \operatorname{dom} B_\pm^*, \tag{5.6}$$

let $M_\pm$ be the Weyl functions of the operators $A_\pm$ corresponding to the boundary triples $(\mathcal{H}^2, \Gamma^{A_\pm}, \Gamma^{B_\pm})$ and let $A_{\pm,j}$ be the restrictions of $B_\pm^*$ to the sets (cf. (2.7))

$$\operatorname{dom} A_{\pm,j} = \ker \Gamma_j^{B_\pm}, \quad j = 1, 2. \tag{5.7}$$

Then the following statements hold.

(i) *$\mathcal{P} \operatorname{dom} A_+^* = \operatorname{dom} B_-^*$, $\mathcal{P} \operatorname{dom} B_+^* = \operatorname{dom} A_-^*$ and*

$$\mathcal{P}A_+^* = B_-^*\mathcal{P}, \quad \mathcal{P}B_+^* = A_-^*\mathcal{P}. \tag{5.8}$$

(ii) *$\mathcal{PT} \operatorname{dom} A_+ = \operatorname{dom} A_-$, $\mathcal{PT} \operatorname{dom} B_+ = \operatorname{dom} B_-$ and*

$$\mathcal{PT}A_+ = A_-\mathcal{PT}, \quad \mathcal{PT}B_+ = B_-\mathcal{PT}. \tag{5.9}$$

(iii) *$\mathcal{PT} \operatorname{dom} A_+^* = \operatorname{dom} A_-^*$, $\mathcal{PT} \operatorname{dom} B_+^* = \operatorname{dom} B_-^*$ and*

$$\mathcal{PT}A_+^* = A_-^*\mathcal{PT}, \quad \mathcal{PT}B_+^* = B_-^*\mathcal{PT}, \quad \mathcal{PT}A_{+,j} = A_{-,j}\mathcal{PT}, \quad j = 1, 2. \tag{5.10}$$

(iv) *The Weyl functions M_+ and M_- of the operators A_+ and A_- are related by*

$$M_+(z) = M_-(\bar{z})^*, \quad z \in \rho(A_{+,2}) = \overline{\rho(A_{-,2})}. \tag{5.11}$$

PROOF. (i) Applying $\mathcal{P}$ to the left and right hand sides of the equalities in (5.5) and using the identity $\mathcal{P}^2 = I$ yields that $A_+\mathcal{P} = \mathcal{P}B_-$ and $B_+\mathcal{P} = \mathcal{P}A_-$. From these identities the statement (i) is immediate.

(ii) Since the dual pairs $(A_\pm, B_\pm)$ are real with respect to $\mathcal{T}_\pm$, one has (4.11). Let $f_+ \in \operatorname{dom} A_+$. Then, by (4.11), $\mathcal{T}_+f_+ \in \operatorname{dom} B_+$ and $B_+\mathcal{T}_+f_+ = \mathcal{T}_+A_+f_+$. By (5.5),

$$\mathcal{PT}_+f_+ \in \operatorname{dom} A_- \quad \text{and} \quad A_-\mathcal{PT}_+f_+ = \mathcal{P}B_+\mathcal{T}_+f_+ = \mathcal{PT}_+A_+f_+.$$

The proofs of the inclusion $\operatorname{dom} A_- \subseteq \mathcal{PT} \operatorname{dom} A_+$ and of the second equality in (5.9) are similar.

(iii) The equalities $\mathcal{PT}A_+^* = A_-^*\mathcal{PT}$, $\mathcal{PT}B_+^* = B_-^*\mathcal{PT}$ follow from (4.13) and item (i). These equalities imply (5.10) since, by (5.6) and (4.12), we have for $f_+ \in \operatorname{dom} B_+^*$, $g_+ \in \operatorname{dom} A_+^*$

(5.12)

$$\begin{aligned}
\Gamma_1^{B_-}(\mathcal{PT}f_+) &= \Gamma_2^{A_+}\mathcal{T}f_+ = \jmath_{\mathcal{H}}\Gamma_1^{B_+}f_+, \qquad & \Gamma_1^{A_-}(\mathcal{PT}g_+) &= \Gamma_2^{B_+}\mathcal{T}g_+ = \jmath_{\mathcal{H}}\Gamma_1^{A_+}g_+,\\
\Gamma_2^{B_-}(\mathcal{PT}f_+) &= \Gamma_1^{A_+}\mathcal{T}f_+ = \jmath_{\mathcal{H}}\Gamma_2^{B_+}f_+, \qquad & \Gamma_2^{A_-}(\mathcal{PT}g_+) &= \Gamma_1^{B_+}\mathcal{T}g_+ = \jmath_{\mathcal{H}}\Gamma_2^{A_+}g_+.
\end{aligned}$$

(iv) In particular, we obtain due to (iii) that

$$\rho(A_{+,1}) = \overline{\rho(A_{-,1})}, \quad \rho(A_{+,2}) = \overline{\rho(A_{-,2})}. \tag{5.13}$$

For every $z \in \rho(A_{+,2})$ and $f_z^+ \in \mathfrak{N}_z(B_+^*)$ we get from (iii) $f_{\bar{z}}^- := \mathcal{PT}f_z^+ \in \mathfrak{N}_{\bar{z}}(B_-^*)$, where we used $\mathcal{PT}zf_z^+ = \bar{z}\mathcal{PT}f_z^+$ which follows from (4.1). By Definition 2.4 the Weyl functions M_+ and M_- satisfy the equalities

$$M_+(z)\Gamma_2^{B_+}f_z^+ = \Gamma_1^{B_+}f_z^+, \quad M_-(\bar{z})\Gamma_2^{B_-}f_{\bar{z}}^- = \Gamma_1^{B_-}f_{\bar{z}}^-.$$

By (5.12)

$$\begin{aligned}
\Gamma_2^{B_-}f_{\bar{z}}^- &= \Gamma_2^{B_-}(\mathcal{PT}f_z^+) = \Gamma_1^{A_+}\mathcal{T}f_z^+ = \jmath_{\mathcal{H}}\Gamma_2^{B_+}f_z^+,\\
\Gamma_1^{B_-}f_{\bar{z}}^- &= \Gamma_1^{B_-}(\mathcal{PT}f_z^+) = \Gamma_2^{A_+}\mathcal{T}f_z^+ = \jmath_{\mathcal{H}}\Gamma_1^{B_+}f_z^+.
\end{aligned}$$

Hence

$$M_-(\bar{z})(\jmath_{\mathcal{H}}\Gamma_2^{B_+}f_z^+) = M_-(\bar{z})(\Gamma_2^{B_-}f_{\bar{z}}^-) = \Gamma_1^{B_-}f_{\bar{z}}^- = \jmath_{\mathcal{H}}\Gamma_1^{B_+}f_z^+ = \jmath_{\mathcal{H}}M_+(z)\Gamma_2^{B_+}f_z^+$$

and thus $M_-(\bar{z}) = \jmath_{\mathcal{H}}M_+(z)\jmath_{\mathcal{H}}$. In view of (4.5), this proves (iv). □

DEFINITION 5.3. *A closed linear operator A in $\mathfrak{H}$ is said to be $\mathcal{PT}$*-symmetric *if for all $f \in \operatorname{dom} A$ we have*

$$\mathcal{PT}f \in \operatorname{dom} A \quad \textit{and} \quad \mathcal{PT}Af = A\mathcal{PT}f.$$

A dual pair (A, B) is said to be $\mathcal{PT}$-symmetric, *if both A and B are $\mathcal{PT}$*-symmetric.

THEOREM 5.4. *Let $\mathcal{P}$ be a parity operator in $\mathfrak{H} = \mathfrak{H}_+ \oplus \mathfrak{H}_-$, let $\mathcal{T}$ be a conjugation in $\mathfrak{H}$, such that* (5.2) *holds and let $(A_\pm, B_\pm)$ be $\mathcal{T}_\pm$-real dual pairs in the Hilbert spaces $\mathfrak{H}_\pm$ intertwined by the parity $\mathcal{P}$. With $\jmath_{\mathcal{H}}$ a conjugation in $\mathcal{H} = \mathbb{C}^d$ let $(\mathcal{H}^2, \Gamma^{A_\pm}, \Gamma^{B_\pm})$ be a $(j_{\mathcal{H}}, \mathcal{T})$-real boundary triples for $(A_\pm, B_\pm)$, such that* (5.6) *holds. Moreover, let the dual pair (A_0, B_0) as in* (4.9) *and let (A, B) be the coupling* *of the dual pairs* (A_+, B_+) *and* (A_-, B_-) *relative to the triples* $(\mathcal{H}^2, \Gamma^{A_+}, \Gamma^{B_+})$ *and* $(\mathcal{H}^2, \Gamma^{A_-}, \Gamma^{B_-})$, *cf.* (3.3), (3.4). *Then the following holds.*

(i) *The dual pair* (A_0, B_0) *is* $\mathcal{PT}$*-symmetric. Moreover, the operators* A_0^*, B_0^* *are* $\mathcal{PT}$*-symmetric.*

(ii) *The dual pair* (A, B) *is* $\mathcal{PT}$*-symmetric. Moreover, the operators* A^*, B^* *are* $\mathcal{PT}$*-symmetric.*

(iii) *The dual pair* (A_1, B_2) *is* $\mathcal{PT}$*-symmetric.*

PROOF. (i) Notice first that, by Lemma 5.2 (i),

$$\mathcal{P}\,\mathrm{dom}\, B_0^* = \mathcal{P}\,\mathrm{dom}\,(B_+^* \oplus B_-^*) = \mathrm{dom}\,(A_-^* \oplus A_+^*) = \mathrm{dom}\, A_0^*. \tag{5.14}$$

We obtain, by (5.4) and (5.8), for $f = \begin{pmatrix} f_+ & f_- \end{pmatrix}^T$, $f_\pm \in \mathrm{dom}\, B_\pm^*$

$$A_0^*\mathcal{P}f = \begin{pmatrix} A_+^* & 0 \\ 0 & A_-^* \end{pmatrix} \mathcal{P} \begin{pmatrix} f_+ \\ f_- \end{pmatrix} = \begin{pmatrix} A_+^*\mathcal{P}f_- \\ A_-^*\mathcal{P}f_+ \end{pmatrix} = \begin{pmatrix} \mathcal{P}B_-^*f_- \\ \mathcal{P}B_+^*f_+ \end{pmatrix} = \mathcal{P}B_0^*f. \tag{5.15}$$

It follows that

$$\mathcal{P}\,\mathrm{dom}\, B_0^* = \mathrm{dom}\, A_0^* \quad \text{and} \quad A_0^*\mathcal{P} = \mathcal{P}B_0^*. \tag{5.16}$$

Next, by (4.13), $\mathcal{PT}\,\mathrm{dom}\, A_0^* = \mathrm{dom}\, A_0^*$ and

$$A_0^*(\mathcal{PT})f = A_0^*\mathcal{P}\begin{pmatrix} \mathcal{T}_+f_+ \\ \mathcal{T}_-f_- \end{pmatrix} = \mathcal{P}B_0^*\begin{pmatrix} \mathcal{T}_+f_+ \\ \mathcal{T}_-f_- \end{pmatrix} = (\mathcal{PT})A_0^*f \tag{5.17}$$

and the operator A_0^* is $\mathcal{PT}$-symmetric. By (5.9), $A_0(\mathcal{PT}) = (\mathcal{PT})A_0$, i.e., the operator A_0 is $\mathcal{PT}$-symmetric. The proofs of the equalities $B_0(\mathcal{PT}) = (\mathcal{PT})B_0$ and $B_0^*(\mathcal{PT}) = (\mathcal{PT})B_0^*$ are similar.

(ii) Due to (3.1) for $g = g_+ \oplus g_- \in \mathrm{dom}\, A^*$ we have $\Gamma_1^{A_+}g_+ = \Gamma_1^{A_-}g_-$ and hence, by (5.12), the following equalities holds

$$\Gamma_1^{A_+}(\mathcal{PT})g_- = \jmath_{\mathcal{H}}\Gamma_1^{A_-}g_- = \jmath_{\mathcal{H}}\Gamma_1^{A_+}g_+ = \Gamma_1^{A_-}(\mathcal{PT})g_+. \tag{5.18}$$

Therefore, $(\mathcal{PT})g = (\mathcal{PT})g_- \oplus (\mathcal{PT})g_+ \in \mathrm{dom}\, A^*$ and, by (5.17),

$$A^*(\mathcal{PT})g = A_0^*(\mathcal{PT})g = (\mathcal{PT})A_0^*g = (\mathcal{PT})A^*g. \tag{5.19}$$

Since the operator $\mathcal{PT} : \mathcal{H} \to \mathcal{H}$ is a bijection, we have $\mathcal{PT}\,\mathrm{dom}\, A^* = \mathrm{dom}\, A^*$ and

$$A^*(\mathcal{PT}) = (\mathcal{PT})A^*. \tag{5.20}$$

By straightforward calculations we derive from (5.20) that $A(\mathcal{PT}) = (\mathcal{PT})A$, i.e., the operator A is $\mathcal{PT}$-symmetric. Similarly, we can show that

$$\mathcal{PT}\,\mathrm{dom}\, B^* = \mathrm{dom}\, B^* \quad \text{and} \quad B^*(\mathcal{PT})f = (\mathcal{PT})B^*f \quad \text{for all} \quad f \in \mathrm{dom}\, B^*. \tag{5.21}$$

This implies the equality $B(\mathcal{PT}) = (\mathcal{PT})B$ and hence B and B^* are $\mathcal{PT}$-symmetric.

(iii) In view of (3.2), (3.3), (3.4) and (5.6), for $f = f_+ \oplus f_- \in \mathrm{dom}\, B^*$ the following equivalences hold

$$f \in \mathrm{dom}\, A_1 \Leftrightarrow \Gamma_1^{B_+}f_+ = -\Gamma_1^{B_-}f_- \Leftrightarrow \Gamma_2^{A_-}\mathcal{P}f_+ = -\Gamma_2^{A_+}\mathcal{P}f_- \Leftrightarrow \mathcal{P}f \in \mathrm{dom}\, B_2.$$

Therefore, by (5.19), $B_2\mathcal{P} = \mathcal{P}A_1$ and hence, by Theorem 4.3,

$$B_2(\mathcal{PT})f = \mathcal{P}A_1\mathcal{T}f = \mathcal{PT}B_2f$$

for all $f \in \mathrm{dom}\, B_2$. Since $A_1 = B_2^*$, this proves also that

$$A_1(\mathcal{PT}) = \mathcal{PT}A_1.$$

□

In the following theorem we consider a dual pair (A, B) from Theorem 5.4 and characterize $\mathcal{PT}$-symmetric proper extensions $\widetilde{A} \in \text{Ext}(A, B)$ of A. Notice, that every proper extension $\widetilde{A}$ of A of rank d, (i.e., factor space $\text{dom}\,\widetilde{A}/\,\text{dom}\,A$ has dimension d) can be represented in the form (2.22): $\widetilde{A} = A_\Theta := B^* \upharpoonright \text{dom}\,A_\Theta$, where Θ is a linear relation in $\mathbb{C}^d$ with the kernel representation

$$\Theta = \ker\begin{pmatrix} C & D \end{pmatrix}, \quad C, D \in \mathbb{C}^{d\times d}, \quad \text{and (2.19) holds.} \tag{5.22}$$

Substitutions of (5.22) in (2.22) shows that $\text{dom}\,A_\Theta$ has the form

$$\text{dom}\,A_\Theta = \{f \in \text{dom}\,B^* | C\Gamma_1^B f + D\Gamma_2^B f = 0\}. \tag{5.23}$$

THEOREM 5.5. *Let in the assumptions of Theorem 5.4 the dual pair (A, B) and the boundary triple $(\mathbb{C}^d \times \mathbb{C}^d, \Gamma^A, \Gamma^B)$ be given by (3.3)–(3.5), let $C, D \in \mathbb{C}^{d\times d}$ and let $A_\Theta \in Ext(A, B)$ be the proper extension of A defined by (5.23). Then A_Θ is $\mathcal{PT}$-symmetric if and only if $\Theta = \jmath_\mathcal{H}\Theta\jmath_\mathcal{H}$, i.e.,* $\ker\begin{pmatrix} C\jmath_\mathcal{H} & D\jmath_\mathcal{H} \end{pmatrix} = \ker\begin{pmatrix} C & D \end{pmatrix}$.

PROOF. By Theorems 4.3 and 5.4, the dual pair (A, B) is $\mathcal{T}$-real and $\mathcal{PT}$-symmetric. It follows from (5.21) that $\mathcal{PT}\,\text{dom}\,B^* = \text{dom}\,B^*$. By (5.12) and (3.2), we obtain for $f \in \text{dom}\,B^*$

$$\begin{pmatrix} \Gamma_1^B(\mathcal{PT}f) \\ \Gamma_2^B(\mathcal{PT}f) \end{pmatrix} = \begin{pmatrix} \Gamma_1^{B_-}\mathcal{PT}f_+ + \Gamma_1^{B_+}\mathcal{PT}f_- \\ \Gamma_2^{B_+}\mathcal{PT}f_- \end{pmatrix} = \begin{pmatrix} \jmath_\mathcal{H}\Gamma_1^{B_+}f_+ + \jmath_\mathcal{H}\Gamma_1^{B_-}f_- \\ \jmath_\mathcal{H}\Gamma_2^{B_+}f_+ \end{pmatrix} = \begin{pmatrix} \jmath_\mathcal{H}\Gamma_1^B f \\ \jmath_\mathcal{H}\Gamma_2^B f \end{pmatrix}$$

Therefore,

$$\Gamma^B\mathcal{PT}f \in \Theta = \ker\begin{pmatrix} C & D \end{pmatrix} \iff \Gamma^B f \in \ker\jmath_\mathcal{H}\Theta\jmath_\mathcal{H} = \ker\begin{pmatrix} C\jmath_\mathcal{H} & D\jmath_\mathcal{H} \end{pmatrix}. \tag{5.24}$$

Now it follows from (5.21) that for $f \in \text{dom}\,A_\Theta$

$$A_\Theta(\mathcal{PT})f = B^*(\mathcal{PT})f = (\mathcal{PT})B^*f = (\mathcal{PT})A_\Theta f.$$

This proves the claim. □

REMARK 5.6. *In the special case of Theorem 5.5 when $d = 1$ and $\jmath_\mathcal{H}$ is the standard complex conjugation in $\mathbb{C}$, the set of all $\mathcal{PT}$-symmetric extensions of A other than the decoupled operator $A_{+,2} \oplus A_{-,2}$ is parameterized by a real parameter $\alpha \in \mathbb{R}$ via*

$$A_\alpha = B^* \upharpoonright_{\text{dom}\,A_\alpha}, \quad \text{dom}\,A_\alpha = \left\{u \in \text{dom}\,B^* \,|\, \Gamma_1^B u = \alpha\Gamma_2^B u\right\}. \tag{5.25}$$

Note that the decoupled operator $A_{+,2} \oplus A_{-,2}$ is also $\mathcal{PT}$-symmetric which follows from Theorem 5.5 with $C = 0$ and $D = 1$.

Next we present a description of $\mathcal{PT}$-symmetric proper extensions of the $\mathcal{PT}$-symmetric operator $A_0 = A_+ \oplus A_-$, see Theorem 4.3.

THEOREM 5.7. *Let in the assumptions of Theorem 5.4 the dual pair (A_0, B_0) and the boundary triple $(\mathbb{C}^{2d} \times \mathbb{C}^{2d}, \Gamma^{A_0}, \Gamma^{B_0})$ be given by (4.9), (4.10) let $\widehat{C}, \widehat{D}$ be $2d \times 2d$-matrices such that (2.19) holds, and let $\widehat{\Theta}$ be a linear relation in $\mathcal{H}^2$, $\mathcal{H} := \mathbb{C}^{2d}$ with the kernel representation $\widehat{\Theta} = \ker\begin{pmatrix} \widehat{C} & \widehat{D} \end{pmatrix}$. Then the extension $(A_0)_{\widehat{\Theta}} = B_0^* \upharpoonright \text{dom}\,(A_0)_{\widehat{\Theta}} \in \text{Ext}(A, B)$ of A_0 with*

$$\text{dom}\,(A_0)_{\widehat{\Theta}} = \{f \in \text{dom}\,B_0^* |\, \widehat{C}\Gamma_1^{B_0} f + \widehat{D}\Gamma_2^{B_0} f = 0\} \tag{5.26}$$

is $\mathcal{PT}$-symmetric if and only if $\widehat{\Theta} = \widehat{\jmath}_\mathcal{H}\widehat{\Theta}\widehat{\jmath}_\mathcal{H}$, i.e.,

$$\ker\begin{pmatrix} \widehat{C}\widehat{\jmath}_\mathcal{H} & \widehat{D}\widehat{\jmath}_\mathcal{H} \end{pmatrix} = \ker\begin{pmatrix} \widehat{C} & \widehat{D} \end{pmatrix} \quad \textit{where } \widehat{\jmath}_\mathcal{H} := \begin{pmatrix} O_d & \jmath_\mathcal{H} \\ \jmath_\mathcal{H} & O_d \end{pmatrix}. \tag{5.27}$$

PROOF. Recall that $\Gamma_1^{B_0}$ and $\Gamma_2^{B_0}$ are given by formulas (4.10).

The invariance of dom B_0^* with respect to $\mathcal{PT}$ is shown in Theorem 5.4 (i). By (5.12) we obtain for $f = f_+ \oplus f_-, g = g_+ \oplus g_- \in \text{dom } B_0^*$

$$\Gamma_1^{B_0}(\mathcal{PT}f) = \begin{pmatrix} \Gamma_1^{B_+}\mathcal{PT}f_- \\ \Gamma_1^{B_-}\mathcal{PT}f_+ \end{pmatrix} = \begin{pmatrix} \mathcal{J}_\mathcal{H}\Gamma_1^{B_-}f_- \\ \mathcal{J}_\mathcal{H}\Gamma_1^{B_+}f_+ \end{pmatrix} = \widehat{\mathcal{J}}_\mathcal{H}\begin{pmatrix} \Gamma_1^{B_+}f_+ \\ \Gamma_1^{B_-}f_- \end{pmatrix} = \widehat{\mathcal{J}}_\mathcal{H}\Gamma_1^{B_0}f$$

and

$$\Gamma_2^{B_0}(\mathcal{PT}f) = \begin{pmatrix} \Gamma_2^{B_+}\mathcal{PT}f_- \\ \Gamma_2^{B_-}\mathcal{PT}f_+ \end{pmatrix} = \begin{pmatrix} \mathcal{J}_\mathcal{H}\Gamma_2^{B_-}f_- \\ \mathcal{J}_\mathcal{H}\Gamma_2^{B_+}f_+ \end{pmatrix} = \widehat{\mathcal{J}}_\mathcal{H}\begin{pmatrix} \Gamma_2^{B_+}f_+ \\ \Gamma_2^{B_-}f_- \end{pmatrix} = \widehat{\mathcal{J}}_\mathcal{H}\Gamma_2^{B_0}f.$$

Therefore, the condition $\mathcal{PT}f \in \text{dom}\,(A_0)_{\widehat{\Theta}}$ is equivalent to

$$\begin{pmatrix} \widehat{C}\widehat{\mathcal{J}}_\mathcal{H} & \widehat{D}\widehat{\mathcal{J}}_\mathcal{H} \end{pmatrix}\begin{pmatrix} \Gamma_1^0 f_+ \\ \Gamma_2^0 f_- \end{pmatrix} = 0. \tag{5.28}$$

Comparison of (5.26) and (5.28) yields (5.27). □

6. $\mathcal{P}$-symmetric and $\mathcal{P}$-selfadjoint extensions of the operator A_0

6.1. Boundary triples for $\mathcal{P}$-symmetric operators. In this section we describe $\mathcal{P}$-symmetric and $\mathcal{P}$-selfadjoint extensions of the operators A_0 and A and compare them with the descriptions of $\mathcal{PT}$-symmetric extensions of A_0 and A from the previous section.

In order to define $\mathcal{P}$-symmetry it is convenient to consider the abstract parity $\mathcal{P}$ from Section 5 as a Gramian which leads to the theory of Krein spaces. Krein spaces are Hilbert spaces equipped with an additional inner product. This additional inner product is a symmetric, non-degenerated sesquilinear form, which is in general not definite. Or, what is the same, the additional inner form is given via a selfadjoint, boundedly invertible Gramian, see [**2, 10**]. Here, $\mathcal{P}$ will be used as the Gramian. More precisely, let $(\mathfrak{H}_\pm, \langle\cdot,\cdot\rangle_{\mathfrak{H}_\pm})$ be two Hilbert spaces and let $\mathcal{P}$ be the abstract parity in $\mathfrak{H} = \mathfrak{H}_+ \oplus \mathfrak{H}_-$. As usual, we define the Hilbert space scalar product in $\mathfrak{H}$ as the sum of the two scalar products in $\mathfrak{H}_\pm$,

$$\langle f, g\rangle_\mathfrak{H} := \langle f_+, g_+\rangle_{\mathfrak{H}_+} + \langle f_-, g_-\rangle_{\mathfrak{H}_-} \quad \text{for } f = f_+ \oplus f_-, g = g_+ \oplus g_- \in \mathfrak{H}_+ \oplus \mathfrak{H}_-.$$

Then the pair $(\mathfrak{H}, [\cdot,\cdot])$ with the inner product given by

$$[f, g] := \langle \mathcal{P}f, g\rangle_\mathfrak{H} \tag{6.1}$$

forms a Krein space. Recall, that a densely defined linear operator A in (a Krein space) $(\mathfrak{H}, [\cdot,\cdot])$ is called *$\mathcal{P}$-symmetric*, if it is symmetric with respect to the new (Krein space) inner product, namely

$$[Af, g] = [f, Ag] \quad \text{for all} \quad f, g \in \text{dom } A.$$

Denote by A^+ the adjoint operator in $(\mathfrak{H}, [\cdot,\cdot])$, i.e., $A^+ = \mathcal{P}A^*\mathcal{P}$. For a $\mathcal{P}$-symmetric operator A one has $A \subseteq A^+$. The operator A is called *$\mathcal{P}$-selfadjoint* if $A = A^+$.

In what follows we will apply the two notions of $\mathcal{P}$-symmetry and $\mathcal{PT}$-symmetry to extensions of the operators A and A_0, defined in Theorem 3.1 and in Theorem 4.3, respectively. The coupling A_1 and the decoupled A_2, cf. Definition 3.2, are examples of operators which are simultaneously $\mathcal{P}$-selfadjoint and $\mathcal{PT}$-symmetric. We have the following chain of operators

$$A_0 = A_+ \oplus A_- \subset A \subset A_j \subset B^* \subset B_0^* = B_+^* \oplus B_-^* \quad j = 1, 2.$$

Note that Definition 2.1 of boundary triples is made for dual pairs. The following definition of a boundary triple for a single operator is from [**12**]. In the case of a Hilbert space symmetric operator it was introduced in [**19**], see also [**16**]. We will use it to describe the extensions of the $\mathcal{P}$-symmetric operators A and A_0.

DEFINITION 6.1. *Let $\mathcal{H}$ be an auxiliary Hilbert space and let Γ_1, Γ_2 be linear operators from* dom A^+ *to $\mathcal{H}$. The triple $(\mathcal{H}, \Gamma_1, \Gamma_2)$ is called a* boundary triple *for the $\mathcal{P}$-symmetric operator A, if:*

(1) *the mapping $\Gamma := \begin{pmatrix} \Gamma_1 \\ \Gamma_2 \end{pmatrix}$ from* dom A^+ *to $\mathcal{H}^2$ is surjective;*
(2) *the following identity holds for every $f, g \in \operatorname{dom} A^+$*

$$[A^+ f, g] - [f, A^+ g] = \langle \Gamma_1 f, \Gamma_2 g \rangle_{\mathcal{H}} - \langle \Gamma_2 f, \Gamma_1 g \rangle_{\mathcal{H}}.$$

Remind [**12**], that given a boundary triple $(\mathcal{H}, \Gamma_1, \Gamma_2)$ for the $\mathcal{P}$-symmetric operator A, the set of all $\mathcal{P}$-selfadjoint extensions $\widehat{A}$ of A can be parametrized by the formula

$$\widehat{A} = A^+ \upharpoonright_{\operatorname{dom} \widehat{A}}, \quad \operatorname{dom} \widehat{A} = \{f \in \operatorname{dom} A^+ | \Gamma f \in \Phi\}, \tag{6.2}$$

where Φ ranges over the set of all selfadjoint relations in $\mathcal{H}$.

In the following theorem we characterize $\mathcal{P}$-selfadjoint extensions $\widehat{A}$ of the operator A_0 of the form (5.26).

THEOREM 6.2. *Let in the assumptions of Theorem 5.4 (A_0, B_0) be a $\mathcal{PT}$-symmetric dual pair, let $\widehat{C}, \widehat{D}$ be 2d $\times$ 2d-matrices such that* (2.19) *holds, and let $\widehat{\Theta} = \ker \begin{pmatrix} \widehat{C} & \widehat{D} \end{pmatrix}$ be a linear relation in $\mathcal{H}^2$, $\mathcal{H} := \mathbb{C}^d$ and let $(A_0)_{\widehat{\Theta}}$ be an extension of A_0 given by* (5.26). *Then the following hold.*

(i) *The operator A_0 is $\mathcal{P}$-symmetric and a boundary triple $(\mathbb{C}^{2d}, \Gamma_1^0, \Gamma_2^0)$ for A_0 can be chosen as*

$$\Gamma_1^0 f = \Gamma_1^{B_0}, \quad \Gamma_2^0 f = J_d \Gamma_2^{B_0}, \quad \textit{where } J_d := \begin{pmatrix} O_d & I_d \\ I_d & O_d \end{pmatrix}. \tag{6.3}$$

(ii) *The extension $(A_0)_{\widehat{\Theta}}$ of the operator A_0 given by* (5.26) *is $\mathcal{P}-$selfadjoint if and only if the linear relation $\Phi := J_d \widehat{\Theta} = \ker \begin{pmatrix} \widehat{C} & \widehat{D} J_d \end{pmatrix}$ is selfadjoint, i.e.,*

$$\widehat{C} J_d \widehat{D}^* - \widehat{D} J_d \widehat{C}^* = O_{2d}. \tag{6.4}$$

PROOF. (i) The $\mathcal{P}$-symmetry of A_0 follows from (5.16) and the fact that (A_0, B_0) forms a dual pair which imply

$$A_0^+ = \mathcal{P} A_0^* \mathcal{P} = B_0^* \supset A_0.$$

For $f = f_+ \oplus f_-, g = g_+ \oplus g_- \in \operatorname{dom} B_0^*$ we obtain with the help of (5.16), (3.12) and (5.6)

$$
\begin{aligned}
[A_0^+ f, g] - [f, A_0^+ g] &= \langle B_0^* f, \mathcal{P} g\rangle - \langle f, \mathcal{P} B_0^* g\rangle = \langle B_0^* f, \mathcal{P} g\rangle - \langle f, A_0^* \mathcal{P} g\rangle \\
&= \langle \Gamma_1^{B+} f_+, \Gamma_1^{A+} \mathcal{P} g_-\rangle_{\mathcal{H}} - \langle \Gamma_2^{B+} f_+, \Gamma_2^{A+} \mathcal{P} g_-\rangle_{\mathcal{H}} \\
&\quad + \langle \Gamma_1^{B-} f_-, \Gamma_1^{A-} \mathcal{P} g_+\rangle_{\mathcal{H}} - \langle \Gamma_2^{B-} f_-, \Gamma_2^{A-} \mathcal{P} g_+\rangle_{\mathcal{H}} \\
&= \langle \Gamma_1^{B+} f_+, \Gamma_2^{B-} g_-\rangle_{\mathcal{H}} - \langle \Gamma_2^{B+} f_+, \Gamma_1^{B-} g_-\rangle_{\mathcal{H}} \\
&\quad + \langle \Gamma_1^{B-} f_-, \Gamma_2^{B+} g_+\rangle_{\mathcal{H}} - \langle \Gamma_2^{B-} f_-, \Gamma_1^{B+} g_+\rangle_{\mathcal{H}} \\
&= \langle \Gamma_1^0 f, \Gamma_2^0 g\rangle_{\mathcal{H}^2} - \langle \Gamma_2^0 f, \Gamma_1^0 g\rangle_{\mathcal{H}^2}.
\end{aligned} \tag{6.5}
$$

So the triple $(\mathcal{H}, \Gamma_1^0, \Gamma_2^0)$ is a boundary triple for the $\mathcal{P}$-symmetric operator A_0.

(ii) The equivalence

$$(A_0)_{\widehat{\Theta}} = (A_0)_{\widehat{\Theta}}^+ \Longleftrightarrow \Phi = \Phi^*$$

is a general fact from [**12**]. The equivalence of the equality $\Phi = \Phi^*$ to (6.4) follows from (2.21). □

REMARK 6.3. *It follows from* (5.19), *that* $A \subseteq B^* = A^+$, *and hence the operator* A *is* $\mathcal{P}$*-symmetric. A boundary triple* $(\mathcal{H}, \Gamma_1, \Gamma_2)$ *for* A^+ *can be chosen as*

$$\mathcal{H} = \mathbb{C}^d, \quad \Gamma_1 f = \Gamma_1^B f := \Gamma_1^{B+} f_+ + \Gamma_1^{B-} f_-, \quad \Gamma_2 f = \Gamma_2^B f := \Gamma_2^{B+} f_+$$

for $f = f_+ \oplus f_- \in \operatorname{dom} B^*$. *For a linear relation* Θ *in* $\mathbb{C}^d$ *with the kernel representation* $\Theta = \ker \begin{pmatrix} C & D \end{pmatrix}$, $C, D \in \mathbb{C}^{d\times d}$, *the extension* A_Θ *of* A *of the form* (5.23) *is* $\mathcal{P}$*-selfadjoint if and only if* $\Theta = \Theta^*$, *i.e.,* (2.21) *holds.*

In the special case when $d = 1$ *the set of all* $\mathcal{P}$*-selfadjoint extensions of* A *different from the decoupled operator* $A_{-,2} \oplus A_{+,2}$ *is parametrized by a real parameter* $\alpha \in \mathbb{R}$ *via*

$$A_\alpha = B^* \restriction \operatorname{dom} A_\alpha, \quad \operatorname{dom} A_\alpha = \left\{u \in \operatorname{dom} B^* | \Gamma_1^B u = \alpha \Gamma_2^B u\right\}. \tag{6.6}$$

Moreover, the set of $\mathcal{PT}$*-symmetric extensions of* A *coincides with the set of its* $\mathcal{P}$*-selfadjoint extensions.*

According to [**9**] boundary condition (5.26) can be rewritten in the following form.

THEOREM 6.4. *Let in the assumptions of Theorem 6.2* $\widehat{\Theta}$ *be a linear relation in* $\mathbb{C}^{2d}$ *such that* $(A_0)_{\widehat{\Theta}}$ *is a* $\mathcal{P}$*-selfadjoint and* $\mathcal{PT}$*-symmetric extensions of* A_0. *Then*

(i) *There exist orthogonal projectors* P_D *and* $P_R = I - P_D$ *in* $\mathbb{C}^{2d}$ *and a selfadjoint matrix* Λ *in* $\mathbb{C}^{2d}$ *such that*

$$\ker \Lambda = P_D \mathcal{H}, \quad \operatorname{ran} \Lambda \subseteq P_R \mathcal{H} \tag{6.7}$$

and $\operatorname{dom}(A_0)_{\widehat{\Theta}}$ *is characterized as the set of* $f \in \operatorname{dom} B_-^* \oplus B_+^*$ *such that*

$$P_D \Gamma_2^0 f = 0, \tag{6.8}$$

$$P_R \Gamma_1^0 f = \Lambda P_R \Gamma_2^0 f. \tag{6.9}$$

(ii) *The extension* $(A_0)_{\widehat{\Theta}}$ *is* $\mathcal{PT}$*-symmetric, if and only if the matrices* P_D *and* Λ *commute with* $\widehat{\mathcal{J}}_{\mathcal{H}}$:

$$P_D \widehat{\mathcal{J}}_{\mathcal{H}} = \widehat{\mathcal{J}}_{\mathcal{H}} P_D, \quad \Lambda \widehat{\mathcal{J}}_{\mathcal{H}} = \widehat{\mathcal{J}}_{\mathcal{H}} \Lambda. \tag{6.10}$$

The condition (6.8) gives the Dirichlet part and the condition (6.9) gives the Robin part of the boundary conditions. As distinct from (5.26) the conditions (6.8)–(6.9) are uniquely defined by the selfadjoint linear relation $\Phi := J_d\widehat{\Theta}$.

PROOF. (i) By Theorem 6.2, the extension $(A_0)_{\widehat{\Theta}}$ of the operator A_0 is $\mathcal{P}$–selfadjoint if and only if the linear relation $\Phi = \ker\begin{pmatrix}\widehat{C} & \widehat{D}J_d\end{pmatrix}$ is selfadjoint. Let P_D be the orthogonal projector onto $\ker\Phi = \ker C$ and let $P_R := I_{2d} - P_D$. As it is known (see e.g. [**9**, Theorem 1.4.4]), the selfadjoint linear relation Φ^{-1} admits the representation

$$\Phi^{-1} = \operatorname{graph}\Lambda \dotplus \begin{pmatrix}0\\ P_D\mathcal{H}\end{pmatrix} = \left\{\begin{pmatrix}x_2\\ \Lambda x_2 + x_1\end{pmatrix} : \begin{array}{c} x_1 \in P_D\mathcal{H}\\ x_2 \in P_R\mathcal{H}\end{array}\right\}, \tag{6.11}$$

where Λ is a selfadjoint matrix Λ in $\mathbb{C}^{2d}$ such that (6.7) holds. For every $f \in \operatorname{dom}(A_0)_{\widehat{\Theta}}$ we have the inclusion $\begin{pmatrix}\Gamma_1^0 f\\ \Gamma_2^0 f\end{pmatrix} \in \Phi$. In view of (6.11) and (6.3) there exist $x_1 \in P_D\mathcal{H}$ and $x_2 \in P_R\mathcal{H}$ such that

$$\begin{pmatrix}\Gamma_2^0 f\\ \Gamma_1^0 f\end{pmatrix} = \begin{pmatrix}x_2\\ \Lambda x_2 + x_1\end{pmatrix}. \tag{6.12}$$

This yields the equations (6.8), (6.9).

(ii) If the extension $(A_0)_{\widehat{\Theta}}$ is $\mathcal{PT}$-symmetric, then, by Theorem 6.2, $\widehat{\Theta}$ is $\widehat{\mathfrak{J}}_{\mathcal{H}}$-invariant and hence the linear relation $\Phi = J_d\widehat{\Theta}$ is also $\widehat{\mathfrak{J}}_{\mathcal{H}}$-invariant. Therefore, the subspaces $P_D\mathcal{H} = \ker\Phi = \ker\widehat{\Theta}$ and $P_R\mathcal{H} = \operatorname{ran}\Phi$ are $\widehat{\mathfrak{J}}_{\mathcal{H}}$-invariant and hence the orthoprojectors P_D and P_R commute with $\widehat{\mathfrak{J}}_{\mathcal{H}}$:

$$P_D\widehat{\mathfrak{J}}_{\mathcal{H}} = \widehat{\mathfrak{J}}_{\mathcal{H}}P_D, \quad P_R\widehat{\mathfrak{J}}_{\mathcal{H}} = \widehat{\mathfrak{J}}_{\mathcal{H}}P_R. \tag{6.13}$$

Moreover, in this case for every $f \in \operatorname{dom}(A_0)_{\widehat{\Theta}}$ we obtain, by (6.9),

$$\begin{pmatrix}\Gamma_1^0 f\\ \Gamma_2^0 f\end{pmatrix} \in \Phi \implies \begin{pmatrix}\widehat{\mathfrak{J}}_{\mathcal{H}}\Gamma_1^0 f\\ \widehat{\mathfrak{J}}_{\mathcal{H}}\Gamma_2^0 f\end{pmatrix} \in \Phi \implies P_R\widehat{\mathfrak{J}}_{\mathcal{H}}\Gamma_1^0 f = \Lambda P_R\widehat{\mathfrak{J}}_{\mathcal{H}}\Gamma_2^0 f.$$

By (6.13), we obtain

$$\widehat{\mathfrak{J}}_{\mathcal{H}}P_R\Gamma_1^0 f = \Lambda\widehat{\mathfrak{J}}_{\mathcal{H}}P_R\Gamma_2^0 f.$$

Comparing this equality with (6.9), we obtain the equality

$$\widehat{\mathfrak{J}}_{\mathcal{H}}\Lambda = \Lambda\widehat{\mathfrak{J}}_{\mathcal{H}}. \tag{6.14}$$

Conversely, if (6.13) and (6.14) hold, then similar reasonings show that $\widehat{\Theta}$ is $\widehat{\mathfrak{J}}_{\mathcal{H}}$-invariant. □

7. Non-Hermitian $\mathcal{PT}$-invariant Hamiltonians

7.1. Dual pairs associated with the Bender-Boettcher Hamiltonian. Here we return to the investigation of the non-Hermitian $\mathcal{PT}$-invariant Hamiltonians presented in the introduction, that is, we study equation (1.1) on the wedge shaped contour Γ, cf. (1.2). By substitution $z(x) := xe^{i\phi\,\operatorname{sgn}\,x}$, one obtains the two differential expressions given by (1.3) and (1.4). Assume that $\mathfrak{a}_\pm$ in (1.4) are in the limit point case at $\pm\infty$. As presented in Section 1, this is the case if and only if the angle ϕ of the wedge satisfies

$$\phi \neq -\frac{N+2}{2N+8}\pi + \frac{2k}{4+N}\pi \quad \text{for } k = 0, \ldots, \left[\frac{N+3}{2}\right]. \tag{7.1}$$

Then, by [**21**, Lemma 1], the differential expressions $\mathfrak{b}_\pm$ in (1.6) are also in the limit point case at $\pm\infty$. Define the operators $A_\pm$ and $B_\pm$ associated with $\mathfrak{a}_\pm$ and $\mathfrak{b}_\pm$ in $L^2(\mathbb{R}_\pm)$ as

$$A_\pm f_\pm := \mathfrak{a}_\pm[f_\pm], \quad B_\pm g_\pm := \mathfrak{b}_\pm[g_\pm] \quad \text{for } f_\pm \in \operatorname{dom} A_\pm,\ g_\pm \in \operatorname{dom} B_\pm,$$

respectively, with the domains

$$\operatorname{dom} A_\pm := \{u_\pm \in L^2(\mathbb{R}_\pm) | \mathfrak{a}_\pm[u_\pm] \in L^2(\mathbb{R}_\pm),\, u'_\pm \in AC_{loc}(\mathbb{R}_\pm),\, u_\pm(0_\pm) = u'_\pm(0_\pm) = 0\},$$

$$\operatorname{dom} B_\pm := \{v_\pm \in L^2(\mathbb{R}_\pm) | \mathfrak{b}_\pm[v_\pm] \in L^2(\mathbb{R}_\pm),\, v'_\pm \in AC_{loc}(\mathbb{R}_\pm),\, v_\pm(0_\pm) = v'_\pm(0_\pm) = 0\}.$$

These operators are called the minimal operators. It follows from [**21**, Proposition 1 and Theorem 3] that the (maximal) operators $A_\pm^*$ and $B_\pm^*$ are generated by differential expressions in $L^2(\mathbb{R}_\pm)$ where the roles of $\mathfrak{a}_\pm$ and $\mathfrak{b}_\pm$ are switched in the sense that the differential expressions $\mathfrak{a}_\pm$ are now related to $B_\pm^*$ and the differential expressions $\mathfrak{b}_\pm$ are related to $A_\pm^*$. More precisely,

$$B_\pm^* f_\pm := \mathfrak{a}_\pm[f_\pm], \quad A_\pm^* g_\pm := \mathfrak{b}_\pm[g_\pm] \quad \text{for } f_\pm \in \operatorname{dom} B_\pm^*,\ g_\pm \in \operatorname{dom} A_\pm^*,$$

with

$$\operatorname{dom} B_\pm^* := \{u_\pm \in L^2(\mathbb{R}_\pm) | \mathfrak{a}_\pm[u_\pm] \in L^2(\mathbb{R}_\pm),\, u'_\pm \in AC_{loc}(\mathbb{R}_\pm)\},$$

$$\operatorname{dom} A_\pm^* := \{v_\pm \in L^2(\mathbb{R}_\pm) | \mathfrak{b}_\pm[v_\pm] \in L^2(\mathbb{R}_\pm),\, v'_\pm \in AC_{loc}(\mathbb{R}_\pm)\}.$$

LEMMA 7.1. *The pairs* (A_-, B_-) *and* (A_+, B_+) *are dual pairs. The triple* $(\mathbb{C}^2, \Gamma^{A_+}, \Gamma^{B_+})$,

$$\Gamma^{B_+} u_+ = \begin{pmatrix} e^{-2i\phi} u'_+(0) \\ u_+(0) \end{pmatrix}, \quad \Gamma^{A_+} v_+ = \begin{pmatrix} v_+(0) \\ e^{2i\phi} v'_+(0) \end{pmatrix}, \quad \begin{array}{l} u_+ \in \operatorname{dom} B_+^*, \\ v_+ \in \operatorname{dom} A_+^*, \end{array} \tag{7.2}$$

is a boundary triple for the dual pair (A_+, B_+). *The triple* $(\mathbb{C}^2, \Gamma^{A_-}, \Gamma^{B_-})$,

$$\Gamma^{B_-} u_- = \begin{pmatrix} -e^{2i\phi} u'_-(0) \\ u_-(0) \end{pmatrix}, \quad \Gamma^{A_-} v_- = \begin{pmatrix} v_-(0) \\ -e^{-2i\phi} v'_-(0) \end{pmatrix}, \quad \begin{array}{l} u_- \in \operatorname{dom} B_-^*, \\ v_- \in \operatorname{dom} A_-^*, \end{array} \tag{7.3}$$

is a boundary triple for the dual pair (A_-, B_-).

PROOF. Integration by parts and [**21**, Proposition 1] show

$$\langle A_\pm u_\pm, v_\pm \rangle = \langle u_\pm, B_\pm v_\pm \rangle, \quad u_\pm \in \operatorname{dom} A_\pm, \quad v_\pm \in \operatorname{dom} B_\pm.$$

This proves the first statement. It follows from [**21**, Proposition 1] that for $u_+ \in \operatorname{dom} B_+^*$ and $v_+ \in \operatorname{dom} A_+^*$

$$\begin{aligned} \langle B_+^* u_+, v_+ \rangle - \langle u_+, A_+^* v_+ \rangle &= -e^{-2i\phi} \int_0^\infty u''_+(x) \overline{v_+(x)} dx + e^{-2i\phi} \int_0^\infty u_+(x) \overline{v''_+(x)} dx \\ &= e^{-2i\phi} (u'_+(0) \overline{v_+(0)} - u_+(0) \overline{v'_+(0)}). \end{aligned}$$

Hence, $(\mathbb{C}^2, \Gamma^{A_+}, \Gamma^{B_+})$ is a boundary triple for the dual pair (A_+, B_+). The statement for the triple (A_-, B_-) is shown in the same way. $\square$

7.2. $\mathcal{PT}$-symmetric extensions of the operator A. The coupling (A, B) of the dual pairs $(A_\pm, B_\pm)$ relative to the boundary triples $(\mathbb{C}^2, \Gamma^{A_\pm}, \Gamma^{B_\pm})$ consists of two operators $A = (B_+^* \oplus B_-^*)|_{\text{dom } A}$, $B = (A_+^* \oplus A_-^*)|_{\text{dom } B}$ with the domains

(7.4)
$$\text{dom } A = \{u_+ \oplus u_- \,|\, u_\pm \in \text{dom } B_\pm^*,\, u_+(0) = u_-(0) = e^{-2i\phi}u_+'(0) - e^{2i\phi}u_-'(0) = 0\},$$

(7.5)
$$\text{dom } B = \{u_+ \oplus u_- \,|\, u_\pm \in \text{dom } A_\pm^*,\, u_+(0) = u_-(0) = e^{2i\phi}u_+'(0) - e^{-2i\phi}u_-'(0) = 0\},$$

see Theorem 3.1. Then the adjoints A^* and B^* are the restrictions of the operators $A_+^* \oplus A_-^*$ and $B_+^* \oplus B_-^*$, respectively, to the domains

$$\text{dom } A^* = \{u_+ \oplus u_- | u_\pm \in \text{dom } A_\pm^*,\, u_+(0) = u_-(0)\}; \tag{7.6}$$

$$\text{dom } B^* = \{u_+ \oplus u_- | u_\pm \in \text{dom } B_\pm^*,\, u_+(0) = u_-(0)\}. \tag{7.7}$$

A boundary triple for (A, B) can be defined on $v \in \text{dom } A^*$ and $u \in \text{dom } B^*$ by

(7.8)
$$\Gamma^A v = \begin{pmatrix} \Gamma_1^A v \\ \Gamma_2^A v \end{pmatrix} := \begin{pmatrix} \Gamma_1^{A_+} v_+ \\ \Gamma_2^{A_+} v_+ + \Gamma_2^{A_-} v_- \end{pmatrix}, \quad \Gamma^B u = \begin{pmatrix} \Gamma_1^B u \\ \Gamma_2^B u \end{pmatrix} := \begin{pmatrix} \Gamma_1^{B_+} u_+ + \Gamma_1^{B_-} u_- \\ \Gamma_2^{B_+} u_+ \end{pmatrix},$$

We define the parity $\mathcal{P}$ and the time reversal $\mathcal{T}$ as in (1.5). It is easy to see that the parity $\mathcal{P}$ and the time reversal $\mathcal{T}$ satisfy (5.2). Due to Theorem 5.5 the operator A is $\mathcal{PT}$-symmetric and $\mathcal{P}$-symmetric in the Krein space $(L^2(\mathbb{R}) = L^2(\mathbb{R}_-) \oplus L^2(\mathbb{R}_+), [\cdot,\cdot])$ with the inner product

$$[\cdot,\cdot] = \langle \mathcal{P}\cdot, \cdot \rangle.$$

The (Krein space) adjoint A^+ of A coincides with the operator $B^* = (B_+^* \oplus B_-^*)|_{\text{dom } B^*}$.

Application of Theorem 5.5 gives a one-parameter family $\{H_\alpha\}_{\alpha \in \mathbb{R}}$ of $\mathcal{PT}$-symmetric and $\mathcal{P}$-selfadjoint extensions of A in the Krein space $(L^2(\mathbb{R}), [\cdot,\cdot])$.

THEOREM 7.2. *Let the angle ϕ satisfies (7.1) and let A be the coupling operator constructed in (7.4). Then the following statements are true.*

(i) *A boundary triple $(\mathbb{C}, \Gamma_1, \Gamma_2)$ for the $\mathcal{P}$-symmetric operator A is given by*

$$\Gamma_1 u = e^{-2i\phi}u_+'(0) - e^{2i\phi}u_-'(0), \quad \Gamma_2 u = u_+(0), \quad u = u_+ \oplus u_- \in \text{dom } B^*.$$

(ii) *The extension H_α of the operator A defined as a restriction of B^* to the domain*

$$\text{dom } H_\alpha = \left\{u_+ \oplus u_- \in \text{dom } B^* : e^{-2i\phi}u_+'(0) - e^{2i\phi}u_-'(0) = \alpha u_+(0)\right\} \tag{7.9}$$

is $\mathcal{P}$-selfadjoint if and only if $\alpha \in \mathbb{R}$.

(iii) *H_α is $\mathcal{PT}$-symmetric if and only if $\alpha \in \mathbb{R}$.*

PROOF. By construction, the dual pairs (A_+, B_+) and (A_-, B_-) are $\mathcal{T}_\pm$-real and the parity operator $\mathcal{P}$ intertwines the operators A_+, B_- and A_-, B_+, that is, (5.5) holds. Moreover, the boundary triples $(\mathbb{C}^2, \Gamma^{A_+}, \Gamma^{B_+})$ and $(\mathbb{C}^2, \Gamma^{A_-}, \Gamma^{B_-})$ are also $(j_\mathbb{C}, \mathcal{T})$-real and satisfy the condition (5.6). Here $j_\mathbb{C}$ stands for the usual complex conjugation in $\mathbb{C}$. Hence, all assumptions in Theorem 5.5 are satisfied and the statements (i)-(iii) in Theorem 7.2 follows directly from Theorem 5.5. $\square$

The boundary conditions in (7.9) are called "δ-type conditions" in the literature, see [**9**]. In [**21**] the extension H_α for the parameter value $\alpha = 0$ was considered, which is specified by "weighted Kirchhoff conditions":

$$\text{dom } H_0 = \left\{ u_+ \oplus u_- | u_\pm \in \text{dom } B_\pm^*,\ u_+(0) - u_-(0) = e^{-2i\phi} u_+'(0) - e^{2i\phi} u_-'(0) = 0 \right\}.$$

7.3. $\mathcal{PT}$-symmetric extensions of A_0. The family $(H_\alpha)_{\alpha\in\mathbb{R}}$, of extensions of A can be also treated via Theorem 6.2 as a family of extensions of the operator $A_0 = A_- \oplus A_+$. A boundary triple for the $\mathcal{P}$-symmetric operator A_0 can be defined on $u \in \text{dom } B_0^*$ by the equalities, see (6.3)

$$\Gamma_1^0 u = \begin{pmatrix} \Gamma_1^{B_+} u_+ \\ \Gamma_1^{B_-} u_- \end{pmatrix} = \begin{pmatrix} e^{-2i\phi} u_+'(0) \\ -e^{2i\phi} u_-'(0) \end{pmatrix},\ \Gamma_2^0 u = \begin{pmatrix} \Gamma_2^{B_-} u_- \\ \Gamma_2^{B_+} u_+ \end{pmatrix} = \begin{pmatrix} u_-(0) \\ u_+(0) \end{pmatrix},\quad u = \begin{pmatrix} u_+ \\ u_- \end{pmatrix} \in \text{dom } B_0^*.$$

By Theorem 6.4, the set of $\mathcal{P}$-selfadjoint and $\mathcal{PT}$-symmetric extensions of A_0 is parametrized via (6.8)–(6.9) by pairs (P_D, Λ) where P_D is an orthogonal projector onto a subspace in $\mathbb{C}^{2d}$ and Λ is a 2×2-matrix which satisfy (6.7) and (6.10).

Let us discuss this formulas in our case when the boundary space is 2-dimensional.

Case 1. Let P_D be the projector on the whole $\mathbb{C}^2$, i.e., $P_D = I_{\mathbb{C}^2}$. Then $\Lambda = 0$ and boundary conditions (6.8)–(6.9) take the form of "Dirichlet boundary conditions":

$$u_+(0) = u_-(0) = 0.$$

The corresponding extension of A_0 is the decoupled operator $A_{+,2} \oplus A_{-,2}$.

Case 2. Let P_D be an 1-dimensional projector onto a $\widehat{\jmath}$-invariant subspace of $\mathbb{C}^2$. Every $\widehat{\jmath}$-invariant subspace $\mathcal{L}$ of $\mathbb{C}^2$ has the form $\mathcal{L} = \text{span} \begin{pmatrix} 1 \\ e^{i\theta} \end{pmatrix}$, $\theta \in [0, 2\pi)$. In particular, for $\theta = \pi$ we have $P_D = \frac{1}{2}\begin{pmatrix} 1 & -1 \\ -1 & 1 \end{pmatrix}$, $P_R = \frac{1}{2}\begin{pmatrix} 1 & 1 \\ 1 & 1 \end{pmatrix}$. Condition (6.7) implies that Λ can be chosen as $\Lambda = \frac{1}{4}\begin{pmatrix} \alpha & \alpha \\ \alpha & \alpha \end{pmatrix}$, $\alpha \in \mathbb{R}$, and boundary conditions (6.8)–(6.9) yield "δ-type conditions":

$$u_-(0) = u_+(0), \quad e^{-2i\phi} u_+'(0) - e^{2i\phi} u_-'(0) = \alpha u_+(0), \quad \alpha \in \mathbb{R}.$$

Case 3. Let the Dirichlet part in boundary conditions is missing, i.e., $P_D = O$. Then (6.7) is fulfilled automatically and $\Lambda = \begin{pmatrix} \lambda_{11} & \lambda_{12} \\ \lambda_{21} & \lambda_{22} \end{pmatrix}$ is an arbitrary selfadjoint 2×2-matrix commuting with $\widehat{\jmath} = \begin{pmatrix} 0 & j \\ j & 0 \end{pmatrix}$. This leads to conditions

$$\lambda_{11} = \lambda_{22} \in \mathbb{R}, \quad \lambda_{21} = \overline{\lambda_{12}}. \tag{7.10}$$

In particular, for $\Lambda = 0$ we obtain the "Neumann boundary conditions":

$$u_+'(0) = u_-'(0) = 0.$$

Setting $\Lambda = \frac{1}{\beta}\begin{pmatrix} 1 & -1 \\ -1 & 1 \end{pmatrix}$, $\beta \in \mathbb{R} \setminus \{0\}$, we obtain the following family $\{T_\beta\}_{\beta \in \mathbb{R}\setminus\{0\}}$ of $\mathcal{PT}$-symmetric extensions of extension of A_0 specified by the boundary conditions

$$e^{-2i\phi}u'_+(0) = e^{2i\phi}u'_-(0) = \frac{1}{\beta}(u_+(0) - u_-(0)), \quad \beta \in \mathbb{R} \setminus \{0\},$$

which are analogues of "δ'-type conditions".

Acknowledgements

The research of Volodymyr Derkach was supported by the Ministry of Education and Science of Ukraine (project # 0121U109525) the German Research Foundation (DFG), grant TR 903/22-1 and a grant of the Volkswagen Foundation.

References

[1] R. Arens, *Operational calculus of linear relations*, Pacific J. Math. **11** (1961), 9–23. MR123188

[2] T. Ya. Azizov and I. S. Iokhvidov, *Linear operators in spaces with an indefinite metric*, Pure and Applied Mathematics (New York), John Wiley & Sons, Ltd., Chichester, 1989. Translated from the Russian by E. R. Dawson; A Wiley-Interscience Publication. MR1033489

[3] T. Ya. Azizov and C. Trunk, *On domains of $\mathcal{PT}$ symmetric operators related to $-y''(x) + (-1)^n x^{2n} y(x)$*, J. Phys. A **43** (2010), no. 17, 175303, 13, DOI 10.1088/1751-8113/43/17/175303. MR2609975

[4] T. Ya. Azizov and C. Trunk, *$\mathcal{PT}$ symmetric, Hermitian and $\mathcal{P}$-self-adjoint operators related to potentials in $\mathcal{PT}$ quantum mechanics*, J. Math. Phys. **53** (2012), no. 1, 012109, 18, DOI 10.1063/1.3677368. MR2919522

[5] C. M. Bender, *PT symmetry in quantum and classical physics*, World Scientific Publishing Co. Pte. Ltd., Hackensack, NJ, 2019. With contributions from Patrick E. Dorey, Clare Dunning, Andreas Fring, Daniel W. Hook, Hugh F. Jones, Sergii Kuzhel, Géza Lévai and Roberto Tateo. MR3889966

[6] C. M. Bender and S. Boettcher, *Real spectra in non-Hermitian Hamiltonians having $\mathcal{PT}$ symmetry*, Phys. Rev. Lett. **80** (1998), no. 24, 5243–5246, DOI 10.1103/PhysRevLett.80.5243. MR1627442

[7] C. M. Bender, D. C. Brody, J.-H. Chen, H. F. Jones, K. A. Milton, and M. C. Ogilvie, *Equivalence of a complex $\mathcal{PT}$-symmetric quartic Hamiltonian and a Hermitian quartic Hamiltonian with an anomaly*, Phys. Rev. D **74** (2006), 025016.

[8] C. M. Bender, D. C. Brody, and H. F. Jones, *Complex extension of quantum mechanics*, Phys. Rev. Lett. **89** (2002), no. 27, 270401, 4, DOI 10.1103/PhysRevLett.89.270401. MR1950305

[9] G. Berkolaiko and P. Kuchment, *Introduction to quantum graphs*, AMS, Providence RI, 2013.

[10] J. Bognar, *Indefinite Inner Product Spaces*, Springer, Berlin, 1974.

[11] B. M. Brown, D. K. R. McCormack, W. D. Evans, and M. Plum, *On the spectrum of second-order differential operators with complex coefficients*, R. Soc. Lond. Proc. Ser. A Math. Phys. Eng. Sci. **455** (1999), no. 1984, 1235–1257, DOI 10.1098/rspa.1999.0357. MR1701776

[12] V. Derkach, *On Weyl function and generalized resolvents of a Hermitian operator in a Kreĭn space*, Integral Equations Operator Theory **23** (1995), no. 4, 387–415, DOI 10.1007/BF01203914. MR1361051

[13] V. A. Derkach, S. Hassi, M. M. Malamud, and H. S. V. de Snoo, *Generalized resolvents of symmetric operators and admissibility*, Methods Funct. Anal. Topology **6** (2000), no. 3, 24–55. MR1903120

[14] V. A. Derkach and M. M. Malamud, *Generalized resolvents and the boundary value problems for Hermitian operators with gaps*, J. Funct. Anal. **95** (1991), no. 1, 1–95, DOI 10.1016/0022-1236(91)90024-Y. MR1087947

[15] V. Derkach, P. Schmitz, and C. Trunk, *$\mathcal{PT}$-symmetric Hamiltonians as couplings of dual pair*, Contributions to mathematics and statistics—essays in honor of Seppo Hassi, Acta Wasaensia, vol. 462, Vaasan Yliopisto, Vaasa, [2021], pp. 55–68. MR4448556

[16] V. I. Gorbachuk and M. L. Gorbachuk, *Boundary value problems for operator differential equations*, Mathematics and its Applications (Soviet Series), vol. 48, Kluwer Academic Publishers Group, Dordrecht, 1991. Translated and revised from the 1984 Russian original, DOI 10.1007/978-94-011-3714-0. MR1154792

[17] S. Hassi and S. Kuzhel, *On J-self-adjoint operators with stable C-symmetries*, Proc. Roy. Soc. Edinburgh Sect. A **143** (2013), no. 1, 141–167, DOI 10.1017/S0308210511001387. MR3023008

[18] H. F. Jones and J. Mateo, *Equivalent Hermitian Hamiltonian for the non-Hermitian $-x^4$ potential*, Phys. Rev. D (3) **73** (2006), no. 8, 085002, 4, DOI 10.1103/PhysRevD.73.085002. MR2221392

[19] A. N. Kočubeĭ, *Extensions of symmetric operators and of symmetric binary relations* (Russian), Mat. Zametki **17** (1975), 41–48. MR365218

[20] H. Langer and C. Tretter, *A Krein space approach to PT-symmetry*, Czechoslovak J. Phys. **54** (2004), no. 10, 1113–1120, DOI 10.1023/B:CJOP.0000044012.95629.b2. MR2109645

[21] F. Leben and C. Trunk, *Operator-based approach to $\mathcal{PT}$-symmetric problems on a wedge-shaped contour*, Quantum Stud. Math. Found. **6** (2019), no. 3, 315–333, DOI 10.1007/s40509-019-00197-3. MR4016656

[22] M. M. Malamud and V. I. Mogilevskii, *Kreĭn type formula for canonical resolvents of dual pairs of linear relations*, Methods Funct. Anal. Topology **8** (2002), no. 4, 72–100. MR1942823

[23] N, Moiseyev, *Non-Hermitian Quantum Mechanics*, Cambridge University Press, 2011.

[24] A. Mostafazadeh, *Pseudo-Hermiticity versus PT symmetry: the necessary condition for the reality of the spectrum of a non-Hermitian Hamiltonian*, J. Math. Phys. **43** (2002), no. 1, 205–214, DOI 10.1063/1.1418246. MR1872494

[25] A. Mostafazadeh, *Pseudo-Hermitian description of PT-symmetric systems defined on a complex contour*, J. Phys. A **38** (2005), no. 14, 3213–3234, DOI 10.1088/0305-4470/38/14/011. MR2132529

[26] A. Mostafazadeh, *Pseudo-Hermitian representation of quantum mechanics*, Int. J. Geom. Methods Mod. Phys. **7** (2010), no. 7, 1191–1306, DOI 10.1142/S0219887810004816. MR2749385

[27] A. R. Sims, *Secondary conditions for linear differential operators of the second order*, J. Math. Mech. **6** (1957), 247–285, DOI 10.1512/iumj.1957.6.56011. MR86224

Volodymyr Derkach: Institut für Mathematik, Technische Universität Ilmenau, PF 100565, 98684 Ilmenau, Germany and Department of Mathematics, Vasyl Stus Donetsk National University, Vinnytsya, Ukraine.

Email address: derkach.v@gmail.com

Carsten Trunk: Institut für Mathematik, Technische Universität Ilmenau, PF 100565, 98684 Ilmenau, Germany.

Email address: carsten.trunk@tu-ilmenau.de

Contemporary Mathematics
Volume **798**, 2024
https://doi.org/10.1090/conm/798/15982

Chebyshev dynamics on two and three intervals and isomonodromic deformations

Vladimir Dragović and Vasilisa Shramchenko

ABSTRACT. We present the dynamics of Chebyshev polynomials on $d = 2$ and $d = 3$ real intervals. This dynamics, for $d = 2$, is associated with isomonodromic deformations and Painlevé VI equations, while for $d = 3$, it is associated with a class of constrained isomonodromic deformations for which we provide the constrained Schlesinger equations. We provide explicit solutions to these equations in terms of differentials on an appropriate family of elliptic curves for $d = 2$ and of genus 2 curves for $d = 3$. From the point of view of the classical algebraic geometry we present and solve the problem of constrained Jacobi inversion for genus 2 curves. We motivate and apply the obtained results to the study of billiards within quadrics in the plane and in 3-dimensional space.

CONTENTS

2020 *Mathematics Subject Classification.* Primary 34M55, 34M56, 33E05, 14H70.

1. Introduction

1.1. Overview. The main aim of this paper is to present a class of isomonodromic deformations of a 2×2 matrix Fuchsian system, the so called *constrained* isomonodromic deformations and to provide explicit families of such deformations in terms of differentials on varying hyperelliptic curves in the case of the smallest possible genus, $g = 2$. The motivation for this study is two-fold. It comes from the theory of integrable billiards within ellipsoids in d-dimensional Euclidean space and from the theory of extremal polynomials on d real intervals. The questions we are going to discuss for $d = 2$ and $d = 3$, $g = d - 1$, along the way in this study are:

- *Given an ellipsoid as a boundary of a billiard system in* $\mathbb{R}^d$ *and a set of* $d - 1$ *caustics tangent to an* n*-periodic trajectory, describe the dynamics of the caustics under the variation of semi-axes of the boundary ellipsoid, provided that the generated billiard trajectories remain periodic with the same period;*
- *Given a set of* d *real intervals which support Chebyshev polynomials (that is polynomial solutions of the Pell equation* (3)*), while keeping one interval fixed and deforming one end-point of each of the remaining* $d - 1$ *intervals, describe the dynamics of the remaining* $d - 1$ *end-points, under the condition that the Pell equation remains solvable during the entire process.*

We will first review our preparatory works [**14**, **15**], where we already answered the above questions in the case $d = 2$ when the underlying curves are elliptic ($g = 1$). The obtained isomonodromic deformations were *unconstrained* for $d = 2, g = 1$. The essential difficulty which comes with $d = 3, g = 2$ with respect to $d = 2, g = 1$ is in the nature of the Jacobi inversion problem. For $g = 1$ there is an isomorphism between an elliptic curve and its Jacobian: the data of an elliptic curve and a marked point on it is equivalent to the data of the Jacobian of the curve and a point on it. Therefore one can consider deformations of the elliptic curve while keeping the association between the two sets of data, and this is precisely what we need for our construction in [**14**] and [**15**]. However, the same thing cannot be done for curves of genus higher than one. The crux of the matter for $d \geq 3, g \geq 2$ is a resolution of the following algebro-geometric problem with a classical flavor. Given a hyperelliptic curve and a marked point on it, the Abel-Jacobi mapping of the marked point gives a marked point on the Jacobian of the hyperelliptic curve. When the curve varies, one can define a variation of the marked point on the Jacobian. However, such a varying marked point is generically not an image of *a single* point from the varying hyperelliptic curve. Its Abel-Jacobi pre-image consists generically of g distinct points. We search here for special, non-generic deformations of genus two hyperelliptic curves where the marked point on the varying Jacobian remains the image of a single point of a varying curve. We will refer to this as *the problem of the constrained variations of the Jacobi inversion for curves of genus* $g = 2$. Such variations are induced by so-called *isoharmonic* deformations, which we introduced recently in [**16**].

1.2. Poncelet theorem. Billiards within quadrics. The classical Poncelet theorem is closely related to algebraic geometry, in particular to the theory of elliptic curves [**19**]. It also has an important mechanical interpretation. *The elliptical billiard* [**28**, **38**] is a dynamical system where a material point of unit mass is moving frictionlessly, with constant velocity inside an ellipse, and obeying the reflection law at the boundary, having congruent angles of incidence and reflection. Any segment of a given elliptical billiard trajectory is tangent to the same conic, confocal with the boundary, called *the caustic* of a given trajectory, [**10**]. If a trajectory becomes closed after n reflections, then the Poncelet Theorem implies that any trajectory of the billiard system, which shares the same caustic curve, is also periodic with the period n.

d-dimensional case. Let us now consider billiards in domains in d-dimensional space bounded by the ellipsoid $\mathcal{Q}_0$ from the confocal family of quadrics:

$$\mathcal{Q}_\lambda \ : \ \frac{x_1^2}{a_1-\lambda}+\cdots+\frac{x_d^2}{a_d-\lambda}=1, \quad \lambda\in\mathbf{R}, \tag{1}$$

with $0<a_1<a_2<\cdots<a_d$ being constants. Any point of the space belongs to exactly d quadrics from the family (1). The parameters $\lambda_1,\ldots,\lambda_d$ of these quadrics are called *the Jacobi coordinates* of the point in $\mathbb{R}^d$ with Cartesian coordinates $(x_1,\ldots,x_d)$ (see [**11**] and references therein), with the convention: $\lambda_1<\lambda_2<\cdots<\lambda_d$. The billiard flow has $d-1$ quadrics from the confocal family (1) as *caustics*, having the parameter λ from (1) equal to $\alpha_1,\ldots,\alpha_{d-1}$. The billiard domain Ω consists of all points of the billiard trajectories within $\mathcal{Q}_0$ with caustics $\mathcal{Q}_{\alpha_1}$, $\ldots$, $\mathcal{Q}_{\alpha_{d-1}}$. In elliptic coordinates, Ω is just a product of d segments in $[0,+\infty)$ where $P_{2d-1}(\lambda)$, defined in (2), is positive. Along any billiard trajectory the elliptic coordinates vary within their segments, and change direction (from increasing to decreasing and vice versa) only at the segments' endpoints. Endpoints of the segments correspond to: i) reflection off the boundary ellipsoid ($\lambda=0$); ii) touching a caustic ($\lambda=\alpha_k, k=1,\ldots,d-1$); iii) crossing a coordinate hyper-plane ($\lambda=a_j, j=1,\ldots,d$). For a given periodic trajectory of period n, denote *the winding numbers*: $m_0=n, m_1,\ldots,m_{d-1}$, where $m_i, i=1,\ldots,d-1$ is equal to the number of turns the Jacobi coordinate λ_i makes in its interval during one period of the billiard trajectory. The billiard dynamics can be analyzed on the Jacobian of the hyperelliptic Riemann surface:

$$\eta^2=P_{2d-1}(\lambda)=(\lambda-a_1)\cdots(\lambda-a_d)(\lambda-\alpha_1)\cdots(\lambda-\alpha_{d-1}); \tag{2}$$

the combination of a boundary $\mathcal{Q}_0$ and $d-1$ caustics allowing for a periodic trajectory of period n corresponds to a point of order n on this hyperelliptic surface.

The connection between ellipsoidal billiards in dimension d and Chebyshev-type extremal polynomials on systems of d intervals was fully established in [**11**]. Exploiting this connection all three Ramirez-Ros conjectures [**35**] were proved there, all sets of caustics which generate periodic trajectories of a given period n of billiards within a given ellipsoid in d-dimensional space were described. In addition, injectivity properties of the billiard frequency maps were proved there and they are going to play important role in the current setting. We use the notation $\{b_1,\ldots,b_{2d-1}\}=\{a_1,\ldots,a_d,\alpha_1,\ldots,\alpha_{d-1}\}$ with the ordering condition $b_1<b_2<\cdots<b_{2d-1}$ and $c_{2d}=0, c_j=1/b_j$ and following [**11**] introduce the polynomial $\mathscr{P}_{2d}(s)=s\prod_{j=1}^{2d-1}(s-c_j)$.

It was proved in [**11**] that the generalized Cayley's condition $C(n,d)$ from [**9**] describing n - periodic trajectories in d-dimensional space with caustics $\alpha_1,\ldots,\alpha_{d-1}$ is satisfied if and only if there exist a pair of real polynomials p_n, q_{n-d} of degrees n and $n-d$ respectively such that the Pell equation holds:

$$p_n^2(s)-\mathscr{P}_{2d}(s)q_{n-d}^2(s)=1. \tag{3}$$

The polynomial p_n is a rescaled extremal polynomial of the Chebyshev type on the system of d intervals $E = [c_{2d}, c_{2d-1}] \cup [c_{2d-2}, c_{2d-3}] \cup \cdots \cup [c_2, c_1]$. Introduce the signature $(\tau_1, \dots, \tau_d)$, where τ_j is the number of zeroes of q_{n-d} in (c_{2j}, c_{2j-1}). The following relations were established [**11**] for the winding numbers: $m_j = m_{j+1} + \tau_j + 1, \quad 1 \le j \le d$.

In this paper we focus our attention to two lowest dimensions, $d = 2$ and $d = 3$ which have the highest significance from the mechanical standpoint and they also allow us to contrast the elliptic versus hyperelliptic settings in the simplest framework of $g = 1$ and $g = 2$. The general situation of arbitrary $d > 2$ and $g > 1$ is presented in [**16**].

1.3. Painlevé VI equations and nonconstrained Schlesinger equations. A classification of all second order ordinary differential equations (with a certain rational structure) for which movable singularities can be poles only, was initiated by Painlevé. It resulted in six new families of equations for which general solutions cannot be expressed in terms of classical special functions, nowadays called the Painlevé equations, see [**18, 25**]. The sixth Painlevé equations, a family parameterized by four constants α, β, γ, and δ, denoted by PVI$(\alpha, \beta, \gamma, \delta)$, are the most general of them as the other five can be obtained from it by different reductions. Equations PVI were discovered by Fuchs [**18**] in the context of isomonodromic deformations of a linear matrix ordinary differential system associated with an elliptic curve.

A direct connection between the Poncelet theorem and Painlevé VI equation was discovered by Hitchin [**23**]. Starting with a point P of order n on the elliptic curve given by the equation $v^2 = u(u-1)(u-x)$, Hitchin [**23**] constructs a function s on the curve with a pole of order n at $P = (u, v)$ and a zero of order n at $P^* = (u, -v)$ obtained from P by the elliptic involution. Recall that a point of order n corresponds to a periodic billiard trajectory of period n. We will call *s the Hitchin function.* Then an algebraic solution of PVI$(1/8; -1/8; 1/8; 3/8)$ is constructed as a (common) u-coordinate of two zeros of the differential $d\log s(Q)$: this u-coordinate becomes a function of x as one considers the moduli space of the elliptic curves of the equation $v^2 = u(u-1)(u-x)$. This was the starting point of the investigations of the authors of the present paper which led to [**14**], [**13**], [**15**]. We briefly present these preliminary results.

Case $g = 1$. Consider a family of elliptic curves $v^2 = u(u-1)(u-x)$ with a varying ramification point $(u, v) = (x, 0)$. The corresponding Jacobians are generated by the lattices $\Lambda(x) =< 2w_1(x), 2w_2(x) >_{\mathbb{Z}}$. Given a point in the Jacobian $z_0 := 2w_1c_1 + 2w_2c_2$, denote by Q_0 its Abel-Jacobi preimage on the elliptic curve. The u-coordinate of Q_0 as a function of x gives the general solution of PVI$(0; 0; 0; 1/2)$ with arbitrary constants c_1 and c_2. We construct in [**14**] a differential on the elliptic curve whose poles correspond to the general solution of PVI$(0; 0; 0; 1/2)$ and the zeros correspond to the general solution of the PVI$(1/8; -1/8; 1/8; 3/8)$ from [**22**]. More precisely, the authors introduced in [**14**] the differential

$$\Omega = \Omega_{Q_0, Q_0^*} - 4\pi i c_2 \omega. \tag{4}$$

Here ω denotes the holomorphic differential on the curve normalized by the condition $\oint_a \omega = 1$; the differentials Ω_{Q_0, Q_0^*} are of the third kind with poles at Q_0 and

Q_0^* with residues $+1$ and -1, respectively, normalized to have vanishing a-period. As a result we have:

THEOREM 1. [**14**] *The two zeros of the differential* Ω (4) *share the same* u*-coordinate projection denoted by* y*. Then* $y = y(x)$ *as the function of the varying branch point* x *provides the general solution of the* $PVI(1/8; -1/8; 1/8; 3/8)$ *depending on two constants* c_1 *and* c_2.

The constants c_1, c_2 play the role of the initial condition for the differential equation. Moreover, the relationship between the poles and zeros of Ω is given by the Okamoto transformation between solutions of $\mathrm{PVI}(0,0,0,\frac{1}{2})$ and $\mathrm{PVI}(\frac{1}{8},-\frac{1}{8},\frac{1}{8},\frac{3}{8})$, see also [**30**].

Note that in the above theorem, the constants c_1, c_2 are not assumed to be rational, as was the case in Hitchin's construction [**23**], where the role of the point Q_0 was played by a point of a finite order on elliptic curves. This leads to another natural question: *is this relationship between Poncelet polygons and Painlevé equations a miraculous isolated event or a tip of an iceberg?*

In [**14**] in addition, solutions of the appropriate rank 2 Schlesinger equations were obtained in terms of geometry of elliptic curves and the above differential Ω from (4).

Case $g > 1$: a generic situation. Nonconstrained Schlesinger systems. The above construction is generalized to hyperelliptic curves in [**13**] as follows. Consider a matrix linear system for a 2×2 matrix function $\Phi(u)$ defined in the Riemann sphere

$$\frac{d\Phi}{du} = A(u)\Phi, \qquad u \in \mathbb{C}P^1 \tag{5}$$

with a matrix $A \in sl(2,\mathbb{C})$ having $2g+1$ simple poles at $u_1, u_2, \ldots, u_{2g+1}$ and at the point at infinity: $A(u) = \sum_{j=1}^{2g+1} A^{(j)}/(u-u_j)$. The corresponding isomonodromic deformation equations form the Schlesinger system of partial differential equations for the residue-matrices $A^{(j)} \in sl(2,\mathbb{C})$ with respect to positions u_i of poles of A as independent variables:

$$\frac{\partial A^{(j)}}{\partial u_k} = \frac{[A^{(k)}, A^{(j)}]}{u_k - u_j}; \qquad \frac{\partial A^{(k)}}{\partial u_k} = -\sum_{j \neq k} \frac{[A^{(k)}, A^{(j)}]}{u_k - u_j}, \tag{6}$$

where $A^{(\infty)} := -A^{(1)} - \cdots - A^{(2g+1)} = const$. If $g = 1$ and $\{u_1, u_2, u_3\} = \{0, 1, x\}$, the Schlesinger system is equivalent to an appropriate PVI equation. The next step was to consider the generic case $g > 1$, when the Schlesinger system is naturally associated with the family of hyperelliptic Riemann surfaces defined by the equation

$$v^2 = (u - u_1) \cdots (u - u_{2g+1}), \tag{7}$$

where the varying branch points $u_1, \ldots, u_{2g+1}$ are given by the positions of the poles in (5). Using the pattern described above in the elliptic case and generalizing it to the hyperelliptic Riemann surfaces, a construction of a meromorphic differential Ω, the hyperelliptic analogue of (4) is proposed in [**13**]: let $c_1, c_2 \in \mathbb{C}^g$ be two constant vectors and consider a point z_0 in the Jacobian of the curve such that

$$z_0 := c_1 + \mathbb{B}c_2. \tag{8}$$

Here $\mathbb{B}$ is the Riemann matrix of the curve and the matrix $(\mathrm{I}\ \mathbb{B})$ generates the lattice Λ of the Jacobian of the curve. The Jacobi inversion of z_0 is a positive divisor $Q_1 + \cdots + Q_g$ of degree g on the hyperelliptic Riemann surface. It is assumed here that the points Q_j *are all distinct* and none of them coincides with a branch point of the hyperelliptic Riemann surface and that no two points of the divisor are paired by the hyperelliptic involution interchanging the sheets of the covering of the u-sphere. A preliminary result from [**13**] is the definition of the following differential: $\Omega(P) = \sum_{j=1}^{g} \Omega_{Q_j Q_j^*}(P) - 4\pi \mathrm{i}\, c_2^t \omega(P)$. Here Q_j^* is the hyperelliptic involution $* : (u, v) \mapsto (u, -v)$ of the point Q_j and $\Omega_{Q_j Q_j^*}$ is the Abelian differential of the third kind with simple poles at Q_j and Q_j^* of residue $+1$ and -1, respectively, normalized by vanishing of all its a-periods; and ω is a vector of holomorphic differentials ω_j normalized by $\oint_{a_j} \omega_i = \delta_{ij}$. Having constructed the differential Ω on one hyperelliptic Riemann surface from the family (7), one starts varying the Riemann surface by allowing small variations of positions of the branch points $\{u_j\}_{j=1}^{2g+1}$. Note that the lattice Λ varies accordingly, however we define the point z_0 by the requirement that the vectors c_1, c_2, its coordinates with respect to the varying lattice Λ, stay constant. In [**13**] a $2g$-parameter family, with vectors c_1 and c_2 as parameters, of solutions to the Schlesinger system (6) is constructed in terms of the differential Ω. As a consequence of this construction and Corollary 6.2.2 in [**25**], the u-coordinates of the pairs of zeros of Ω, understood as the $2g-1$ functions of the branch points $\{u_j\}_{j=1}^{2g+1}$, give solutions [**13**] to the multidimensional Garnier system of the form given in [**25**]. From the billiard dynamics perspective, the last construction has in its background the billiard ordered games (originally introduced in [**9**]) having g confocal quadrics as boundaries, as indicated in [**13**].

In order to derive the isomonodromic deformations which would correspond to the canonical billiard *within one quadric* in $\mathbb{R}^d$ for $d > 2$ we need to study deformations of the Chebyshev polynomials on d intervals. As a preparation for this, we introduce *the Akhiezer functions* as a generalization of the Hitchin function s from [**23**].

1.4. Akhiezer's function as a generalized Hitchin's function. In the recent years there is an extensive study of isoperiodic deformations, see for example [**20**], where the period-preserving flows in nonlinear Schrödinger systems were studied. Our construction related to PVI equation and described above is similar to the set-up of isoperiodic deformations. In the elliptic case, we deformed the Riemann surface by moving the branch points but the periods of the differential remain constant (note that $\oint_a \Omega = -4\pi \mathrm{i} c_2$ and $\oint_b \Omega = 4\pi \mathrm{i} c_1$). Moreover, from the billiard perspective, the underlying elliptical billiards deform in a way that both the caustic and the boundary change simultaneously to preserve the period of the trajectories. In other words, they deform, preserving the rotational number which is rational in the periodic case. The above mentioned generalization of the Hitchin construction to arbitrary c_1 and c_2 from [**14**] can be seen as a deformation of the boundary and the caustic which preserves the rotational number, which is not necessary rational any more. Thus, one can say that our approach to PVI $(1/8; -1/8; 1/8; 3/8)$ was based on the study of *the iso-rotational/iso-frequency/iso-winding deformations*, which was naturally related to *isoperiodic deformations.*

A geometric construction of four-periodic and six-periodic families of Poncelet polygons inscribed in a given circle with caustics belonging to a confocal family of conics was recently presented in [**12**].

In the present paper, we investigate the following deformations of Pell's equations (3) over $d = 2$ and $d = 3$ intervals. We will keep the first interval fixed, and in the remaining $d-1$ intervals we will deform one end-point of each of the remaining $d-1$ intervals, as $d-1$ independent variables. We consider the second end-point of each of these intervals as $d-1$ dependent variables, under the assumption that the deformed systems of d intervals still allow for solutions of Pell's equation and the signature (see Section 1.2) is preserved. The role of the generalization of the Hitchin function s will be played by the Akhiezer function. This is a function defined on the hyperelliptic curve $\mathscr{C}_{2d}$ associated with the polynomial $\hat{\mathscr{P}}_{2d}$ from a solution to the Pell equation (3). We now introduce the Akhiezer function together with the related differential on $\mathscr{C}_{2d}$:

$$\mathbb{A} = p_n - \sqrt{\mathscr{P}_{2d}}\, q_{n-d}, \qquad \Omega_{\mathbb{A}} = \frac{1}{n} d \log \mathbb{A}. \tag{9}$$

A goal of this paper is to derive equations governing the outlined dynamics of the intervals supporting the Pell equation and then to solve the dynamics in $d-1$ times in terms of the differential $\Omega_{\mathbb{A}}$. The proposed dynamics deforms the Pell equation (3) in a way that the corresponding winding numbers (or frequencies) are preserved. Thus it can be seen as an iso-winding or iso-frequency deformation. This is equivalent to the preservation of the signature and the degree of the polynomial.

This dynamics provides the special deformation of the curve $\mathscr{C}_{2d}$ mentioned at the beginning of the introduction in Section 1.1. The obtained deformations allow for solutions of the constrained variations of the Jacobi inversion for the curves $\mathscr{C}_{2d}$, that is we have a family of curves $\mathscr{C}_{2d}$ and a point z_0 defined by (8) in the Jacobian of each curve from the family, such that there exists a point Q_0 on each curve from the family the Abel-Jacobi image of which is z_0. This solution to the constrained variations of the Jacobi inversion gives rise to new equations, the constrained Schlesinger equations (15) - (17) (see Section 2). We provide their explicit solution in terms of the differential $\Omega_{\mathbb{A}}$ (see Theorem 5).

2. Schlesinger and constrained Schlesinger systems

Let us consider a linear non-resonant Fuchsian system for the function $\Phi(u) \in M_{2\times 2}(\mathbb{C})$ with Fuchsian singularities in the set $B := \{0, 1, x_1, x_2, \nu\}$ and at ∞ :

$$\frac{d\Phi}{du} = \Big(\frac{A_0}{u} + \frac{A_1}{u-1} + \frac{A_{x_1}}{u-x_1} + \frac{A_{x_2}}{u-x_2} + \frac{A_\nu}{u-\nu}\Big)\Phi, \qquad u \in \mathbb{C} \setminus B, \tag{10}$$

where A_{a_j} are matrices independent of u. This system is non-resonant if the differences of eigenvalues of the residue matrices A_{a_j} with $a_j \in B$ are all non-integer. A fundamental matrix of solutions of (10) is multi-valued in $\mathbb{C} \setminus B$: it transforms under analytical continuation along the generators of $\pi_1(\mathbb{C} \setminus B)$ by left multiplication by *monodromy matrices.* In other words, we have the *monodromy representation* $\rho : \pi_1(\mathbb{C} \setminus B) \to GL(2, \mathbb{C})$.

Equivalently, the meromorphic 1-form

$$A(u)du := \sum_{a_j \in B} \frac{A_{a_j}}{u - a_j} du \tag{11}$$

can be seen as the connection form of a flat meromorphic connection with simple poles at $B \cup \infty$ in the trivial rank two vector bundle over $\mathbb{C}P^1$. In this case the monodromy representation is the holonomy representation of the connection.

More precisely, in a non-resonant case, a fundamental matrix of solutions behaves as follows close to a singularity $a_j \in B$:

$$\Phi(u) \underset{u \sim a_j}{=} \left(G_{a_j} + \mathcal{O}(u - a_j)\right)(u - a_j)^{T_{a_j}} C_{a_j} \tag{12}$$

with some matrices G_{a_j}, T_{a_j}, C_{a_j} independent of u. The residue matrices A_{a_j} are then given by $A_{a_j} = G_{a_j} T_{a_j} G_{a_j}^{-1}$ and the monodromies of Φ are $M_{a_j} = C_{a_j}^{-1} e^{2\pi i T_{a_j}} C_{a_j}$.

The classical isomonodromic deformation of system (10) is a problem of finding the residue matrices A_{a_j} with $a_j \in B$ as functions of x_1, x_2, ν such that all monodromies of solution matrices Φ stay constant under small variations of the positions of Fuchsian singularities at x_1, x_2, ν. This is equivalent to requiring the matrices T_{a_j} and C_{a_j} to be constant under small variations of x_1, x_2, ν. The Schlesinger system is a system of partial differential equations for the residues A_{a_j} as functions of x_1, x_2, ν solutions of which define an isomonodromic system (10), that is a system of the form (10) whose solutions have constant monodromy matrices.

In this paper, we propose to look at the *constrained isomonodromic deformations*, that is we assume that x_1, x_2 vary independently whereas the third variable position of Fuchsian singularity ν is a function of x_1 and x_2. We thus look for the residue matrices A_{a_j} with $a \in B$ as functions of x_1, x_2 only, such that all monodromies of solutions of (10) be constant. This leads to a generalization of the Schlesinger system which we call *constrained Schlesinger system* as follows.

Assuming the matrices T_{a_j} and C_{a_j} constant, we can differentiate (12) with respect to x_i to get the following behaviour of the derivatives of Φ

$$\Phi_{x_i}\Phi^{-1} \underset{u \sim x_i}{=} -\frac{A_{x_i}}{u - x_i} + \mathcal{O}(1), \qquad \Phi_{x_i}\Phi^{-1} \underset{u \sim \nu}{=} -\frac{A_\nu}{u - \nu}\frac{\partial \nu}{\partial x_i} + \mathcal{O}(1). \tag{13}$$

and see that $\Phi_{x_i}\Phi^{-1}$ has two simple poles, at $u = x_i$ and at $u = \nu$, and no other singularities. Assuming a normalization of Φ which implies that $\Phi_{x_i}\Phi^{-1}$ vanishes at infinity, we see that a fundamental matrix solution Φ of the Fuchsian system (10) satisfies the following system

$$\Phi_{x_i}\Phi^{-1} = -\frac{A_{x_i}}{u - x_i} - \frac{A_\nu}{u - \nu}\frac{\partial \nu}{\partial x_i}, \quad i = 1, 2. \tag{14}$$

The constrained Schlesinger system arises as a compatibility condition of systems (10) and (14). Let us work out the details. Differentiating (10) with respect to x_i and writing a_j for an arbitrary point of the set B , we have

$$\begin{aligned}\Phi_{ux_i} &= \Big(\sum_{a_j \in B} \frac{\partial_{x_i} A_{a_j}}{u - a_j} + \frac{A_{x_i}}{(u - x_i)^2} + \frac{A_\nu}{(u - \nu)^2}\frac{\partial \nu}{\partial x_i}\Big)\Phi \\ &\quad - \sum_{a_j \in B} \frac{A_{a_j}}{u - a_j}\left(\frac{A_{x_i}}{u - x_i} + \frac{A_\nu}{u - \nu}\frac{\partial \nu}{\partial x_i}\right)\Phi.\end{aligned}$$

On the other hand, differentiating (14) with respect to u, we get

$$\Phi_{x_i u} = \Big(\frac{A_{x_i}}{(u - x_i)^2} + \frac{A_\nu}{(u - \nu)^2}\frac{\partial \nu}{\partial x_i}\Big)\Phi - \left(\frac{A_{x_i}}{u - x_i} + \frac{A_\nu}{u - \nu}\frac{\partial \nu}{\partial x_i}\right)\sum_{a_j \in B}\frac{A_{a_j}}{u - a_j}\Phi.$$

Equating the expressions for $\Phi_{x_i u}\Phi^{-1}$ and $\Phi_{ux_i}\Phi^{-1}$ obtained from the two above equations, the following system with $i = 1, 2$ arrises as the equality of the respective leading coefficients in the principal parts of Laurent series at the five simple poles at $u \in B$:

$$\partial_{x_i} A_{a_j} = \frac{[A_{x_i}, A_{a_j}]}{x_i - a_j} + \frac{[A_\nu, A_{a_j}]}{\nu - a_j}\frac{\partial \nu}{\partial x_i}, \tag{15}$$

$$\partial_{x_i} A_\nu = \frac{[A_{x_i}, A_\nu]}{x_i - \nu} - \sum_{\substack{a_j \in B \\ a \neq \nu}} \frac{[A_\nu, A_{a_j}]}{\nu - a_j}\frac{\partial \nu}{\partial x_i}, \tag{16}$$

$$\partial_{x_i} A_{x_i} = - \sum_{\substack{a_j \in B \\ a \neq x_i}} \frac{[A_{x_i}, A_{a_j}]}{x_i - a_j} + \frac{[A_\nu, A_{x_i}]}{\nu - x_i}\frac{\partial \nu}{\partial x_i}. \tag{17}$$

We have thus obtained the constrained Schlesinger system (15) - (17) for the residue matrices A_{aj} . Note that the last equation (17) may be replaced by the condition $A_\infty := -\sum_{k=1}^N A_k = const$, where A_∞ is the residue at infinity of the 1-form $A(u)du$ (11). In the particular case of $\partial_{x_i}\nu = 0$, equations (15) - (17) coincide with the classical Schlesinger system.

3. Extremal polynomials

Modern algebraic approximation theory, especially extremal polynomials and continued fraction theory was established by Chebyshev along with his Saint-Petersburg school in the second half of the XIX century, and their followers in the modern times. Highlights of these studies included the discovery of the Chebyshev polynomials, their generalizations to the systems of intervals, the points of alternance and the theorem of alternance, continued fractions, to mention a few. We need all of this in the present work.

Chebyshev's motivation for these studies was his interest in practical problems: in mechanism theory to estimate the error of the mechanisms which transform linear motion into circular.

3.1. One interval case: classical Chebyshev polynomials. It is well known (see [**5**] for example) that the Chebyshev polynomials $T_n(x)$, $n = 0, 1, 2, \dots$ defined by the recursion

$$T_0(x) = 1, \quad T_1(x) = x, \quad T_{n+1}(x) + T_{n-1}(x) = 2xT_n(x), \tag{18}$$

for $n = 1, 2, \dots$ can be parameterized by

$$T_n(x) = \cos n\phi, \quad x = \cos\phi, \tag{19}$$

or, alternatively,

$$T_n(x) = \frac{1}{2}\left(v^n + \frac{1}{v^n}\right), \quad x = \frac{1}{2}\left(v + \frac{1}{v}\right). \tag{20}$$

Denote $\hat{L}_0 = 1$ and $\hat{L}_n = 2^{n-1}$ for $n = 1, 2, \dots$. Then the Chebyshev theorem states that the polynomials $\hat{T}_n(x) = T_n(x)/\hat{L}_n$ are characterized as solutions of the following minmax problem:

Find the monic polynomial of degree n which minimizes the uniform norm on the interval $I = [-1, 1]$.

As shown by Chebyshev, the polynomials T_n satisfy the polynomial Pell equation, i.e. there exist a polynomial Q_{n-1} of degree $n-1$ such that

$$T_n^2(x) - (x-1)(x+1)Q_{n-1}^2(x) = 1. \tag{21}$$

Given an arbitrary interval I_1, there is an orientation preserving affine transformation which maps it onto the interval $[-1,1]$. By composing the Chebyshev polynomials T_n with this affine transformation, one gets polynomials of degree n which satisfy the Pell equation corresponding to I_1 similar to (21) and the obtained polynomials solve the minmax problem on I_1. Thus, the interval $I = [-1,1]$ can be replaced by an arbitrary real interval I_1 and the entire previous considerations hold with trivial adjustments.

3.2. Two intervals case: the Zolotarev polynomials. However, the situation when one interval is replaced by the union of two real intervals is much more delicate. Consider the set $\{\lambda_1, \lambda_2, \lambda_3, \lambda_4\} \subset \mathbb{R}$ with the ordering $\lambda_1 < \lambda_2 < \lambda_3 < \lambda_4$ and consider the union of two intervals,

$$I_2 = [\lambda_1, \lambda_2] \cup [\lambda_3, \lambda_4].$$

Then the polynomial Pell equation

$$p_n^2(\lambda) - (\lambda-\lambda_1)(\lambda-\lambda_2)(\lambda-\lambda_3)(\lambda-\lambda_4)q_{n-2}^2(\lambda) = 1, \tag{22}$$

does not necessarily have solutions among polynomials p_n and q_{n-2} of degrees n and $n-2$, respectively, for arbitrary $\lambda_1 < \lambda_2 < \lambda_3 < \lambda_4$. Thus, as pointed out in [**37**], the above Pell equation should be understood as equation for "everything", i.e. for the polynomials p_n, q_{n-2} *and for the coefficient* $(\lambda-\lambda_1)(\lambda-\lambda_2)(\lambda-\lambda_3)(\lambda-\lambda_4)$. Note that affine transformations in the λ-plane map solutions of a Pell equation into solutions of a Pell equation of the same form.

Zolotarev [**39**] in 1868 described solutions of the polynomial Pell equation (22) with the coefficient $\mathscr{P}(\lambda) = (\lambda-\lambda_1)(\lambda-\lambda_2)(\lambda-\lambda_3)(\lambda-\lambda_4)$ being a polynomial of degree 4. The solutions were given in terms of elliptic functions, see formulas (4) in [**6**]. There are two essential parameters involved in the parametrization of solutions along with the degree n of the polynomial: the modulus κ and an integer $m = 1, \dots, n-1$. We will call solutions of Pell equations (22), *the Zolotarev polynomials*, the union of two intervals $I_2 = [\lambda_1, \lambda_2] \cup [\lambda_3, \lambda_4]$ *the support* of the Zolotarev polynomial, and the interval (λ_2, λ_3) will be called *the gap interval.* An alternative description of the supports of Zolotarev polynomials, the unions of two intervals $I_2 = [\lambda_1, \lambda_2] \cup [\lambda_3, \lambda_4]$ which allow for solutions of the Pell equations, can be given as follows: *they are the polynomial inverse images of the segment* $I = [-1,1]$. The segments $[\lambda_1, \lambda_2]$ and $[\lambda_3, \lambda_4]$ are composed of $n-m$ and m pre-images of I, respectively. A Zolotarev polynomial has one critical point in the gap interval, $\hat{\gamma} \in (\lambda_2, \lambda_3)$, which is referred to as *the gap critical point.* It is natural to associate with a Zolotarev polynomial the elliptic curve of the equation determined by the four endpoints of the support of the polynomial:

$$\mu^2 = (\lambda-\lambda_1)(\lambda-\lambda_2)(\lambda-\lambda_3)(\lambda-\lambda_4).$$

There is an alternative approach to the Zolotarev polynomials. Namely, one can consider the question of finding the monic polynomial of certain degree n which minimizes the maximum norm on the union of two intervals. Denote such a *minimal* polynomial by $\hat{p}_n$ and its norm L_n. The fact that a polynomial p_n is a solution of the Pell equation (22) on the union of intervals $[\lambda_1,\lambda_2]\cup[\lambda_3,\lambda_4]$ is equivalent to the following conditions:

(i) $p_n = \hat{p}_n/(\pm L_n)$;
(ii) the set $[\lambda_1,\lambda_2]\cup[\lambda_3,\lambda_4]$ is the maximal subset of $\mathbb{R}$ for which $\hat{p}_n$ is the minimal polynomial in the above sense.

It was this extremal problem on two intervals that motivated Akhiezer to study the Zolotarev polynomials in [**1**–**3**] (see also [**4**, **5**]). For some most recent results on Zolotarev polynomials see [**15**]. The polynomials p_n are also called *generalized Chebyshev polynomials* on two intervals $[\lambda_1,\lambda_2]\cup[\lambda_3,\lambda_4]$, with an appropriate normalization.

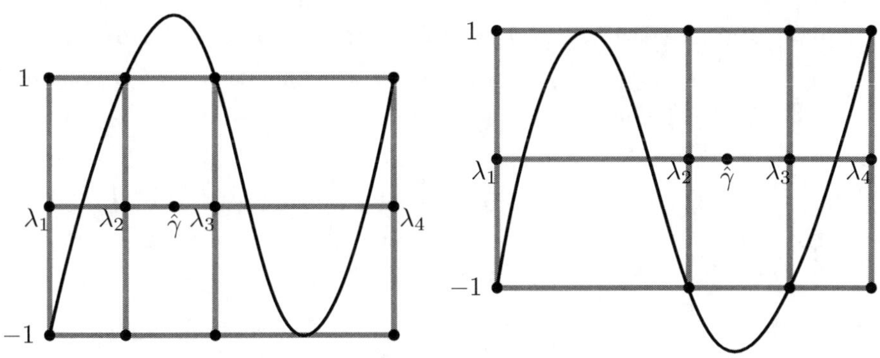

FIGURE 1. Zolotarev polynomials of degree 3 on the union of two intervals $[\lambda_1,\lambda_2]\cup[\lambda_3,\lambda_4]$; the critical point on the gap-interval (λ_2,λ_3) is denoted by $\hat{\gamma}$.

It follows from the above, that an arbitrary pair of intervals is not necessarily the support of a Zolotarev polynomial. We can use affine transformations to fix one of the intervals in an arbitrary position, or to fix the left end-point of the first interval and the right end-point of the second interval.

We will use the classical Jacobi elliptic functions sn, cn, and dn and the theta-functions θ_i, $i=0,1,2,3$ (see e.g. [**5**]). Given a constant κ distinct from $0,1$, define

$$K=\int_0^1\frac{dt}{\sqrt{(1-t^2)(1-\kappa^2t^2)}}.$$

Following [**1**–**3**], let us consider the union of two intervals $E_{n,m}=[-1,\alpha_{n,m}]\cup[\beta_{n,m},1]$, where

$$\alpha_{n,m}=1-2\,\mathrm{sn}^2\left(\frac{m}{n}K\right),\quad \beta_{n,m}=2\,\mathrm{sn}^2\left(\frac{n-m}{n}K\right)-1.$$

Define

$$\hat{a}_n(x,m,\kappa_{n,m})=L_{n,m}\left(v_{n,m}^n(u)+\frac{1}{v_{n,m}^n(u)}\right),$$

where

$$v_{n,m}(u)=\frac{\theta_1\left(u-\frac{m}{n}K\right)}{\theta_1\left(u+\frac{m}{n}K\right)},\quad x=\frac{\mathrm{sn}^2(u)\,\mathrm{cn}^2(\frac{m}{n}K)+\mathrm{cn}^2(u)\,\mathrm{sn}^2(\frac{m}{n}K)}{\mathrm{sn}^2(u)-\mathrm{sn}^2(\frac{m}{n}K)},$$

and

$$L_{n,m} = \frac{1}{2^{n-1}} \left(\frac{\theta_0(0)\theta_3(0)}{\theta_0(\frac{m}{n}K)\theta_3(\frac{m}{n}K)} \right)^{2n}, \quad \kappa^2 = \frac{2(\beta_{n,m} - \alpha_{n,m})}{(1 - \alpha_{n,m})(1 + \beta_{n,m})}.$$

Relationships between the Riemann theta functions θ_i, $i = 0, 1, 2, 3$, and the Jacobi functions are as follows, see e.g. [**5**]:

$$\mathrm{sn}(w) = \frac{\theta_3(0)\theta_1(w)}{\theta_2(0)\theta_0(w)}, \ \mathrm{cn}(w) = \frac{\theta_0(0)\theta_2(w)}{\theta_2(0)\theta_0(w)}, \ \mathrm{dn}(w) = \frac{\theta_0(0)\theta_3(w)}{\theta_3(0)\theta_0(w)}.$$

Akhiezer [**1**–**3**] proved the following result:

THEOREM 2 (Akhiezer). *Let n and m be natural numbers such that $n > m$.*

(a) *The function $\hat{a}_n(x, m, \kappa)$ is a polynomial of degree n in x with the leading coefficient 1 and the second coefficient equal to $-n\tau_1^{(n,m)}$, where*

$$\tau_1^{(n,m)} = -1 + 2\frac{\mathrm{sn}(\frac{m}{n}K)\,\mathrm{cn}(\frac{m}{n}K)}{\mathrm{dn}(\frac{m}{n}K)} \left(\frac{1}{sn(\frac{2m}{n}K)} - \frac{\theta_0'(\frac{m}{n}K)}{\theta_0(\frac{m}{n}K)} \right).$$

(b) *The maximum of the modulus of $\hat{a}_n$ on the union of the two intervals $[-1, \alpha_{n,m}] \cup [\beta_{n,m}, 1]$ is $L_{n,m}$.*

(c) *The function $\hat{a}_n$ takes values $\pm L_{n,m}$ with alternating signs at $n - m + 1$ consecutive points of the interval $[-1, \alpha]$ and at $m + 1$ consecutive points of the interval $[\beta, 1]$. In addition*

$$\hat{a}_n(\alpha_{n,m}, m, \kappa) = \hat{a}_n(\beta_{n,m}, m, \kappa) = (-1)^m L_{n,m},$$

and for any $x \in (\alpha_{n,m}, \beta_{n,m})$, it holds:

$$(-1)^m \hat{a}_n(x, m, \kappa) > L_{n,m}.$$

(d) *The polynomials $\hat{a}_n(x, m, \kappa)/L_{n,m}$ are the Zolotarev polynomials for $E_{n,m} = [-1, \alpha_{n,m}] \cup [\beta_{n,m}, 1]$ with the norm $L_{n,m} = ||\hat{a}_n(x, m, \kappa)||_{E_{n,m}}$ and*

$$E_{n,m} = \hat{a}_n^{-1}[-L_{n,m}, L_{n,m}].$$

(e) *Outside $E_{n,m}$ the derivative of the polynomial $\hat{a}_n(x, m, \kappa)$ with respect to x has only one zero $c_{n,m}$. It belongs to $[\alpha_{n,m}, \beta_{n,m}]$ and*

(23) $$c_{n,m} = \frac{\alpha_{n,m} + \beta_{n,m}}{2} - \tau_1^{(n,m)}.$$

(f) *Let F be a monic polynomial of degree n in x, such that:*
  (i) *$max|F(x)| = L_{n,m}$ for $x \in [-1, \alpha_{n,m}] \cup [\beta_{n,m}, 1]$;*
  (ii) *$F(x)$ takes values $\pm L_{n,m}$ with alternating signs at $n - m + 1$ consecutive points of the interval $[-1, \alpha_{n,m}]$ and at $m + 1$ consecutive points of the interval $[\beta_{n,m}, 1]$.*

  Then $F(x) = \hat{a}_n(x, m, \kappa)$.

Note that, in particular, Akhiezer's theorem implies the existence of a Zolotarev polynomial of degree n for any natural n.

3.3. d intervals case. Let us sketch the theory of extremal polynomials on d real intervals. We consider polynomials p_n of degree n satisfying the Pell equation (3). Here $\mathscr{P}_{2d}$ is a monic polynomial vanishing at the ends of the d finite real intervals $[c_{2d}, c_{2d-1}] \cup [c_{2d-2}, c_{2d-3}] \cup \cdots \cup [c_2, c_1]$ and q_{n-d} is a polynomial of degree $n-d$. In other words, p_n is a solution of the Pell equation, if there exist polynomials $\mathscr{P}_{2d}$ and q_{n-d} such that (3) holds. Solutions p_n of the Pell equation are also called the (generalized) Chebyshev polynomials; they satisfy the following extremality conditions [**6**]. By dividing out the leading coefficient of p_n, we obtain a monic polynomial $\hat{p}_n$ which has the least possible deviation from zero over the above set of intervals among all monic polynomials of the same degree and of the same signature. The signature takes into account the number of oscillations of the polynomial on each interval, i.e. this is a d-vector of integers, where the j-th component counts the number of zeros of q_{n-d} in the j-th interval. The corresponding set of intervals is then the maximal subset of the real line on which $\hat{p}_n$ has the above minimal deviation property.

LEMMA 1 (Theorem 2.12 from [**33**]). *Let p_n, p_n^* be two polynomials of degree n, which solve Pell's equations. Denote by*

$$\mathcal{I}_d = \cup_{j=0}^{d-1} [c_{2(d-j)}, c_{2(d-j)-1}] \quad \text{and} \quad \mathcal{I}_d^* = \cup_{j=0}^{d-1} [c^*_{2(d-j)}, c^*_{2(d-j)-1}]$$

respectively the sets $\{x \mid |p_n(x)| \leq 1\}$ and $\{x \mid |p_n^*(x)| \leq 1\}$. Suppose that:

(i) *at least one of the intervals from $\mathcal{I}_d$ coincides with one of the intervals from $\mathcal{I}_d^*$;*

(ii) *for each $j \in \{0, \dots, d-1\}$, $c_{2(d-j)} = c^*_{2(d-j)}$ or $c_{2(d-j)-1} = c^*_{2(d-j)-1}$;*

(iii) *in each pair of the corresponding intervals*

$$[c_{2(d-j)}, c_{2(d-j)-1}] \quad \text{and} \quad [c^*_{2(d-j)}, c^*_{2(d-j)-1}]$$

the polynomials p_n, p_n^* have the same number of extreme points.

Then the polynomials p_n, p_n^* coincide up to an overall sign and sets $\mathcal{I}_d$ and $\mathcal{I}_d^*$ coincide.

We will use the above lemma in two special cases $d = 2$ and $d = 3$. For $d = 2$ it says that if the interval $[c_4, c_3]$ is fixed and one end-point of the second interval is fixed, say the smaller one, c_2, then there exists at most one point $c_1 > c_2$ such that the Pell equation has a solution of a given degree and with the given numbers of extremal points at each of the intervals $[c_4, c_3]$ and $[c_2, c_1]$. Using Theorem 2.7 from [**33**] and the above lemma we have the following result which we formulate only for $d = 2$ and $d = 3$.

LEMMA 2. *[A corollary of Theorem 2.7 [**33**]]*

(a) *Let $\mathcal{I}_4^0 = [c_4^0, c_3^0] \cup [c_2^0, c_1^0]$ be a union of two disjoint intervals which admits a solution p_n^0 of degree n of Pell's equation. Then there is an ϵ-neighborhood of c_2^0 such that there is a smooth function $c_1 = c_1(c_2)$ which for every c_2 from that neighborhood, satisfies $c_1(c_2) > c_2$ and such that $c_1(c_2^0) = c_1^0$ and the union of two intervals $\mathcal{I}_4 = [c_4^0, c_3^0] \cup [c_2, c_1(c_2)]$ admits a solution p_n of degree n of Pell's equation with the same number of extremal points in the first and the second interval as p_n^0 has in the corresponding intervals.*

(b) *Let $\mathcal{I}_6^0 = [c_6^0, c_5^0] \cup [c_4^0, c_3^0] \cup [c_2^0, c_1^0]$ be a union of three disjoint intervals which admits a solution p_n^0 of degree n of Pell's equation. Then there is an ϵ-neighborhood of c_4^0, c_2^0 such that there are smooth functions c_3, c_1 of*

c_4, c_2 which for every c_4, c_2 from that neighborhood, satisfy $c_3(c_4, c_2) > c_4$, $c_1(c_4, c_2) > c_2$ and $c_1(c_4^0, c_2^0) = c_1^0$, $c_3(c_4^0, c_2^0) = c_3^0$ and are such that the union of three intervals $\mathcal{I}_6 = [c_6^0, c_5^0] \cup [c_4, c_3(c_4, c_2)] \cup [c_2, c_1(c_4, c_2)]$ admits a solution p_n of degree n of Pell's equation with the same number of extremal points in the first and the second interval as p_n^0 has in the corresponding intervals.

The internal extremal points of p_n, the ones contained in the interior of the union of the intervals, are also smooth functions of the variable end-points of the intervals.

3.4. Hyperelliptic surfaces associated with extremal polynomials. Given polynomials p_n, q_{n-d} and $\mathscr{P}_{2d} = \prod_{j=1}^{2d}(\lambda - \lambda_j)$ satisfying Pell equation (3) with real $\lambda_1 < \lambda_2 < \cdots < \lambda_{2d}$, consider an algebraic curve defined by

$$\mathscr{C} = \{(\mu, \lambda) \in \mathbb{C}^2 \mid \mu^2 = \mathscr{P}_{2d}(\lambda)\}. \tag{24}$$

Let us denote by $\hat{\mathscr{C}}$ the corresponding compact Riemann surface of genus $g = d-1$ obtained by adding two points at infinity. The added points are denoted by P_{∞^+} and P_{∞^-} and distinguished by the following conditions: $\mu \sim \lambda^d$ at P_{∞^+} and $\mu \sim -\lambda^d$ at P_{∞^-}.

Let us now fix a canonical homology basis on $\hat{\mathscr{C}}$ consisting of $2g$ closed cycles $\{\mathbf{a_j}, \mathbf{b_j}\}_{j=1}^g$. This choice of the basis in $H_1(\hat{\mathscr{C}})$ determines g holomorphic Abelian differentials $\omega_1, \dots, \omega_g$ normalized by the condition $\oint_{\mathbf{a_j}} \omega_k = \delta_{jk}$ as well as the Riemann $g \times g$ matrix $\mathbb{B}$ and the Jacobian $\mathscr{J}(\hat{\mathscr{C}}) = \mathbb{C}^g/\{\mathbb{Z}^g + \mathbb{B}\mathbb{Z}^g\}$ of the surface $\hat{\mathscr{C}}$. The *Abel map* $\mathscr{A}_{\infty^-}$ based at the point P_{∞^-} is a map from the surface to its Jacobian defined by:

$$\mathscr{A}_{\infty^-}(P) = \int_{P_{\infty^-}}^P \omega\,, \tag{25}$$

where $\omega := (\omega_1, \dots, \omega_g)^T$. According to the Abel theorem, a divisor $\mathscr{D}$ of degree zero is a principal divisor, that is a divisor of a non-zero meromorphic function, on the surface $\hat{\mathscr{C}}$ if and only if $\mathscr{A}_{\infty^-}(\mathscr{D}) \equiv 0$ in $\mathscr{J}(\hat{\mathscr{C}})$. A point $P \in \hat{\mathscr{C}}$ is said to be of order $n \in \mathbb{N}$ if $n\mathscr{A}_{\infty^-}(P) \equiv 0$. In other words, a point $P \in \hat{\mathscr{C}}$ is of order n if and only if the divisor $nP - nP_{\infty^-}$ is principal on $\hat{\mathscr{C}}$, that is if and only if there exists a meromorphic function on $\hat{\mathscr{C}}$ with a zero of order n at P and a pole of order n at P_{∞^-} and no other zeros or poles. We will use the following theorem, see e.g. [**6**].

THEOREM 3. *Let $\mathscr{P}_{2d}$ be a monic polynomial of degree $2d$ with $2d$ distinct real roots $\lambda_1 < \cdots < \lambda_{2d}$ and consider the compact Riemann surface $\hat{\mathscr{C}}$ corresponding to the algebraic curve (24). Let $P_{\pm\infty} \in \hat{\mathscr{C}}$ be the points added at infinity specified, respectively, by the conditions $\mu \sim \pm\lambda^d$.*

The union $E = [\lambda_1, \lambda_2] \cup \cdots \cup [\lambda_{2d-1}, \lambda_{2d}]$ of d intervals determined by the zeros of the polynomial $\mathscr{P}_{2d}$ admits a solution (p_n, q_{n-d}) of the Pell equation (3) where p_n, q_{n-d} are polynomials of degrees n and $n-d$, respectively, if and only if the point P_{∞^+} is a point of order n on $\hat{\mathscr{C}}$, in the sense that $n\mathscr{A}_{\infty^-}(P_{\infty^+}) \equiv 0$.

Proof. The Akhiezer function (9) is a function on the surface $\hat{\mathscr{C}}$ and can be written in the form

$$\mathbb{A}(P) = p_n(\lambda) - \mu q_{n-d}(\lambda), \quad P = (\lambda, \mu) \in \hat{\mathscr{C}}\,, \tag{26}$$

where λ and μ are naturally extended to the points $P_{\infty^\pm}$. Since $\mu \sim -\lambda$ at P_{∞^-}, the Akhiezer function has a pole of order n at P_{∞^-}. To see that it has a zero of order n at P_{∞^+}, consider the function

$$\mathbb{A}^*(P) = p_n(\lambda) + \mu q_{n-d}(\lambda),$$

which satisfies $\mathbb{A}^*(P) = \mathbb{A}(P^*)$, where $P^* = (\lambda, -\mu)$ if $P = (\lambda, \mu)$, that is P^* is obtained from P by the hyperelliptic involution on $\hat{\mathscr{C}}$. Since p_n, q_{n-d} and $\mu^2 = \mathscr{P}_{2d}$ is a solution to the Pell equation (3), we have

$$\mathbb{A}(P)\mathbb{A}^*(P) = p_n^2(\lambda) - \mu q_{n-d}^2(\lambda) = 1$$

and thus $\mathbb{A}^*(P)$ has a zero of order n at the point where the Akhiezer function has a pole of order n, that is at P_{∞^-}. Due to the relations $\mathbb{A}^*(P) = \mathbb{A}(P^*)$ and $P_{\infty^+} = P^*_{\infty^-}$, we conclude that the Akhiezer function has a zero of order n at P_{∞^+}.

This argument also shows that every pole of $\mathbb{A}$ is paired by the hyperelliptic involution to a zero of $\mathbb{A}$. Since from expression (26) it follows that the only pole of $\mathbb{A}$ is at infinity, we conclude that $\mathbb{A}$ does not have zeros in the affine part of the curve (away from points at infinity). Thus the divisor of $\mathbb{A}$ is $nP_{\infty^+} - nP_{\infty^-}$ and thus P_{∞^+} is a point of order n on $\hat{\mathscr{C}}$. □

4. Elliptic curves associated with Zolotarev polynomials

In this section we use the theory of extremal polynomials on 2 intervals, the Zolotarev polynomials, to construct solutions to the Painlevé-VI$(\frac{1}{8}; -\frac{1}{8}; \frac{1}{8}; \frac{3}{8})$ equation in terms of elliptic functions from Theorem 2. In the subsequent sections, this construction is generalized to the case of extremal polynomials on 3 intervals.

4.1. Akhiezer function and the associated differential. Let $\hat{\mathscr{C}}_1$ be the genus one curve defined as the projective compactification of the algebraic curve (24) with $d = 2$. In other words, $\hat{\mathscr{C}}_1$ is the compactification of the curve $\mathscr{C}_1$ defined by

$$\mathscr{C}_1 = \{(\mu, \lambda) \in \mathbb{C}^2 \mid \mu^2 = (\lambda - \lambda_1)(\lambda - \lambda_2)(\lambda - \lambda_3)(\lambda - \lambda_4)\} \tag{27}$$

and p_n, q_{n-2} are polynomials satisfying the Pell equation

$$p_n^2(\lambda) - \mu^2 q_{n-2}^2(\lambda) = 1. \tag{28}$$

Recall that $\lambda_1 < \lambda_2 < \lambda_3 < \lambda_4$ and p_n is the Zolotarev polynomial of degree n.

As explained in Section 3.4, the Akhiezer function

$$\mathbb{A}(P) = p_n(\lambda) - \mu q_{n-2}(\lambda) \tag{29}$$

from (9) defined for $P \in \hat{\mathscr{C}}_1$ is the function that has a pole of order n at P_{∞^-}, a zero of order n at P_{∞^+} and no other zeros or poles. Therefore, the differential $\Omega_{\mathbb{A}}$ (9) is a differential of the third kind having two simple poles with residue 1 at P_{∞^+} and with residue -1 at P_{∞^-} and no other singularities. As a consequence, $\Omega_{\mathbb{A}}$ has two zeros on the compact surface $\hat{\mathscr{C}}_1$ and since $\Omega_{\mathbb{A}}(P^*) = -\Omega_{\mathbb{A}}(P)$ due to the form of the divisor of the Akhiezer function, the zeros of $\Omega_{\mathbb{A}}$ are paired by the involution exchanging the sheets, that is they have the same projection on the λ-plane.

The following lemma can be found in the literature on Pell equations and approximation theory, see, for example, [**21**] and [**7**].

LEMMA 3. *Let $p_n(\lambda)$ be the Zolotarev polynomial of order n and $\Omega_{\mathbb{A}}$ be the Akhiezer differential* (9) *defined on the associated compact Riemann surface $\hat{\mathscr{C}}_1$. The λ-coordinate of the zeros of $\Omega_{\mathbb{A}}$ coincides with the gap critical point of the Zolotarev polynomial p_n, that is the unique critical point of p_n occurring in the interval (λ_2, λ_3). Moreover, denoting the gap critical point by $\hat{\gamma}$, we have*

$$\Omega_{\mathbb{A}}(P) = -\frac{(\lambda - \hat{\gamma})d\lambda}{\mu}, \qquad P = (\lambda, \mu).$$

Proof. The gap critical point $\hat{\gamma}$ satisfies $p_n'(\hat{\gamma}) = 0$ and $\hat{\gamma} \in (\lambda_2, \lambda_3)$. Let us first show that

$$p_n'(\lambda) = nq_{n-2}(\lambda)(\lambda - \hat{\gamma}). \tag{30}$$

Note first that all zeros of q_{n-2} belong to the support $[\lambda_1, \lambda_2] \cup [\lambda_3, \lambda_4]$ of the Zolotarev polynomial p_n due to the Pell equation (28) and the fact that $|p_n(\lambda)| > 1$ for λ outside of the support of p_n.

Now, differentiating Pell's equation (28), we obtain

$$p_n(\lambda)p_n'(\lambda)d\lambda = \mu q_{n-2}(\lambda)(q_{n-2}(\lambda)d\mu + \mu q_{n-2}'(\lambda)d\lambda). \tag{31}$$

From the Pell equation, we see that p_n and q_{n-2} do not have common zeros. Thus (31) implies that q_{n-2} divides p_n'. Since p_n' has $n-1$ zeros, and since $p_n = 1$ at the zeros of p_n' belonging to the support, there exist a zero of p_n' which is not a zero of q_{n-2} and thus lies outside of the support of the Zolotarev polynomial. Denote this zero by γ. Given that all oscillations of p_n occur between λ_1 and λ_4, we conclude that $\hat{\gamma}$ belongs to the gap interval (λ_2, λ_3) and thus, because μ^2 is monic, p_n and q_{n-2} have equal leading coefficients and we have

$$p_n'(\lambda) = nq_{n-2}(\lambda)(\lambda - \hat{\gamma}). \tag{32}$$

Plugging this into (31) we get

$$np_n(\lambda)(\lambda - \hat{\gamma})d\lambda = q_{n-2}(\lambda)\mu d\mu + \mu^2 q_{n-2}'(\lambda)d\lambda. \tag{33}$$

Now, from definitions (9) and (29) and using (32) and (33) in the second line, we obtain

$$\begin{aligned}\Omega_{\mathbb{A}}(P) \quad &= \frac{1}{n}\frac{d\mathbb{A}(P)}{\mathbb{A}(P)} = \frac{1}{n}\frac{p_n'(\lambda)d\lambda - q_{n-2}(\lambda)d\mu - \mu q_{n-2}'(\lambda)d\lambda}{p_n(\lambda) - \mu q_{n-2}(\lambda)} \\ &= \frac{q_{n-2}(\lambda)(\lambda - \hat{\gamma})d\lambda - p_n(\lambda)(\lambda - \hat{\gamma})d\lambda/\mu}{p_n(\lambda) - \mu q_{n-2}(\lambda)} = -\frac{(\lambda - \hat{\gamma})d\lambda}{\mu}. \qquad \square\end{aligned}$$

The form of $\Omega_{\mathbb{A}}$ from Lemma 3 in terms of λ and μ shows that the differential vanishes at two points λ-coordinate of which is $\hat{\gamma}$. Let us denote these points by $P_{\hat{\gamma}}$ and $P_{\hat{\gamma}}^*$. Let us now rewrite the differential $\Omega_{\mathbb{A}}$ in yet another form using the function theory of Riemann surfaces introduced in Section 3.4. Recall that the points at infinity of the surface $\hat{\mathscr{C}}_1$ satisfy $n\mathscr{A}_{\infty^-}(P_{\infty^+}) \equiv 0$, the condition that we rewrite in the next lemma in a more explicit form in terms of the Riemann matrix τ of the surface.

LEMMA 4. *Let $\{\mathbf{a}, \mathbf{b}\}$ be a canonical basis of cycles in $H_1(\hat{\mathscr{C}}_1)$ and ω the corresponding normalized holomorphic differential. Let $\tau = \oint_{\mathbf{b}} \omega$ be the Riemann matrix* *of* $\hat{\mathscr{C}_1}$. *Let the integers* k *and* l *be such that for the point* P_{∞^+} *of order* n *on* $\hat{\mathscr{C}_1}$ *we have*

$$n\int_{P_{\infty^-}}^{P_{\infty^+}}\omega = k\tau + l\,. \tag{34}$$

Let $\Omega_{\infty^\pm}$ *be a differential of the third kind with simple poles at* P_{∞^+} *and* P_{∞^-} *with residues* 1 *and* -1, *respectively, with vanishing* **a**-*period:* $\oint_{\mathbf{a}}\Omega_{\infty^\pm}=0$. *Then the differential* $\Omega_{\mathbb{A}}$ *from Lemma 3 has the form*

$$\Omega_{\mathbb{A}} = \Omega_{\infty^\pm} - 2\pi \mathrm{i}\frac{k}{n}\omega.$$

Proof. Given the description of the Akhiezer function $\mathbb{A}$ as a function on the surface $\hat{\mathscr{C}_1}$ having a pole of order n at P_{∞^-} and a zero of order n at P_{∞^+}, this function can be written in therm of the Jacobi theta-function θ_1 as follows:

$$\mathbb{A}(P) = \frac{\theta_1^n\left(\int_{P_{\infty^+}}^{P}\omega\right)}{\theta_1^n\left(\int_{P_{\infty^-}}^{P}\omega\right)}\exp\left\{-2k\pi\mathrm{i}\int_{P_{\infty^-}}^{P}\omega\right\}, \qquad P\in\hat{\mathscr{C}_1}, \tag{35}$$

where k is the integer from (34). This formula for $\mathbb{A}$ can be justified using (34) and the following properties of the theta-function:

$$\theta_1(0;\tau)=0, \qquad \theta_1(z+1;\tau)=-\theta_1(z;\tau), \qquad \theta_1(z+\tau;\tau) = -\mathrm{e}^{-\pi\mathrm{i}\tau-2\pi\mathrm{i}z}\theta_1(z;\tau)\,.$$

To complete the proof it remains to differentiate (35):

$$\Omega_{\mathbb{A}}(P) = \frac{1}{n}\frac{d\mathbb{A}(P)}{\mathbb{A}(P)} = d_P\log\frac{\theta_1\left(\int_{P_{\infty^+}}^{P}\omega\right)}{\theta_1\left(\int_{P_{\infty^-}}^{P}\omega\right)} - 2\pi\mathrm{i}\frac{k}{n}\omega(P) = \Omega_{\infty^\pm}(P) - 2\pi\mathrm{i}\frac{k}{n}\omega(P). \qquad \square$$

To summarize this section, we have the following representations for the differential of the third kind on $\hat{\mathscr{C}_1}$ which has two simple poles at P_{∞^+} and P_{∞^-} and two simple zeros at the two points $P_{\hat{\gamma}}$ and $P_{\hat{\gamma}}^*$ on $\hat{\mathscr{C}_1}$ having the λ-coordinate equal to the gap critical point $\hat{\gamma}\in(\lambda_2,\lambda_3)$ of the Zolotarev polynomial p_n and no other zeros or poles:

$$\Omega_{\mathbb{A}}(P) = \frac{1}{n}\frac{d\mathbb{A}(P)}{\mathbb{A}(P)} = -\frac{(\lambda-\hat{\gamma})d\lambda}{\mu} = \Omega_{\infty^\pm}(P) - 2\pi\mathrm{i}\frac{k}{n}\omega(P).$$

The Akhiezer function $\mathbb{A}$ on $\hat{\mathscr{C}_1}$ and the associated differential $\Omega_{\mathbb{A}}$ coincide with the function $s(P)$ and the differential $d\log s$ from paper [**23**] establishing the link between Poncelet trajectories and algebraic solutions of the equation Painlevé-VI$(1/8;-1/8;1/8;3/8)$, or, in other words, between points of a finite order on an elliptic curve and algebraic solutions of PVI$(1/8;-1/8;1/8;3/8)$.

4.2. From the Akhiezer function to Painlevé-VI$\left(\frac{1}{8};-\frac{1}{8};\frac{1}{8};\frac{3}{8}\right)$. Here we are going to make a connection between the differential $\Omega_{\mathbb{A}}$ and solutions of the Painlevé-VI equation with constants $\left(\frac{1}{8};-\frac{1}{8};\frac{1}{8};\frac{3}{8}\right)$ using the results of [**14**]. In order to do this, we need to perform a Möbius transformation in the λ-sphere of the curve $\hat{\mathscr{C}_1}$ of Section 4.1 bringing the branch points $\lambda_1,\lambda_2,\lambda_4$ to the points $0,1,\infty$, that is we need to make the following change of variables

$$u=\phi(\lambda) = \frac{\lambda-\lambda_1}{\lambda-\lambda_4}\frac{\lambda_2-\lambda_4}{\lambda_2-\lambda_1}\,. \tag{36}$$

Let us denote by $\hat{\mathscr{C}}_x$ the compact genus one surface corresponding to the algebraic cuve

(37)
$$\mathscr{C}_x = \{(u,v) \in \mathbb{C}^2 \mid v^2 = u(u-1)(u-x)\} \quad \text{with} \quad x = \phi(\lambda_3) = \frac{\lambda_3 - \lambda_1}{\lambda_3 - \lambda_4}\frac{\lambda_2 - \lambda_4}{\lambda_2 - \lambda_1}$$

obtained from (27) by the above change of variables. The Möbius transformation ϕ extends naturally to the points of the surface $\hat{\mathscr{C}}_1$, we denote this extension also by ϕ. The canonical homology cycles $\mathbf{a}$ and $\mathbf{b}$ are transformed into a canonical basis of cycles $\phi \circ \mathbf{a}$ and $\phi \circ \mathbf{b}$ on the new surface $\hat{\mathscr{C}}_x$ and the holomorphic differential on $\hat{\mathscr{C}}_x$ normalized with respect to the basis $\phi \circ \mathbf{a}$, $\phi \circ \mathbf{b}$ can be obtained from the holomorphic normalized differential ω on $\hat{\mathscr{C}}_1$ by the pullback $(\phi^{-1})^*\omega$.

Let us denote by $Q_0 \in \hat{\mathscr{C}}_x$ the image by ϕ of the point $P_{\infty^+} \in \hat{\mathscr{C}}_1$. For the u-coordinate of Q_0, denoted by y_0, we thus have

$$y_0 = \frac{\lambda_2 - \lambda_4}{\lambda_2 - \lambda_1} \quad \text{and} \quad x = \frac{\lambda_3 - \lambda_1}{\lambda_3 - \lambda_4} y_0 \,. \tag{38}$$

Let $\mathscr{A}_\infty$ be the Abel map on $\hat{\mathscr{C}}_x$ based at the point $P_\infty = \hat{\phi}(\lambda_4, 0)$. Equation (34) implies

$$n\int_{P_{\infty^-}}^{P_{\infty^+}} \omega = n\int_{Q_0^*}^{Q_0} (\phi^{-1})^*\omega = n(\mathscr{A}_\infty(Q_0) - \mathscr{A}_\infty(Q_0^*)) = k\tau + l\,. \tag{39}$$

The space of holomorphic differentials on a genus one curve $\hat{\mathscr{C}}_x$ is one dimensional and thus the differential $(\phi^{-1})^*\omega$ is proportional to the du/v. Therefore, our holomorphic normalized differential satisfies $(\phi^{-1})^*\omega(P^*) = -(\phi^{-1})^*\omega(P)$ for two points $P = (u,v)$ and $P^* = (u,-v)$. Since the base of the Abel map $\mathscr{A}_\infty$ is a ramification point, we have $\mathscr{A}_\infty(Q_0) = -\mathscr{A}_\infty(Q_0^*)$ and thus (39) implies

$$\mathscr{A}_\infty(Q_0) = \frac{k}{2n}\tau + \frac{l}{2n}\,, \tag{40}$$

that is Q_0 is a point of order $2n$ on $\hat{\mathscr{C}}_x$. Let us denote by Ω_{Q_0,Q_0^*} the differential of the third kind on $\hat{\mathscr{C}}_x$ with simple poles at Q_0 and Q_0^* of residues 1 and -1, respectively, normalized by the vanishing $\phi \circ \mathbf{a}$-period. We have $\Omega_{Q_0,Q_0^*} = (\phi^{-1})^*\Omega_{\pm\infty}$ where $\Omega_{\pm\infty}$ is the normalized differential of the third kind on $\hat{\mathscr{C}}_1$ from Lemma 4. Let us now denote by Ω the differential on $\hat{\mathscr{C}}_x$ obtained by the pullback of $\Omega_{\mathbb{A}}$:

$$\Omega := (\phi^{-1})^*\Omega_{\mathbb{A}} = \Omega_{Q_0,Q_0^*} - 4\pi \mathrm{i}\frac{k}{2n}(\phi^{-1})^*\omega.$$

Due to results of Lemmas 3 and 4, the differential Ω has two simple zeros at the points of $\hat{\mathscr{C}}_x$ that have u-coordinate equal to $\phi(\hat{\gamma})$, that is equal to the image by the Möbius transformation (36) of the gap critical point of the Zolotarev polynomial p_n from Section 4.1.

On the other hand, it is shown in [**14**] that the u-coordinate of the zero of Ω as a function of x from (37) satisfies the Painlevé-VI equation with constants $(\frac{1}{8}, -\frac{1}{8}, \frac{1}{8}, \frac{3}{8})$. Denoting $y := \phi(\hat{\gamma})$ we conclude that $y(x)$ is a solution to $\mathrm{PVI}(\frac{1}{8}, -\frac{1}{8}, \frac{1}{8}, \frac{3}{8})$, where

$$y := \phi(\hat{\gamma}) = \frac{\hat{\gamma} - \lambda_1}{\hat{\gamma} - \lambda_4}\frac{\lambda_2 - \lambda_4}{\lambda_2 - \lambda_1} \quad \text{and} \quad x = \phi(\lambda_3) = \frac{\lambda_3 - \lambda_1}{\lambda_3 - \lambda_4}\frac{\lambda_2 - \lambda_4}{\lambda_2 - \lambda_1}\,. \tag{41}$$

Let us outline the way this result is proven in [**14**]. Note that y_0 from (38), the u-coordinate of the point Q_0 defined by (40), as a function of the variable branch point x, satisfies the Painlevé-VI equation with constants $(0,0,0,\frac{1}{2})$ and is a particular case of the Picard solution [**34**].

The Picard general solution $y_0(x)$ to PVI$(0,0,0,\frac{1}{2})$ and the general solution $y(x)$ to PVI$(\frac{1}{8},-\frac{1}{8},\frac{1}{8},\frac{3}{8})$ are related by the following Okamoto transformation [**32**], see also [**14**]:

$$y(x) = y_0 + \frac{y_0(y_0-1)(y_0-x)}{x(x-1)y_0' - y_0(y_0-1)}. \tag{42}$$

On the other hand, the differential Ω can also be written in terms of the coordinate u as follows:

$$\Omega(P) = \frac{du(P)}{I_0 v\, \omega(Q_0)}\left[\frac{1}{u(P)-y_0} - \sigma\right] - 4\pi \mathrm{i}\frac{k}{2n}\frac{du}{I_0 v}, \tag{43}$$

where σ is a normalization constant ensuring that expression (43) has the prescribed a-period, $\omega(Q_0)$ stands for $(I_0 y_0(y_0-1)(y_0-x))^{-1/2}$, and I_0 is another normalization constant. From here, we obtain the relationship between $y_0 = u(Q_0)$ and the u-coordinate y of the two zeros of Ω :

$$\frac{1}{y-y_0} = \sigma + 4\pi\mathrm{i}\frac{k}{2n}\omega(Q_0). \tag{44}$$

Then, by differentiating (40), it is proven in [**14**], see also Remark 2 in Section 7.2.1 below, that the derivative y_0' entering (42) is given by

$$\frac{dy_0}{dx} = \frac{y_0(y_0-1)(y_0-x)}{x(x-1)}\left[4\pi\mathrm{i}\,\frac{k}{2n}\,\omega(Q_0) - \frac{1}{x-y_0} + \sigma\right]. \tag{45}$$

Using (45) to express the right hand side of (44), one sees that the relationship between the position y_0 of poles of Ω and the position y of its zeros coincides with the Okamoto transformation (42).

REMARK 1. In this section we used the point of a finite order Q_0 obtained by a Möbius transformation of the point at infinity on the curve associated with a Zolotarev polynomial. A similar construction is done in [**23**] using a point of finite order on an elliptic curve given by the Cayley condition for existence of Poncelet trajectories. We also used implicitly that (40) is preserved under the variations of the complex structure of the curve produced by variation of x. Given that for a genus one surface the Abel map is biholomorphic and the surface is isomorphic to its Jacobian, the use of the Zolotarev polynomials in this subsection is not essential (it will become essential in the next subsection however). In fact, we may as well define a point Q_0 on $\hat{\mathscr{C}}_x$ by the condition $\mathscr{A}_\infty(Q_0) = c_1 + c_2\tau$ requiring the complex constants c_1 and c_2 to be independent of x. All results of this section concerning the relationship with the Painlevé-VI will remain valid, see [**14**].

However, in Section 5 we are going to generalize the construction of this section to surfaces of genus two and then it is much more difficult to define a point on a varying surface by fixing the coordinates of its Abel image in the Jacobian and requiring those coordinates to be constant through the variation of the surface. In the case of genus two, we will construct a point of a finite order on the surface by the process we used here, that is by starting with a surface associated with an extremal polynomial as in Section 4. Due to Theorem 3, this surface has a point of a finite order at infinity. We will then use Lemmas 1 and 2 from Section 3.3 allowing to deform the surface while keeping the association with a generalized Chebyshev polynomial and thus while keeping the point at infinity to be of finite order n, where n is equal to the degree of the polynomial.

4.3. Solutions of the PVI$(\frac{1}{8},-\frac{1}{8},\frac{1}{8},\frac{3}{8})$ in terms of elliptic functions. In Section 4.2, we obtained a solution $y(x)$ of PVI$(\frac{1}{8},-\frac{1}{8},\frac{1}{8},\frac{3}{8})$ in terms of the gap critical point $\hat{\gamma}$ of a Zolotarev polynomial with support $[\lambda_1,\lambda_2]\cup[\lambda_3,\lambda_4]$. On the other hand, the theorem of Akhiezer 2 gives a complete parameterizations for the Zolotarev polynomials with support of the form $[-1,\alpha_{n,m}]\cup[\beta_{n,m},1]$ and their gap critical points in terms of elliptic functions sn, cn and dn. In this section, we combine these two results to obtain solutions to PVI$(\frac{1}{8},-\frac{1}{8},\frac{1}{8},\frac{3}{8})$ in terms of elliptic functions.

THEOREM 4. *Let $n>m$ be natural numbers and let the functions $\alpha_{n,m}$, $\beta_{n,m}$ and $\tau_1^{(n,m)}$ be those defined in Theorem 2. Then the following functions $y_{n,m}(x)$ are solutions to the Painlevé-VI equation with constants $(\frac{1}{8},-\frac{1}{8},\frac{1}{8},\frac{3}{8})$:*

$$y_{n,m}(x)=\frac{2+\alpha_{n,m}+\beta_{n,m}-2\tau_1^{(n,m)}}{2-\alpha_{n,m}-\beta_{n,m}+2\tau_1^{(n,m)}}\cdot\frac{1-\alpha_{n,m}}{1+\alpha_{n,m}},$$

where

$$x=\frac{\beta_{n,m}+1}{\beta_{n,m}-1}\cdot\frac{\alpha_{n,m}-1}{\alpha_{n,m}+1}.$$

Proof. In (44) we have obtained a solution $y(x)$ to PVI$(\frac{1}{8},-\frac{1}{8},\frac{1}{8},\frac{3}{8})$ in terms of the gap critical point $\hat{\gamma}$ of a Zolotarev polynomial with support being the set $[\lambda_1,\lambda_2]\cup[\lambda_3,\lambda_4]$. The theorem of Akhiezer, Theorem 2 gives expression (23) for the gap critical point of a Zolotarev polynomial corresponding to the set $[-1,\alpha_{n,m}]\cup[\beta_{n,m},1]$. Thus, it only remains to use the following identifications in (44): $\lambda_1=-1$, $\lambda_2=\alpha_{n,m}$, $\lambda_3=\beta_{n,m}$, $\lambda_4=1$, and $\hat{\gamma}=c_{n,m}$ from (23). □

In Section 6 of [**23**], some particular examples of solutions from Theorem 4 are written in an alternative form.

Note that the Painlevé-VI equation encodes the isomonodromy condition for the Fuchsian system (5) with four Fuchsian singularities, at $0,1,x,\infty$ and is thus equivalent to the corresponding Schlesinger system (6), see [**25**]. The representation of solutions of the PVI$(\frac{1}{8},-\frac{1}{8},\frac{1}{8},\frac{3}{8})$ as zeros of the differential Ω (43) led to solutions to the appropriate Schlesinger system in terms of Ω in [**14**]. In the rest of the present paper, we adapt the approach of [**14**] to differentials on hyperelliptic surfaces of genus two associated with solutions of the Pell equation on the union of three real intervals. This leads to solutions of the constrained Schlesinger system from Section 2.

5. Chebyshev family of genus two curves, $d=3$

Consider a family of the following hyperelliptic curves

(46) $$\mathscr{C}=\{(u,v)\in\mathbb{C}^2\mid v^2=u(u-1)(u-\nu)(u-x_1)(u-x_2)\}$$

where x_1 and x_2 are independent variables and ν is a function of x_1 and x_2. The projection on the first coordinate $f_u:\mathscr{C}\to\mathbb{C}$ defined by $f_u(u,v)=u$ represents a curve of the family as a two-fold ramified covering of the u-plane. This covering is ramified over the *branch points*, or critical values of f_u, that is points of the set

$\{0, 1, \nu, x_1, x_2\}$. In what follows, it will be convenient to denote the set of branch points by B and an arbitrary point from this set by a_j, that is $B = \{a_1, \ldots, a_5\} = \{0, 1, \nu, x_1, x_2\}$.

The family (46) is thus parameterized by the values of two branch points x_1 and x_2. The corresponding compact projective curves $\hat{\mathscr{C}}$ are of genus two; the compactification is achieved by adding one (singular) point at infinity. The point at infinity is a branch point of the extended function $\hat{f}_u : \hat{\mathscr{C}} \to \mathbb{C}P^1$. The two-fold covering defined by $\hat{f}_u$ and denoted by $(\mathscr{C}, \hat{f}_u)$ is a Riemann surface which we are going to work with. By abuse of notation, we denote the Riemann surface corresponding in this way to the curve (46) also by $\mathscr{C}$.

In order to introduce the standard atlas for the complex structure of the Riemann surface $\mathscr{C}$, let us fix some notation. The pre-image by $\hat{f}_u$ of a branch point a_j is denoted by P_{a_j}. This is the point of the algebraic curve with (u, v)-coordinates being, respectively, a_j and 0, we can write $P_{a_j} = (a_j, 0)$. The pole of $\hat{f}_u$, the point at infinity of the Riemann surface $\mathscr{C}$, is denoted by P_∞.

We use the standard local charts defining the complex structure on $\mathscr{C}$ given as follows. For a point $P \in C$ being in a neighbourhood of

- $P \sim P_{a_j}$ with $a_j \in \{0, 1, \nu, x_1, x_2\}$, the local parameter is $\xi_j(P) = \sqrt{f_u(P) - a_j}$;
- $P \sim P_\infty$, the local parameter is $\xi_\infty(P) = 1/\sqrt{f_u(P)}$;
- $P \sim Q = (u, v)$ with $u \in \mathbb{C} \setminus \{0, 1, \nu, x_1, x_2\}$, the local parameter is $\xi(P) = f_u(P)$.

Note that the Riemann surfaces corresponding to two distincts sets $\{x_1, x_2\}$ are in general non isomorphic. We thus have a family of Riemann surfaces parametrized by values of two variables, x_1 and x_2.

Another tool we need is a canonical homology basis on our compact orientable genus two surfaces $\mathscr{C}$. Let us choose one such basis $\mathbf{a_1}, \mathbf{a_2}; \mathbf{b_1}, \mathbf{b_2}$ on one of the surfaces $\mathscr{C}_0$ corresponding to the branch points $\{0, 1, \nu(x_1^0, x_2^0), x_1^0, x_2^0\}$. Now, consider the images by $\hat{f}_u$ of the four basis cycles. From now on, we assume that the independent branch points x_1 and x_2 vary in neighbourhoods of x_1^0 and x_2^0, respectively, and that these neighbourhoods are small enough to allow for canonical homology bases to be chosen on the surfaces of the family in such a way that the images of the basis cycles by $\hat{f}_u$ be independent of x_1 and x_2, that is be the same for all surfaces in the family.

Given the choice of a canonical homology basis, we can introduce two linearly independent holomorphic differentials ω_1, ω_2 satisfying the normalization conditions

$$\oint_{\mathbf{a_j}} \omega_k = \delta_{jk}. \tag{47}$$

The $\mathbf{b}$-periods of ω_1, ω_2 form the Riemann matrice $\mathbb{B}$ of the surface, $\mathbb{B}_{jk} = \oint_{\mathbf{b_j}} \omega_k$, which is a symmetric matrix with positive definite imaginary part. The four vectors of the $\mathbf{a}$- and $\mathbf{b}$-periods of the vector $\omega := (\omega_1, \omega_2)^T$ define the following lattice in $\mathbb{C}^2$

$$\Lambda := \left\{ \begin{pmatrix} m_1 \\ m_2 \end{pmatrix} + \mathbb{B} \begin{pmatrix} n_1 \\ n_2 \end{pmatrix} \,\middle|\, m_i, n_i \in \mathbb{Z} \right\}. \tag{48}$$

The *Jacobian* of the surface $\mathscr{C}$ is the quotient $\mathscr{J}(\mathscr{C}) = \mathbb{C}^2/\Lambda$. The *Abel map* based at the point P_∞ is the following map from the surface to its Jacobian:

$$\mathscr{A}_\infty(P) = \int_{P_\infty}^{P} \omega\,, \tag{49}$$

where as before $\omega := (\omega_1, \omega_2)^T$.

Let us impose one more condition on our family of hyperelliptic surfaces. Let us assume the existence of a point Q_0 on each surface $\mathscr{C}$ of the family, that is a point $Q_0 = Q_0(x_1, x_2)$ on the surface corresponding to the branch points $\{0, 1, \nu(x_1, x_2), x_1, x_2\}$ such that

$$\int_{P_\infty}^{Q_0} \omega = c_1 + \mathbb{B}c_2 \tag{50}$$

where the $\mathbb{R}^2$ vectors c_1 and c_2 are independent of x_1, x_2, that is c_1 and c_2 are the same for all curves of the family.

Note that condition (50) can be seen as a condition for the function $\nu(x_1, x_2)$. Next, we show that such a family of surfaces does exist for some $c_1, c_2 \in \mathbb{Q}^2$ and is given by the dynamics of generalized Chebyshev polynomials.

Let us denote by $\mathscr{T}$ the family of Riemann surfaces (46) described here endowed with a consistent choice of canonical homology basis through the whole family and satisfying condition (50); let us refer to the family $\mathscr{T}$ as a *Chebyshev family of Riemann surfaces.*

5.1. Existence of Chebyshev families of genus two surfaces. We need to show existence of a family of compact surfaces $\mathscr{C} = \mathscr{C}_{x_1,x_2}$ of genus 2 corresponding to the curves of the form (46) parametrized by x_1, x_2 with $\nu = \nu(x_1, x_2)$ such that for any x_1, x_2 in some small set $X \subset \mathbb{R}$, there exists a point $Q_0 = Q_0(x_1, x_2) \in \mathscr{C}_{x_1,x_2}$ smoothly depending on $x_1, x_2 \in X$ and satisfying (50) with some constants c_1, c_2 independent of x_1, x_2.

Let us start with a generalized Chebyshev polynomial $p_n(\lambda)$ satisfying the Pell equation (3) on the union of three intervals $[0, 1] \cup [\lambda_1, \lambda_2] \cup [\lambda_3, \lambda_4]$, that is with $d = 3$ and the coefficient being $\mathscr{P}_6(\lambda) = \lambda(\lambda - 1)\prod_{j=1}^4(\lambda - \lambda_j)$.

Due to Theorem 3, the point at infinity $P_{\infty+}$ on the compact surface corresponding to equation $\mu^2 = \mathscr{P}_6(\lambda)$ is of the order n, that is $n\mathscr{A}_{\infty^-}(P_{\infty^+}) \equiv 0$. Here we use the notation of Section 3.4.

Lemmas 1 and 2(b) from Section 3.4 imply that we can consider λ_1 and λ_3 as independent variables and define the functions $\lambda_2 = \lambda_2(\lambda_1, \lambda_3)$ and $\lambda_4 = \lambda_4(\lambda_1, \lambda_3)$ by the condition that the set $[0, 1] \cup [\lambda_1, \lambda_2(\lambda_1, \lambda_3)] \cup [\lambda_3, \lambda_4(\lambda_1, \lambda_3)]$ is a support of the Pell equation, that is there exist polynomials $p_n(\lambda)$ and $q_{n-3}(\lambda)$ of degrees n and $n - 3$, respectively, such that

$$p_n^2(\lambda) - \mathscr{P}_{\lambda_1,\lambda_3}(\lambda)q_{n-3}^2(\lambda) = 1$$

with

$$\mathscr{P}_{\lambda_1,\lambda_3}(\lambda) = \lambda(\lambda - 1)\prod_{j=1}^{4}(\lambda - \lambda_j) \quad \text{where} \quad \lambda_k = \lambda_k(\lambda_1, \lambda_3),\ k = 2, 4.$$

Using Theorem 3 again, we conclude that the compact surface, denote it by $\mathscr{C}_{\lambda_1,\lambda_3}$, corresponding to the equation $\mu^2 = \mathscr{P}_{\lambda_1,\lambda_3}(\lambda)$ has a point of order n at infinity:

$n\mathscr{A}_\infty\text{-}(P_{\infty^+}) \equiv 0$. We thus have a continuous family of compact surfaces $\mathscr{C}_{\lambda_1,\lambda_3}$ each of which has a point of order n at infinity.

Now, to bring this situation to the one described in Section 5, we need to bring the point of order n from infinity to a finite place, that is to the affine part of the algebraic curve. To this end, let us apply the following Möbius transformation in the λ-sphere:

$$u := \phi(\lambda) = \frac{\lambda(1-\lambda_4)}{\lambda - \lambda_4}. \tag{51}$$

Let us denote the images of the λ_j as follows

$$x_1 := \phi(\lambda_1), \qquad x_2 := \phi(\lambda_3), \qquad \nu := \phi(\lambda_2),$$

and denote also by ϕ the natural extension of this Möbius transformation to the surface $\mathscr{C}_{\lambda_1,\lambda_3}$ and by $\mathscr{C}_{x_1,x_2}$ the compact surface obtained in the image of ϕ. We see that ϕ sends the ramification point $P_{\lambda_4} = (\lambda_4, 0)$ to the point at infinity denoted by $P_\infty \in \mathscr{C}_{x_1,x_2}$. The points $\lambda = 0$ and $\lambda = 1$ are also branch points of the covering $\mathscr{C}_{x_1,x_2} \to \mathbb{C}P^1 \ni u$. Thus the compact surface $\mathscr{C}_{x_1,x_2}$ corresponds to the algebraic curve (46) from Section 5, where we can assume that x_1 and x_2 are independent variables and $\nu = \nu(x_1, x_2)$ is a smooth function. In other words, we have a family of surfaces $\mathscr{C}_{x_1,x_2}$ corresponding to the family (46) with variable x_1 and x_2.

Let us denote by $Q_0 \in \mathscr{C}_{x_1,x_2}$ the image by the Möbius transformation of the point of the order n at infinity: $Q_0 := \phi(P_{\infty^+})$. The argument from Section 4.2 leading to (40) can be used to show that Q_0 is a point of order $2n$ on $\mathscr{C}_{x_1,x_2}$ where the Abel map $\mathscr{A}_\infty$ is based at the point $P_\infty \in \mathscr{C}_{x_1,x_2}$. In other words, the vectors c_1 and c_2 from (50) have rational components and are such that $2nc_1, 2nc_2 \in \mathbb{Z}^2$. In particular, being rational, c_1, c_2 stay constant through the continuous variation of λ_1, λ_3, or, equivalently, of x_1, x_2. We thus have a continuous family of Riemann surfaces $\mathscr{C}_{x_1,x_2}$ for some sufficiently small variations of $x_1, x_2 \in \mathbb{R}$ as desired.

6. Differentials on the surfaces of the Chebyshev family

The space of holomorphic differentials on the surfaces (46) is two-dimensional and there are two standard ways to choose a basis for it. One standard basis is written explicitly in terms of the coordinates u and v of the curve; it consists of the differentials ϕ and $u\phi$, where

$$\phi = \frac{du}{v}\,. \tag{52}$$

Another standard basis consists of the holomorphic normalized differentials ω_1 and ω_2 from (47) in Section 5.

We need to introduce one more basis of holomorphic differentials, and in order to do so, we need to define what we call the *value* of a differential at a given point of the Riemann surface. We define the value $\psi(R)$ of a differential ψ at a fixed point R of a Riemann surface (46), at which ψ is non-singular, to be the constant term in the Taylor series expansion of ψ in the powers of the appropriate standard local parameter ξ, ξ_j or ξ_∞ at R as defined in Section 5. More precisely,

$$\psi(R) = \frac{\psi(P)}{d\xi(P)}\Big|_{P=R}. \tag{53}$$

Let us calculate the value of the differential ϕ at the special point Q_0 and and at a ramification point P_{a_k}. Denoting the u-coordinate of Q_0 by q_0, that is $u(Q_0) = q_0$, formula (53) gives

$$\phi(Q_0) = \frac{1}{\sqrt{q_0(q_0-1)(q_0-\nu)(q_0-x_1)(q_0-x_2)}} \tag{54}$$

and

$$\phi(P_{a_k}) = \frac{2}{\sqrt{\prod_{a_j\in B\setminus\{a_k\}}(a_k-a_j)}}\,, \tag{55}$$

where the evaluation at Q_0, a regular point of the covering $(\mathscr{C}, \hat{f}_u)$, is done with respect to the local parameter $\xi = u - q_0$ and the evaluation at a ramification point P_{a_k} is done using $\xi = \zeta_k$ from Section 5. One can verify that ϕ vanishes at P_∞ to the order $2g-2$, and thus $\phi(P_\infty) = 0$.

Let us assume from now on that the divisor $P_\nu + Q_0$ is non-special for all the curves of the Chebyshev family, that is that there is no holomorphic differential vanishing at P_ν and Q_0. In this case, we can introduce one more basis of holomorphic differentials, which we will use along with the first two. This basis is given by differentials v_1 and v_2 defined by the following conditions:

$$v_1(P_\nu) = 1, \qquad v_1(Q_0) = 0, \tag{56}$$

$$v_2(P_\nu) = 0, \qquad v_2(Q_0) = 1\,, \tag{57}$$

where $\nu = \nu(x_1, x_2)$ and P_ν is the corresponding ramification point, Q_0 is the point satisfying (50). Explicitly, these differentials can be written as

$$v_1(P) = \frac{\phi(P)(u(P)-q_0)}{\phi(P_\nu)(\nu-q_0)}, \qquad v_2(P) = \frac{\phi(P)(u(P)-\nu)}{\phi(Q_0)(q_0-\nu)} \tag{58}$$

where $P = (u, v)$ is a point of the curve from the family (46)and $u(P) = f_u(P)$, the u-coordinate of the point P.

An important tool for us is the following symmetric bidifferential, that we call the *Riemann bidifferential* W. It can be defined for any Riemann surface with a fixed basis of **a**- and **b**-cycles by its three characteristic properties:

- Symmetry with respect the two arguments, W(P,Q)=W(Q,P);
- No singularity except for a second order pole along the diagonal $P = Q$ with biresidue 1: for ξ being a local parameter near $P = Q$, the bidifferential has the following local expansion:

$$W(P,Q) \underset{P\sim Q}{=} \left(\frac{1}{(\xi(P)-\xi(Q))^2} + \mathscr{O}(1)\right) d\xi(P)d\xi(Q);$$

- Normalization by vanishing of all **a**-periods: $\oint_{\mathbf{a_k}} W(P,Q) = 0$, where the integral can be taken with respect to either of the arguments.

Here P and Q are two arbitrary points on the Riemann surface. Alternatively, the Riemann bidifferential can be defined by an explicit formula using a theta-function with nonsingular odd characteristics $[\alpha, \beta]$, that is $\alpha, \beta \in \frac{1}{2}\mathbb{Z}^2$ such that $4(\alpha_1\beta_1 + \alpha_2\beta_2)$ is an odd integer and the gradient of $\theta_{[\alpha,\beta]}$ at zero does not vanish, $(\partial_{z_1}\theta_{[\alpha,\beta]}(0), \partial_{z_2}\theta_{[\alpha,\beta]}(0)) \neq 0$, as follows:

$$W(P,Q) = d_P d_Q \theta_{[\alpha,\beta]}\left(\int_Q^P \omega\right),$$

where $\omega = (\omega_1, \omega_2)^T$ is the vector of holomorphic normalized differentials (47). Here the theta-function is defined using the Riemann matrix $\mathbb{B}$ of the Riemann surface in question. As a consequence of this definition we have: $\oint_{\mathbf{b_k}} W(P,Q) = 2\pi\mathrm{i}\,\omega_k(P)$.

By fixing one of the arguments of the bidifferential, we obtain a differential with respect to the other argument. We can then evaluate the obtained differential at a given point with respect to the standard local parameter according to (53). In this way, we can talk about differentials of the form $W(P, P_{a_j})$ or $W(P, P_\infty)$.

Let us express $W(P, P_{a_j})$ and $W(P, P_\nu)$ using the two bases of holomorphic differentials, ω_1, ω_2 and v_1, v_2, introducing the normalization coefficients α_k and γ_k as follows:

$$W(P, P_{a_j}) = \frac{1}{u(P) - a_j}\frac{\phi(P)}{\phi(P_{a_j})} - \sum_{k=1}^{2} \alpha_k^{(j)} \omega_k(P) \tag{59}$$

$$W(P, P_\nu) = \frac{1}{u(P) - \nu}\frac{\phi(P)}{\phi(P_\nu)} - \sum_{k=1}^{2} \gamma_k v_k(P)\,. \tag{60}$$

The existence of a special point Q_0 (50) on every curve from the Chebyshev family allows us to introduce the following Abelian differential of the third kind

$$\Omega(P) = \int_{Q_0^*}^{Q_0} W(P,Q) - 4\pi\mathrm{i}c_2^T\omega(P), \tag{61}$$

where Q_0^* is the point obtained from Q_0 by the hyperelliptic involution, that is if Q_0 and Q_0^* are the two points on the curve (46) with u-coordinate being $u(Q_0) = u(Q_0^*) = q_0$. The constant vector c_2 in (61) is the one defined by the point Q_0 through relation (50) and $\omega = (\omega_1, \omega_2)^T$.

One can also write a more explicit expression for Ω similar to (60) by introducing coefficients δ_1 and δ_2 depending on the curve such that

$$\Omega(P) = \frac{1}{u(P) - q_0}\frac{\phi(P)}{\phi(Q_0)} + \sum_{i=1}^{2} \delta_i v_i(P)\,. \tag{62}$$

As one can check, the right hand side of (62) is a differential having simple poles at $P = Q_0$ and $P = Q_0^*$ with the correct residues. The constants δ_1 and δ_2 are defined by the condition that the a-period of this differential vanish.

Note that (61) on the surface $\mathscr{C}_{x_1,x_2}$ from Section 5.1 can also be written in the form similar to (4) by introducing notation Ω_{Q_0,Q_0^*} for the differential of the third kind having two simple poles at Q_0 and at Q_0 with residues 1 and -1, respectively:

$$\Omega = \Omega_{Q_0,Q_0^*} - 4\pi\mathrm{i}c_2^T\omega. \tag{63}$$

As in the genus one case, this differential is the pullback by ϕ^{-1} of the differential $\Omega_{\mathbb{A}}$ from (9), in other words, the following analogue of Lemma 4 holds.

LEMMA 5. *Let Ω be the differential defined by* (63) *on a surface $\mathscr{C}_{x_1,x_2}$ obtained in Section* 5.1 *by the Möbius transformation* (51) *from the surface $\mathscr{C}_{\lambda_1,\lambda_3}$ associated with a Chebyshev polynomial p_n. Then Ω is the pullback by the inverse Möbius transformation of Section* 5.1 *of the differential $\Omega_{\mathbb{A}}$ obtained by the derivative of the logarithm of the Akhiezer function* (9) *with $d = 3$ and $\mathscr{P}_6 = \mathscr{P}_{\lambda_1,\lambda_3}$:*

$$\Omega = (\phi^{-1})^*\Omega_{\mathbb{A}}. \tag{64}$$

Proof. Let us denote again by $\Omega_{\infty^\pm}$ the differential of the third kind on the surface $\mathscr{C}_{\lambda_1,\lambda_3}$ having two simple poles, at two points at infinity P_{∞^+} and P_{∞^-} with residues 1 and -1, respectively, and normalized by vanishing of its **a**-periods. Let us also denote by $\omega_{\lambda_1,\lambda_3}$ the vector of normalized holomorphic differential on $\mathscr{C}_{\lambda_1,\lambda_3}$. Then we have

$$\Omega = \left(\phi^{-1}\right)^*\Omega_{\infty^\pm} - 4\pi \mathrm{i} c_2^T \left(\phi^{-1}\right)^*\omega_{\lambda_1,\lambda_3}$$

and due to the existence of the Akhiezer function

$$(65) \qquad n\int_{P_{\infty^-}}^{P_{\infty^+}} \omega_{\lambda_1,\lambda_3} = n\int_{Q_0^*}^{Q_0} \omega = \mathbb{B}k + l \quad \Rightarrow \quad \mathscr{A}_\infty(Q_0) = \frac{1}{2n}\mathbb{B}k + \frac{1}{2n}l\,,$$

where $k, l \in \mathbb{Z}^2$ and $\mathbb{B}$ is the Riemann matrix of the two surfaces $\mathscr{C}_{\lambda_1,\lambda_3}$ and $\mathscr{C}_{x_1,x_2}$ related by the Möbius transformation (51). The implication in (65) is analogous to the passage from (39) to (40). Comparing to (49) and (50), we obtain that $c_2 = \frac{1}{2n}k$? Thus it remains to prove that

$$(66) \qquad \Omega_{\mathbb{A}} = \Omega_{\infty^\pm} - \frac{2\pi \mathrm{i}}{n}k_2^T\omega_{\lambda_1,\lambda_3}.$$

This can be done similarly to the proof of Lemma 4 by representing the Akhiezer function in a way analogous to (35). However, in genus two, we need to use the prime form E, see [**17**], instead of the theta-function θ_1. The prime form $E(P,Q)$ for two points $P, Q \in \mathscr{C}_{\lambda_1,\lambda_3}$ is anti-symmetric, $E(P,Q) = E(Q,P)$, has a simple zero at $P = Q$ and transforms as follows as one of its arguments makes a full loop along a cycle from a canonical homology basis **a**, **b**. Denoting such transformation of an argument, for example, by $P \mapsto P + a_j$ for the homology cycle a_j, we have

$$E(P + a_j, Q) = E(P,Q),$$

$$E(P + b_j, Q) = E(P,Q)\exp\{-\pi \mathrm{i}\mathbb{B}_{jj} - 2\pi \mathrm{i}\,(\int_Q^P \omega_{\lambda_1,\lambda_3})_j\}, \quad j = 1, 2.$$

These properties allow us to write, similarly to (35):

$$\mathbb{A}(P) = \frac{E^n(P, P_{\infty^+})}{E^n(P, P_{\infty^-})}\exp\left\{-2k^T\pi \mathrm{i}\int_{P_{\infty^-}}^P \omega_{\lambda_1,\lambda_3}\right\}, \qquad P \in \mathscr{C}_{\lambda_1,\lambda_3}.$$

Differentiating the logarithm of this expression, we see that $\frac{1}{n}d\log\mathbb{A}$ coincides with (66). □

7. Variational formulas

For our construction, we need to know how the quantities we have defined on the Riemann surfaces of curves (46) vary under small variations of the independent branch points x_1, x_2. The local parameters from Section 5 change under such variations entailing the dependence on x_1, x_2 of the complex structure and thus of all the meromorphic functions and differentials defined on the Riemann surfaces. This is what we study in this section; we will also derive an expression for the derivative of the dependent branch point ν with respect to x_1, x_2.

7.1. Rauch variation. In this subsection, we consider a family of hyperelliptic curves of genus 2 with all ramification points being independent of one another. More precisely, we consider Riemann surfaces obtained by compactification and desingularization of the algebraic curves given by

$$\mathscr{L} = \{(u,v) \in \mathbb{C}^2 \mid v^2 = \prod_{j=1}^{5}(u-a_j)\}\,. \tag{67}$$

Note that everything presented in this subsection extends without change for the case of hyperelliptic curves of any genus. As before, we consider the meromorphic function $f_u : \mathscr{L} \to \mathbb{C}P^1$ which is the projection on the u-coordinate, that is applied to a point $P = (u,v) \in \mathscr{L}$, it gives $f_u(P) = u = u(P)$. We keep the notation $\mathscr{L}$ for the Riemann surface corresponding to the algebraic curve (67) and f_u for the natural extension of the above function to the Riemann surface. As before, we denote by P_{a_j} the point of the curve with coordinates $P_{a_j} = (a_j, 0)$ and P_∞ the point at infinity of $\mathscr{L}$ added to compactify the algebraic curve.

The function f_u represents the surface $\mathscr{L}$ as a two-fold ramified covering of the u-$\mathbb{C}P^1$. By definition, a ramified covering is a pair $(\mathscr{L}, f_u)$ of a Riemann surface and a meromorphic function on it. Our two-fold coverings are ramified over the set $B := \{a_j\}_{j=1}^5$ and over the point at infinity, the points P_{a_j} and P_∞ are the corresponding ramification points and their images by f_u are called branch points. Let us denote by $\mathscr{H}_2$ the set of all such two-fold ramified coverings of genus 2 indexed by the set B. This set is an example of a *Hurwitz space*. The elements of the set B play the role of coordinates on the Hurwitz space $\mathscr{H}_2$.

We consider the space $\mathscr{H}_2$ locally, that is we assume that the set of branch points B as a subset of $\mathbb{C}^5$ belongs to a small neighbourhood D of a fixed set $B^0 \in \mathbb{C}^5 \setminus \Delta$, where

$$\Delta = \{a_1, \ldots, a_5 \mid \text{ there exists } j \neq k\,,\ 1 \leq j\,, k \leq 5 \text{ such that } a_j = a_k\}$$

is a set of diagonals. We assume that D is small enough to allow for a choice of a canonical homology basis on $\mathscr{L}$ such that the images by f_u of the elements of this basis are fixed for all $B \in D$.

Recall that, in this section, we consider the branch points a_j to be varying independently of each other in a way so that $B \in D$.

We can now consider all functions and differentials defined on a Riemann surface $\mathscr{L}$ as objects defined on the Hurwitz space $\mathscr{H}_2$ as follows. Recall that a set B of branch points completely defines the Riemann surface of the corresponding curve from (67); the holomorphic atlas being given by the local parameters from Section (5). A meromorphic differential ψ on a surface $\mathscr{L}$ thus depends on point of the surface $\mathscr{L}$ and on the branch points $\{a_j\}_{j=1}^5$ defining the surface $\mathscr{L}$ in the family (67), that is we can write $\psi = \psi(P; a_1, \ldots, a_5)$ instead of $\psi(P)$. The dependence of differentials on the branch points a_j is captured by the *Rauch variational formulas* that we now summarize, see [**17, 27**].

The Rauch variation of a branch point a_j is performed assuming that the point P on a varying surface is fixed by the requirement that its u-coordinate stays fixed under the variation: $u(P) = \text{const}$. We denote the corresponding derivative as follows:

$$\frac{\partial^{\text{Rauch}}}{\partial a_k}\psi(P) := \frac{\partial}{\partial a_k}\Big|_{u(P)=const}\psi(P) \tag{68}$$

where ψ is an Abelian differential defined on the Riemann surface of the curve (67). In the case of the Riemann bidifferential we need to require that both $u(P)$ and $u(Q)$ stay fixed. We have the following Rauch variational formula for the W, see [**27**]:

$$\frac{\partial^{\text{Rauch}}}{\partial a_k} W(P,Q) := \frac{\partial}{\partial a_k}\Big|_{\substack{u(P)=const\\ u(Q)=const}} W(P,Q) = \frac{1}{2} W(P, P_{a_j}) W(P_{a_j}, Q).$$

This formula also determines the Rauch variation of the holomorphic normalized differentials and the Riemann matrix due to relations $\omega_j(P) = \oint_{\mathbf{b_j}} W(P,Q)/(2\pi \mathrm{i})$ and $\mathbb{B}_{jk} = \oint_{\mathbf{b_k}} \omega_j$. We have

(69)

$$\frac{\partial^{\text{Rauch}} \omega_j(P)}{\partial a_k} = \frac{1}{2}\omega_j(P_{a_k}) W(P, P_{a_k}), \qquad \frac{\partial^{\text{Rauch}} \mathbb{B}_{ij}}{\partial a_k} = \frac{\pi \mathrm{i}}{2} \omega_j(P_{a_k}) \omega_i(P_{a_k}).$$

To obtain Rauch derivatives of the differential of the third kind Ω defined by (61), we need to assume for the moment that the point $Q_0 \in \mathscr{L}$ has a fixed projection q_0 on the u-sphere, $u(Q_0) = q_0$, independent of the branch points. From definition (61) of Ω and the above derivatives of W and ω, we derive the following Rauch variational formula

$$\frac{\partial^{\text{Rauch}} \Omega(P)}{\partial a_k} = \frac{1}{2} \Omega(P_{a_k}) W(P, P_{a_k}). \tag{70}$$

Evaluating (70) at P_{a_j} for $k \neq j$ we have

$$\frac{\partial^{\text{Rauch}} \Omega(P_{a_j})}{\partial a_k} = \frac{1}{2} \Omega(P_{a_k}) W(P_{a_j}, P_{a_k}). \tag{71}$$

Note that (70) does not allow to obtain the Rauch derivative of $\Omega(P_{a_j})$ with respect to a_j. We derive an expression for this derivative from the following lemma.

LEMMA 6. *Let $a_k \in B$ be an arbitrary branch point of the curve $\mathscr{L}$ (67) and $Q_0, Q_0^* \in \mathscr{L}$ be two regular points u-coordinate of which, denoted by $q_0 = u(Q_0) = u(Q_0^*)$, is fixed and independent of the branch points. Let Ω be the differential of the third kind having simple poles at the points Q_0 and Q_0^* defined by (61) in which $c_2 \in \mathbb{C}^2$ is a constant vector and $\omega = (\omega_1, \omega_2)^T$ is a vector of holomorphic normalized differentials. The following variational formula holds:*

$$\frac{\partial \Omega(P_{a_k})}{\partial a_k} = -\frac{1}{2} \sum_{\substack{a_j \in B\\ a_j \neq a_k}} \Omega(P_{a_j}) W(P_{a_k}, P_{a_j}) - 2W(Q_0, P_{a_k}).$$

Proof. Let ε be a complex number with small absolute value and consider two ramified coverings $f_u : \mathscr{L} \to \mathbb{C}P^1$ sending a point $P = (u, v)$ to $f_u(P) = u$ and $f_u^\varepsilon : \mathscr{L} \to \mathbb{C}P^1$ sending a point $P = (u, v)$ to $f_u^\varepsilon(P) = u + \varepsilon$. The covering f_u is ramified over the points a_j and the covering f_u^ε is ramified over the points $a_j + \varepsilon$ for $a_j \in B$. The standard local parameters induced by both coverings as defined in Section 5 coincide and the evaluation of differentials defined by (53) is the same in the standard charts induced by both coverings. Note that all the differentials we are working with are defined on the Riemann surface $\mathscr{L}$ which is the same for both coverings.

On the other hand, the quantity $\Omega(P_{a_k})$ is a function of q_0 and the branch points $a_1, \ldots, a_5$ of the covering f_u. At the same time, considering $\mathscr{L}$ as the surface of the covering f_u^ε, the same quantity can be seen as $\Omega^\varepsilon(P_{a_k})$ which is a function of $q_0+\varepsilon$ and the branch points $a_1+\varepsilon,\dots,a_5+\varepsilon$ of f_u^ε. Since we have $\Omega^\varepsilon(P_{a_k})=\Omega(P_{a_k})$, we conclude that $\frac{d}{d\varepsilon}\Omega^\varepsilon(P_{a_k})=0$ and thus

$$0=\frac{d}{d\varepsilon}\Big|_{\varepsilon=0}\Omega^\varepsilon(P_{a_k})=\left(\sum_{a_j\in B}\frac{\partial\Omega^\varepsilon(P_{a_k})}{\partial(a_j+\varepsilon)}+\frac{\partial\Omega^\varepsilon(P_{a_k})}{\partial(q_0+\varepsilon)}\right)\Big|_{\varepsilon=0}=\sum_{a_j\in B}\frac{\partial\Omega(P_{a_k})}{\partial a_j}+\frac{\partial\Omega(P_{a_k})}{\partial q_0}.$$

From here we deduce the expression for the derivative of $\Omega(P_{a_k})$ with respect to a_k. Noting that the partial derivative $\frac{\partial\Omega(P_{a_j})}{\partial a_k}$ for $k\neq j$ coincides with the Rauch derivative (68), (71), we obtain

$$\frac{\partial\Omega(P_{a_k})}{\partial a_k}=-\sum_{\substack{a_j\in B\\ a_j\neq a_k}}\frac{\partial^{\mathrm{Rauch}}\Omega(P_{a_k})}{\partial a_j}-\frac{\partial\Omega(P_{a_k})}{\partial q_0}.$$

From definition (61) of Ω, we see that

$$\frac{\partial\Omega(P_{a_k})}{\partial q_0}=W(P_{a_k},Q_0)-W(P_{a_k},Q_0^*)=2W(P_{a_k},Q_0)\,, \tag{72}$$

where in the last equality we used the anti-invariance of $W(P_{a_k},\cdot)$ with respect to the hyperelliptic involution, $W(P_{a_j},Q_0)=-W(P_{a_j},Q_0^*)$, which follows, for example, from (59) and (52). With this and (71) for the Rauch derivatives, we finish the proof. □

7.2. Variation in the Chebyshev family of curves.

7.2.1. *Variation of dependent branch point and q_0 in the Chebyshev family.* Let ω as before denote the vector of holomorphic normalized differentials, $\omega=(\omega_1,\omega_2)^T$ and recall that on a Chebyshev family of curves, there exists a section of points Q_0 satisfying condition (50). Given that the normalized holomorphic differentials are linear combination of the basis given by $\phi,u\phi$ with ϕ from (52), the vector ω changes sign under the hyperelliptic involution, that is $\omega(P^*)=-\omega(P)$ and thus (50) can be equivalently rewritten as

$$\frac{1}{2}\int_{Q_0^*}^{Q_0}\omega=c_1+\mathbb{B}c_2\,. \tag{73}$$

Differentiating this relation with respect to the independent branch points x_1 and x_2 we have

$$\frac{1}{2}\int_{Q_0^*}^{Q_0}\frac{\partial^{\mathrm{Rauch}}\omega}{\partial x_i}+\frac{1}{2}\frac{\partial\nu}{\partial x_i}\int_{Q_0^*}^{Q_0}\frac{\partial^{\mathrm{Rauch}}\omega}{\partial\nu}+\frac{1}{2}\left(\omega(Q_0)-\omega(Q_0^*)\right)\frac{\partial q_0}{\partial x_i}=\frac{\partial^{\mathrm{Rauch}}\mathbb{B}}{\partial x_i}c_2\,,$$

which, by plugging in the Rauch derivatives (69), using $\omega(Q_0^*)=-\omega(Q_0)$ and definition (61) of the differential Ω, becomes

$$\frac{1}{4}\omega(P_{x_i})\Omega(P_{x_i})+\frac{1}{4}\omega(P_\nu)\Omega(P_\nu)\frac{\partial\nu}{\partial x_i}+\omega(Q_0)\frac{\partial q_0}{\partial x_i}=0\,. \tag{74}$$

The vector ω has two components, and thus this is a linear system of two equations for the two unknown functions $\frac{\partial\nu}{\partial x_i}$ and $\frac{\partial q_0}{\partial x_i}$, solving which by Cramer's rule we obtain

$$\frac{\partial \nu}{\partial x_i} = -\frac{1}{4}\frac{\Omega(P_{x_i})}{\Omega(P_\nu)}\frac{\begin{vmatrix} \omega_1(P_{x_i}) & \omega_1(Q_0) \\ \omega_2(P_{x_i}) & \omega_2(Q_0) \end{vmatrix}}{\begin{vmatrix} \omega_1(P_\nu) & \omega_1(Q_0) \\ \omega_2(P_\nu) & \omega_2(Q_0) \end{vmatrix}}$$

and

$$\omega_k(Q_0)\frac{\partial q_0}{\partial x_i} = -\frac{1}{4}\omega_k(P_{x_i})\Omega(P_{x_i}) - \frac{1}{4}\omega_k(P_\nu)\Omega(P_\nu)\frac{\partial \nu}{\partial x_i}, \qquad i,k = 1,2. \tag{75}$$

Note that the last two expressions are invariant under replacing the basic differentials ω_i by any of their linear combinations. Given the form of the equations, the basis given by holomorphic differentials v_1 and v_2 satisfying relations (56), (57) becomes very useful. In the basis $\{v_1, v_2\}$ in the space of holomorphic differentials, the variational formulas for ν and q_0 have the following simpler form

$$\frac{\partial \nu}{\partial x_i} = -\frac{\Omega(P_{x_i})}{\Omega(P_\nu)}v_1(P_{x_i}) \tag{76}$$

and

$$\frac{\partial q_0}{\partial x_i} = -\frac{1}{4}\Omega(P_{x_i})v_2(P_{x_i}) \tag{77}$$

for $i = 1, 2$.

REMARK 2. For genus one curves considered in Section 4, formula (77) becomes

$$\frac{dy_0}{dx} = -\frac{1}{4}\Omega(P_x)\frac{\omega(P_x)}{\omega(Q_0)}$$

which is equivalent to (45) from Section 4, see [**14**].

7.2.2. *Variation of differentials ϕ and Ω.* Having obtained expressions for derivatives of ν and q_0 with respect to independent variables x_1 and x_2, we can obtain the variation of the holomorphic differential ϕ evaluated at various points of the curves by the straightforward differentiation of expressions (54) and (55). In this way, for $a_j \neq x_i$ we have

$$\frac{\partial \phi(P_{a_j})}{\partial x_i} = \frac{\phi(P_{a_j})}{2}\left(\frac{1}{a_j - x_i} + \frac{1}{a_j - \nu}\frac{\partial \nu}{\partial x_i}\right), \tag{78}$$

$$\frac{\partial \phi(P_\nu)}{\partial x_i} = \frac{\phi(P_\nu)}{2}\left(\frac{1}{\nu - x_i} - \sum_{a_j \neq \nu}\frac{1}{\nu - a_j}\frac{\partial \nu}{\partial x_i}\right), \tag{79}$$

$$\frac{\partial \phi(Q_0)}{\partial x_i} = \frac{\phi(Q_0)}{2}\left(\sum_{a_j}\frac{1}{a_j - q_0}\frac{\partial q_0}{\partial x_i} - \frac{1}{x_i - q_0} - \frac{1}{\nu - q_0}\frac{\partial \nu}{\partial x_i}\right), \tag{80}$$

where $i = 1, 2$ and $\frac{\partial \nu}{\partial x_i}$ and $\frac{\partial q_0}{\partial x_i}$ are given by (76) and (77), respectively.

In order to obtain variation of similar quantities related to Ω (61), we can use the Rauch variational formulas and Lemma 6 from Section 7.1. Note that the Rauch variation allows us to obtain derivatives $\partial_{x_i}\Omega(P_{a_j})$ only if a_j is independent of x_i, that is different from x_i and ν.

PROPOSITION 1. *Let Ω be the differential of the third kind defined by* (61) *on the genus 2 curves from a Chebyshev family* (46). *Let $\Omega(P_{a_j})$ be its evaluation at the ramification point P_{a_j} with respect to the standard local parameter $\xi(P) = \sqrt{u(P) - a_j}$ as defined by* (53). *Let $i \in \{1, 2\}$. For $a_j \neq \nu$ and $a_j \neq x_i$ we have:*

$$\frac{\partial \Omega(P_{a_j})}{\partial x_i} = \frac{\Omega(P_{x_i})}{2} \frac{\phi(P_{x_i})(x_i - q_0)}{\phi(P_{a_j})(a_j - q_0)} \left(\frac{1}{x_i - a_j} - \frac{1}{\nu - a_j} \right).$$

Proof. Let us differentiate $\Omega(P_{a_j})$ using its representation (61) and the Rauch variational formulas. We obtain

$$\begin{aligned} \frac{\partial \Omega(P_{a_j})}{\partial x_i} &= \frac{\partial^{\text{Rauch}}}{\partial x_i} \Omega(P_{a_j}) + \frac{\partial}{\partial q_0} \Omega(P_{a_j}) \frac{\partial q_0}{\partial x_i} + \frac{\partial^{\text{Rauch}}}{\partial \nu} \Omega(P_{a_j}) \frac{\partial \nu}{\partial x_i} \\ &= \frac{1}{2} W(P_{a_j}, P_{x_i}) \Omega(P_{x_i}) + 2W(P_{a_j}, Q_0) \frac{\partial q_0}{\partial x_i} + \frac{1}{2} W(P_{a_j}, P_\nu) \Omega(P_\nu) \frac{\partial \nu}{\partial x_i} \end{aligned}$$

where we used expression (71) for the Rauch derivatives and (72) for the derivative with respect to q_0. Evaluating (59) at P_{x_i} and P_ν, we can replace $W(P_{a_j}, P_{x_i})$ and $W(P_{a_j}, P_\nu)$ by their more explicit expressions:

$$\begin{aligned} \frac{\partial \Omega(P_{a_j})}{\partial x_i} = \frac{1}{2} &\left(\frac{1}{x_i - a_j} \frac{\phi(P_{x_i})}{\phi(P_{a_j})} - \sum_{k=1}^{2} \alpha_k^{(j)} \omega_k(P_{x_i}) \right) \Omega(P_{x_i}) \\ &+ 2 \left(\frac{1}{q_0 - a_j} \frac{\phi(Q_0)}{\phi(P_{a_j})} - \sum_{k=1}^{2} \alpha_k^{(j)} \omega_k(Q_0) \right) \frac{\partial q_0}{\partial x_i} \\ &+ \frac{1}{2} \left(\frac{1}{\nu - a_j} \frac{\phi(P_\nu)}{\phi(P_{a_j})} - \sum_{k=1}^{2} \alpha_k^{(j)} \omega_k(P_\nu) \right) \Omega(P_\nu) \frac{\partial \nu}{\partial x_i}. \end{aligned}$$

This expression simplifies a lot due to relation (75) for the derivative of q_0. Following this simplification, replacing $\partial_{x_i} q_0$ from (77) and $\partial_{x_i} \nu$ from (76), and also writing explicitly v_1 and v_2 as in (58), we are left with

$$\frac{\partial \Omega(P_{a_j})}{\partial x_1} = \frac{\Omega(P_{x_i})}{2} \frac{\phi(P_{x_i})}{\phi(P_{a_j})} \left(\frac{1}{x_i - a_j} - \frac{x_i - \nu}{(q_0 - a_j)(q_0 - \nu)} - \frac{x_i - q_0}{(\nu - a_j)(\nu - q_0)} \right),$$

which is equivalent to the claim of the proposition. □

To obtain further variational formulas, we need first to establish the auxiliary relations of the next lemma.

LEMMA 7. *Let W, Ω, ϕ be differentials defined in Section 6 on a compactified hyperelliptic curve of genus 2 from the family given by equation* (46) *with a choice of a canonical homology basis. Let the evaluations of differentials at given points be defined by* (53). *Then the following relations hold*

$$\frac{1}{2} \sum_{a_j \in B} \Omega(P_{a_j}) \phi(P_{a_j}) + 2\phi(Q_0) = 0; \tag{81}$$

$$\frac{1}{2} \sum_{\substack{a_j \neq \nu \\ a_j \in B}} \frac{\Omega(P_{a_j}) \phi(P_{a_j})}{a_j - \nu} + \operatorname*{res}_{P = P_\nu} \frac{\Omega(P) \phi(P)}{(u(P) - \nu) du(P)} + 2 \frac{\phi(Q_0)}{q_0 - \nu} = 0; \tag{82}$$

$$\frac{1}{2}\sum_{\substack{a_j\neq\nu\\a_j\in B}}\Omega(P_{a_j})W(P_{a_j},P_\nu)+\operatorname*{res}_{P=P_\nu}\frac{\Omega(P)W(P,P_\nu)}{du(P)}+2W(Q_0,P_\nu)=0\,. \tag{83}$$

Proof. Equalities (81) and (82) are the vanishing of the sum of residues of the differentials $\frac{\Omega(p)\phi(p)}{du}$ and $\frac{\Omega(p)\phi(p)}{(u-\nu)du}$, respectively. The vanishing of the sum of residues of the differential $\frac{\Omega(p)W(p,P_\nu)}{du}$ gives (83). □

PROPOSITION 2. *In the settings of Proposition 1, we have for the derivatives of $\Omega(P_\nu)$ with respect to independently varying branch points x_1 and x_2:*

$$\frac{\partial\Omega(P_\nu)}{\partial x_i}=\frac{\Omega(P_{x_i})}{2}\frac{\phi(P_{x_i})(x_i-q_0)}{\phi(P_\nu)(\nu-q_0)}\left(\frac{1}{x_i-\nu}-\sum_{\substack{a_j\neq\nu\\a_j\in B}}\frac{\Omega(P_{a_j})\phi(P_{a_j})}{\Omega(P_\nu)\phi(P_\nu)}\frac{(a_j-q_0)}{(\nu-q_0)(\nu-a_j)}\right).$$

Proof. Differentiating $\Omega(P_\nu)$ with respect to x_i with the help of Rauch variational formulas (70), Lemma 6 and (72), and using (77) for $\partial_{x_i}q_0$ and (76) for $\partial_{x_i}\nu$, we get

$$\frac{\partial\Omega(P_\nu)}{\partial x_i}=\frac{\partial^{\mathrm{Rauch}}\Omega(P_\nu)}{\partial x_i}+\frac{\partial\Omega(P_\nu)}{\partial q_0}\frac{\partial q_0}{\partial x_i}+\frac{\partial\Omega(P_\nu)}{\partial\nu}\frac{\partial\nu}{\partial x_i}=\frac{1}{2}\Omega(P_{x_i})W(P_{x_i},P_\nu)$$

$$-\frac{1}{2}W(Q_0,P_\nu)\Omega(P_{x_i})v_2(P_{x_i})+\left(\frac{1}{2}\sum_{a_j\neq\nu}\Omega(P_{a_j})W(P_{a_j},P_\nu)+2W(Q_0,P_\nu)\right)$$

$$\times\frac{\Omega(P_{x_i})}{\Omega(P_\nu)}v_1(P_{x_i})\,.$$

Now we can replace the last two terms using the sum of residues (83):

$$\begin{aligned}\frac{\partial\Omega(P_\nu)}{\partial x_i}&=\frac{1}{2}\Omega(P_{x_i})W(P_{x_i},P_\nu)-\frac{1}{2}W(q_0,P_\nu)\Omega(P_{x_i})v_2(P_{x_i})\\&\quad-\frac{\Omega(P_{x_i})}{\Omega(P_\nu)}v_1(P_{x_i})\operatorname*{res}_{P=P_\nu}\frac{\Omega(P)W(P,P_\nu)}{du}\,.\end{aligned}$$

Replacing $W(\cdot,P_\nu)$ with the expression from (60), we obtain:

$$\frac{\partial\Omega(P_\nu)}{\partial x_i}=\frac{1}{2}\Omega(P_{x_i})\left(\frac{1}{x_i-\nu}\frac{\phi(P_{x_i})}{\phi(P_\nu)}-\sum_{k=1}^{2}\gamma_k v_k(P_{x_i})\right)$$

$$-\frac{1}{2}\left(\frac{1}{q_0-\nu}\frac{\phi(Q_0)}{\phi(P_\nu)}-\sum_{k=1}^{2}\gamma_k v_k(Q_0)\right)\Omega(P_{x_i})v_2(P_{x_i})-\frac{\Omega(P_{x_i})}{\Omega(P_\nu)}v_1(P_{x_i})\operatorname*{res}_{p=P_\nu}\frac{\Omega(P)W(P,P_\nu)}{du}\,.$$

This expression simplifies due to relations $v_1(Q_0)=0$, $v_2(Q_0)=1$ from (56), (57), in particular the terms proportional to γ_2 cancel out. Note also that, due to (60), and (56), (57) we have

$$\frac{1}{\phi(P_\nu)}\operatorname*{res}_{P=P_\nu}\frac{\Omega(P)\phi(P)}{(u-\nu)du}-\operatorname*{res}_{P=P_\nu}\frac{\Omega(P)W(P,P_\nu)}{du}=\operatorname*{res}_{P=P_\nu}\frac{\Omega(P)}{du}\sum_{k=1}^{2}\gamma_k v_k(P)=\frac{1}{2}\Omega(P_\nu)\gamma_1\,. \tag{84}$$

The first residue in (84) can be expressed from (82) and then using (81) for $\phi(Q_0)$. Thus (84) implies

$$\frac{\sum_{a_j}\Omega(P_{a_j})\phi(P_{a_j})}{2\phi(P_\nu)(q_0-\nu)}-\frac{1}{2\phi(P_\nu)}\sum_{a_j\neq\nu}\frac{\Omega(P_{a_j})\phi(P_{a_j})}{a_j-\nu}=\operatorname*{res}_{P=P_\nu}\frac{\Omega(P)W(P,P_\nu)}{du}+\frac{1}{2}\Omega(P_\nu)\gamma_1\,,$$

which, together with explicit expressions (58) for v_1, v_2, allows to rewrite our expression for the derivative in the form

$$\frac{\partial\Omega(P_\nu)}{\partial x_i}=\frac{\Omega(P_{x_i})\phi(P_{x_i})}{2\phi(P_\nu)}$$
$$\times\left(\frac{1}{x_i-\nu}-\frac{x_i-\nu}{(q_0-\nu)^2}+\frac{x_i-q_0}{\nu-q_0}\left(\sum_{a_j\neq\nu}\frac{\frac{\Omega(P_{a_j})\phi(P_{a_j})}{\Omega(P_\nu)\phi(P_\nu)}}{a_j-\nu}-\frac{\sum_{a_j}\frac{\Omega(P_{a_j})\phi(P_{a_j})}{\Omega(P_\nu)\phi(P_\nu)}}{q_0-\nu}\right)\right).$$

which is equivalent to the claim of the proposition. □

Proposition 3. *Let Ω be the differential of the third kind defined by (61) on the genus 2 curves from a Chebyshev family (46). Let $\Omega(P_{x_i})$ be its evaluation at the ramification point P_{x_i} with respect to the standard local parameter $\xi(P)=\sqrt{u(P)-x_i}$ and let $\Omega(P_\infty)$ be its evaluation at the ramification point P_∞ with respect to the standard local parameter $\xi_\infty(P)=u(P)^{-1/2}$ as defined by (53). Let $i\in\{1,2\}$. The following variational formula holds:*

$$\frac{\partial\Omega(P_\infty)}{\partial x_i}=\frac{1}{4}\Omega(P_{x_i})\phi(P_{x_i})(x_i-q_0)(\nu-x_i)\,.$$

Proof. Evaluating at $P=P_\infty$ the Rauch formula (70) with $a_k=\nu$ and $a_k=x_i$, we obtain the Rauch variation of $\Omega(P_\infty)$. Together with (72), this gives

$$\frac{\partial\Omega(P_\infty)}{\partial x_i}=\frac{1}{2}W(P_{x_i},P_\infty)\Omega(P_{x_i})+\frac{1}{2}W(P_\nu,P_\infty)\Omega(P_\nu)\frac{\partial\nu}{\partial x_i}+2W(P_\infty,Q_0)\frac{\partial q_0}{\partial x_i}\,.$$

Note that evaluating the bidifferential W at infinity with respect to one of the arguments, we get the following differential:

$$W(P,P_\infty)=-\frac{1}{2}u^2\phi(P)+\frac{1}{2}\sum_{k=1}^{g}I_k\omega_k(P)\,, \tag{85}$$

where $I_k:=\oint u^2\phi(P)$ and $P=(u,v)\in\mathscr{C}$. Using this expression for $W(P,P_\infty)$ with $P=P_{x_i}$, $P=P_\nu$, and $P=Q_0$, we have, for example, $W(P_{x_i},P_\infty)=-\frac{x_i^2\phi(P_{x_i})}{2}+\frac{1}{2}\sum_{k=1}^{g}I_k\omega_k(P_{x_i})$. Using this in the above expression for ∂ and then (75) for $\omega_k(Q_0)\frac{\partial q_0}{\partial x_i}$, we get

$$\frac{\partial\Omega(P_\infty)}{\partial x_i}=-\frac{\Omega(P_{x_i})x_i^2\phi(P_{x_i})}{4}-\frac{\Omega(P_\nu)\nu^2\phi(P_\nu)}{4}\frac{\partial\nu}{\partial x_i}-q_0^2\phi(Q_0)\frac{\partial q_0}{\partial x_i}\,.$$

It remains to plug in expressions (77) and (76) for derivatives $\frac{\partial q_0}{\partial x_i}$ and $\frac{\partial\nu}{\partial x_i}$, as well as expressions (58) for v_1 and v_2 to prove the proposition. □

8. Solutions to constrained Schelsinger system

Let Ω be the differential (61) defined on a Chebyshev family of Riemann surfaces corresponding to the algebraic curves of equation (46) and let ϕ be the corresponding basic holomorphic differential (52). Let us consider the following differential on the u-sphere:

$$A^{12}(u)du = \frac{t}{2}\frac{\Omega(P)}{\phi(P)}\frac{(u-q_0)^2du}{v^2} \tag{86}$$

with some arbitrary complex parameter t, where $u = u(P)$ in the right hand side. From expression (62) for Ω, we see that differential (86) as a differential on the u-sphere has a simple pole at $u = a_j$ for every $a_j \in B$ and no other singularities. It is the residues of this differential that are going to play an important role for us. Thus for any a_j we define

$$A^{12}_{a_j} := \operatorname*{res}_{u=a_j} A^{12}(u)du = \frac{t}{4}\Omega(P_{a_j})\phi(P_{a_j})(a_j-q_0)^2\,, \tag{87}$$

where as before, the evaluations of differentials at given points be defined by (53). Note that

$$\sum_{a_j} A^{12}_{a_j} = 0 \tag{88}$$

as a sum of residues of the differential $A^{12}(u)du$.

Let us now rewrite expressions (76) and (77) in terms of the new quantities $A^{12}_{a_j}$:

$$\frac{\partial\nu}{\partial x_i} = -\frac{A^{12}_{x_i}(\nu-q_0)}{A^{12}_{\nu}(x_i-q_0)} \tag{89}$$

and

$$\frac{\partial q_0}{\partial x_i} = \frac{A^{12}_{x_i}(x_i-\nu)}{t\phi(Q_0)(\nu-q_0)(x_i-q_0)^2}\,. \tag{90}$$

It will be sometimes convenient to write a longer expression for $\partial_{x_i}q_0$ maintaining the splitting of the formulas into two parts, grouping together the terms containing the derivative $\partial_{x_i}\nu$. We get such an expression from (75) by replacing ω_k with ϕ :

$$\frac{\partial q_0}{\partial x_i} = -\frac{A^{12}_{x_i}}{t\phi(Q_0)(x_i-q_0)^2} - \frac{\partial\nu}{\partial x_i}\frac{A^{12}_{\nu}}{t\phi(Q_0)(\nu-q_0)^2}\,. \tag{91}$$

The results of Propositions 1-3 in terms of quantities $A^{12}_{a_j}$ take the form:

$$\frac{\partial\Omega(P_{a_j})}{\partial x_i} = \frac{\Omega(P_{a_j})}{2}\frac{A^{12}_{x_i}(a_j-q_0)}{A^{12}_{a_j}(x_i-q_0)}\left(\frac{1}{x_i-a_j}-\frac{1}{\nu-a_j}\right), \tag{92}$$

$$\frac{\partial\Omega(P_{\nu})}{\partial x_i} = \frac{\Omega(P_{\nu})}{2}\frac{A^{12}_{x_i}(\nu-q_0)}{A^{12}_{\nu}(x_i-q_0)}\left(\frac{1}{x_i-\nu}-\sum_{a_j\neq\nu}\frac{A^{12}_{a_j}(\nu-q_0)}{A^{12}_{\nu}(a_j-q_0)}\frac{1}{(\nu-a_j)}\right), \tag{93}$$

$$\frac{\partial\Omega(P_{\infty})}{\partial x_i} = \frac{A^{12}_{x_i}}{t}\frac{\nu-x_i}{x_i-q_0}\,. \tag{94}$$

Let us introduce

(95)
$$\beta_{a_j} = A^{12}_{a_j}\left(\frac{1}{\phi(Q_0)(a_j-q_0)^2} + X_{a_j}\right), \qquad \text{where} \quad X_{a_j} = -\frac{1}{2}\frac{\sum_k \frac{1}{a_k-q_0}}{(a_j-q_0)\phi(Q_0)} - \Omega(P_\infty).$$

In the future calculations we will need the following relation which follows directly from the definition of β_{a_j} :

$$\frac{\beta_{a_j}}{A^{12}_{a_j}} - \frac{\beta_{a_i}}{A^{12}_{a_i}} = \frac{(a_i-a_j)(a_i-q_0+a_j-q_0)}{\phi(Q_0)(a_j-q_0)^2(a_i-q_0)^2} + X_{a_j} - X_{a_i} \tag{96}$$
$$= \frac{a_i-a_j}{\phi(Q_0)(a_j-q_0)(a_i-q_0)}\left(\frac{a_i-q_0+a_j-q_0}{(a_j-q_0)(a_i-q_0)} - \frac{1}{2}\sum_k \frac{1}{a_k-q_0}\right).$$

THEOREM 5. *Consider a Chebyshev family of Riemann surfaces corresponding to the algebraic curves of equation* (46) *introduced in Section* 5 *and the differentials* Ω (61) *and* ϕ (52) *defined on the surfaces of the family. The traceless matrices*

$$A_{a_j} = \begin{pmatrix} A^{11}_{a_j} & A^{12}_{a_j} \\ A^{21}_{a_j} & -A^{11}_{a_j} \end{pmatrix}, \qquad a_j \in B$$

where the coefficients $A^{12}_{a_j}$ *are given by* (87) *and*

$$A^{11}_{a_j} = -\frac{1}{4} - \frac{1}{t}\beta_{a_j}, \qquad A^{21}_{a_j} = -\frac{t\beta_{a_j} + 2\beta^2_{a_j}}{2t^2 A^{12}_{a_j}} \tag{97}$$

with β_{a_j} *defined by* (95) *solve the constrained Schlesinger system* (15)–(17)

9. Proof of Theorem 5

The proof of Theorem 5 which we give is a long but rather straightforward calculation, involving however some non-trivial relations between the objects on Riemann surfaces that we introduced. Let us first establish some useful identities.

LEMMA 8. *Let* Ω, ϕ *be differentials defined in Section* 6 *on a compactified hyperelliptic curve of genus 2 from the family given by equation* (46) *with a choice of a canonical homology basis. Let the evaluations of differentials at given points be defined by* (53) *and* u *stand for* $u(P)$. *Then the following relations hold*

$$\sum_{a_j \in B} \frac{A^{12}_{a_j}}{a_j - q_0} = 0, \tag{98}$$

$$\sum_{a_j \in B} \frac{A^{12}_{a_j}}{(a_j - q_0)^3} = -t \operatorname*{res}_{P=Q_0} \frac{\Omega(P)\phi(P)}{(u-q_0)du}, \tag{99}$$

$$\sum_{a_j \in B} \frac{A^{12}_{a_j}}{(a_j - q_0)^4} = -t \operatorname*{res}_{P=Q_0} \frac{\Omega(P)\phi(P)}{(u-q_0)^2du}. \tag{100}$$

Proof. Equality (98) is the statement that the sum of residues of the differential $\frac{\Omega(p)\phi(p)(u-q_0)}{du}$ on our Riemann surface vanishes. Equalities (99) and (100) are obtained by equating to zero the sum of residues of the differentials $\frac{\Omega(P)\phi(P)}{(u-q_0)du}$ and $\frac{\Omega(P)\phi(P)}{(u-q_0)^2du}$, respectively. $\square$

Let us also note that identity (81) can be rewritten in terms of the quantities $A^{12}_{a_j}$ as

$$\sum_{a_j\in B}\frac{A^{12}_{a_j}}{(a_j-q_0)^2}=-t\phi(Q_0)\,. \tag{101}$$

9.1. Proof of Theorem 5 for the (12)**-component.** The (12)-component of the matrix constrained Schlesinger system (15)–(17) gives a set of equations for the derivatives of $A^{12}_{a_j}$. There are two essentially different cases, $a_j\neq\nu$ and $a_j=\nu$. In this section we prove that derivatives $\partial_{x_i}A^{12}_{a_j}$ for the $A^{12}_{a_j}$ defined by (87) satisfy the corresponding equations of system (15)–(16) in both cases. The two cases are considered separately, in the two proposition of the section.

PROPOSITION 4. *Let a_j be different from ν and x_i. Then the functions A^{12} defined by* (87) *and A^{11} defined by* (97) *with β_{a_k} given by* (95) *satisfy for $i=1,2$:*

$$\frac{\partial A^{12}_{a_j}}{\partial x_i}=2\frac{A^{11}_{x_i}A^{12}_{a_j}-A^{12}_{x_i}A^{11}_{a_j}}{x_i-a_j}+2\frac{\partial\nu}{\partial x_i}\frac{A^{11}_{\nu}A^{12}_{a_j}-A^{12}_{\nu}A^{11}_{a_j}}{\nu-a_j}\,. \tag{102}$$

Proof. Replacing the $A^{11}_{a_j}$ by their expressions in terms of β_{a_j}, we have (102) in the form:

$$\begin{aligned}\frac{\partial A^{12}_{a_j}}{\partial x_i}&=\frac{1}{2}\frac{A^{12}_{x_i}-A^{12}_{a_j}}{x_i-a_j}+\frac{2}{t}\frac{A^{12}_{x_i}A^{12}_{a_j}}{x_i-a_j}\left(\frac{\beta_{a_j}}{A^{12}_{a_j}}-\frac{\beta_{x_i}}{A^{12}_{x_i}}\right)\\&\quad+\frac{\partial\nu}{\partial x_i}\left(\frac{1}{2}\frac{A^{12}_{\nu}-A^{12}_{a_j}}{\nu-a_j}+\frac{2}{t}\frac{A^{12}_{\nu}A^{12}_{a_j}}{\nu-a_j}\left(\frac{\beta_{a_j}}{A^{12}_{a_j}}-\frac{\beta_{\nu}}{A^{12}_{\nu}}\right)\right),\end{aligned} \tag{103}$$

which, due to relation (96) for $\frac{\beta_{a_j}}{A^{12}_{a_j}}$ and expression (89) for $\partial_{x_i}\nu$ is equivalent to

$$\begin{aligned}\frac{\partial A^{12}_{a_j}}{\partial x_i}&=\frac{1}{2}\frac{A^{12}_{x_i}-A^{12}_{a_j}}{x_i-a_j}-\frac{1}{2}\frac{A^{12}_{x_i}(\nu-q_0)}{(x_i-q_0)(\nu-a_j)}+\frac{A^{12}_{x_i}A^{12}_{a_j}}{2A^{12}_{\nu}}\frac{\nu-q_0}{(x_i-q_0)(\nu-a_j)}\\&\quad+\frac{2}{t}\frac{A^{12}_{x_i}A^{12}_{a_j}}{(x_i-a_j)}\left(\frac{(x_i-a_j)(x_i-q_0+a_j-q_0)}{\phi(Q_0)(a_j-q_0)^2(x_i-q_0)^2}+X_{a_j}-X_{x_i}\right)\\&\quad-\frac{2}{t}\frac{A^{12}_{x_i}A^{12}_{a_j}(\nu-q_0)}{(x_i-q_0)(\nu-a_j)}\left(\frac{(\nu-a_j)(\nu-q_0+a_j-q_0)}{\phi(Q_0)(a_j-q_0)^2(\nu-q_0)^2}+X_{a_j}-X_{\nu}\right).\end{aligned}$$

Let us now rewrite this in the following way.

$$\begin{aligned}\frac{\partial A^{12}_{a_j}}{\partial x_i}&=-\frac{1}{2}\frac{A^{12}_{a_j}}{x_i-a_j}+\frac{A^{12}_{x_i}A^{12}_{a_j}}{2A^{12}_{\nu}}\frac{\nu-q_0}{(x_i-q_0)(\nu-a_j)}+\frac{1}{2}\frac{A^{12}_{x_i}(a_j-q_0)(\nu-x_i)}{(x_i-a_j)(x_i-q_0)(\nu-a_j)}\\&\quad+\frac{2}{t}\frac{A^{12}_{x_i}A^{12}_{a_j}(\nu-x_i)}{\phi(Q_0)(a_j-q_0)(x_i-q_0)^2(\nu-q_0)}+\frac{2}{t}\frac{A^{12}_{x_i}A^{12}_{a_j}}{(x_i-q_0)}\\&\quad\times\left((X_{\nu}-X_{a_j})\frac{\nu-q_0}{\nu-a_j}-(X_{x_i}-X_{a_j})\frac{x_i-q_0}{x_i-a_j}\right).\end{aligned}$$

Due to definition (95) for X_{a_j}, the last term vanishes and we have

(104)

$$\frac{\partial A^{12}_{a_j}}{\partial x_i} = -\frac{1}{2}\frac{A^{12}_{a_j}}{x_i - a_j} + \frac{A^{12}_{x_i}A^{12}_{a_j}}{2A^{12}_{\nu}}\frac{\nu - q_0}{(x_i - q_0)(\nu - a_j)} + \frac{1}{2}\frac{A^{12}_{x_i}(a_j - q_0)(\nu - x_i)}{(x_i - a_j)(x_i - q_0)(\nu - a_j)} + \frac{2}{t}\frac{A^{12}_{x_i}A^{12}_{a_j}(\nu - x_i)}{\phi(Q_0)(a_j - q_0)(x_i - q_0)^2(\nu - q_0)}.$$

We want to prove that (104) holds for the functions $A^{12}_{a_j}$. Let us calculate the derivative in the left hand side using the expression for $A^{12}_{a_j}$ in terms of differentials:

$$\frac{\partial A^{12}_{a_j}}{\partial x_i} = \frac{t}{4}\partial_{x_i}\left\{\Omega(P_{a_j})\phi(P_{a_j})(a_j - q_0)^2\right\}$$

Using the result of Proposition 1 for derivative of $\Omega(P_{a_j})$, then (78) for derivative of $\phi(P_{a_j})$, as well as (90) and (89) for derivatives of q_0 and ν, we have

(105)

$$\frac{\partial A^{12}_{a_j}}{\partial x_i} = \frac{1}{2}\frac{A^{12}_{x_i}(a_j - q_0)}{x_i - q_0}\left(\frac{1}{x_i - a_j} - \frac{1}{\nu - a_j}\right) + \frac{A^{12}_{a_j}}{2}\left(\frac{1}{a_j - x_i} - \frac{A^{12}_{x_i}(\nu - q_0)}{A^{12}_{\nu}(x_i - q_0)(a_j - \nu)}\right) - \frac{2}{t}\frac{A^{12}_{a_j}A^{12}_{x_i}(x_i - \nu)}{\phi(Q_0)(a_j - q_0)(\nu - q_0)(x_i - q_0)^2}.$$

As is easy to see, this coincides with (104). □

PROPOSITION 5. *The function* $A^{12}_{a_j}$ *defined by* (87) *and functions* $A^{11}_{a_j}$ *defined by* (97) *with* β_{a_j} *given by* (95) *satisfy for* $i = 1, 2$

$$\frac{\partial A^{12}_{\nu}}{\partial x_i} = 2\frac{A^{11}_{x_i}A^{12}_{\nu} - A^{12}_{x_i}A^{11}_{\nu}}{x_i - \nu} - 2\frac{\partial \nu}{\partial x_i}\sum_{\substack{a_j \in B\\ a_j \neq \nu}}\frac{A^{11}_{\nu}A^{12}_{a_j} - A^{12}_{\nu}A^{11}_{a_j}}{\nu - a_j}. \tag{106}$$

Proof. Let us first rewrite (106) in terms of the quantities β (95) and $A^{12}_{a_j}$. We obtain

$$\frac{\partial A^{12}_{\nu}}{\partial x_i} = \frac{1}{2}\frac{A^{12}_{x_i} - A^{12}_{\nu}}{x_i - \nu} + \frac{2}{t}\frac{A^{12}_{x_i}A^{12}_{\nu}}{x_i - \nu}\left(\frac{\beta_{\nu}}{A^{12}_{\nu}} - \frac{\beta_{x_i}}{A^{12}_{x_i}}\right) - \frac{\partial \nu}{\partial x_i}\sum_{a_j \neq \nu}\left(\frac{1}{2}\frac{A^{12}_{\nu} - A^{12}_{a_j}}{\nu - a_j} + \frac{2}{t}\frac{A^{12}_{\nu}A^{12}_{a_j}}{\nu - a_j}\left(\frac{\beta_{a_j}}{A^{12}_{a_j}} - \frac{\beta_{\nu}}{A^{12}_{\nu}}\right)\right), \tag{107}$$

which, due to relation (96) for $\frac{\beta_{a_j}}{A^{12}_{a_j}}$ is equivalent to

(108)

$$\frac{\partial A^{12}_{\nu}}{\partial x_i} = \frac{1}{2}\frac{A^{12}_{x_i} - A^{12}_{\nu}}{x_i - \nu} + \frac{2}{t}A^{12}_{x_i}A^{12}_{\nu}\frac{X_{\nu} - X_{x_i}}{x_i - \nu} + \frac{2}{t}A^{12}_{x_i}A^{12}_{\nu}\frac{x_i - q_0 + \nu - q_0}{\phi(Q_0)(\nu - q_0)^2(x_i - q_0)^2}$$
$$-\frac{\partial \nu}{\partial x_i}\sum_{a_j \neq \nu}\left(\frac{1}{2}\frac{A^{12}_{\nu} - A^{12}_{a_j}}{\nu - a_j} + \frac{2}{t}A^{12}_{\nu}A^{12}_{a_j}\frac{X_{a_j} - X_{\nu}}{\nu - a_j} + \frac{2}{t}A^{12}_{\nu}A^{12}_{a_j}\frac{\nu - q_0 + a_j - q_0}{\phi(Q_0)(a_j - q_0)^2(\nu - q_0)^2}\right).$$

Now we can compute the derivative $\partial_{x_i}A^{12}_{\nu}$, using (93) following from Proposition 2 for $\partial_{x_i}\Omega(P_{\nu})$ and expressions (79) and (90) for $\partial_{x_i}\phi(P_{\nu})$ and $\partial_{x_i}q_0$:

$$\frac{\partial A^{12}_\nu}{\partial x_i} = \frac{t}{4}\partial_{x_i}\left\{\Omega(P_\nu)\phi(P_\nu)(\nu-q_0)^2\right\}$$

$$= \frac{A^{12}_{x_i}(\nu-q_0)}{2(x_i-q_0)}\left(\frac{1}{x_i-\nu} - \sum_{a_j\neq\nu}\frac{A^{12}_{a_j}}{A^{12}_\nu}\left(\frac{1}{\nu-a_j}+\frac{1}{a_j-q_0}\right)\right)$$

$$+\frac{A^{12}_\nu}{2}\left(\frac{1}{\nu-x_i} - \frac{\partial\nu}{\partial x_i}\sum_{a_j\neq\nu}\frac{1}{\nu-a_j}\right) + \frac{2A^{12}_\nu}{\nu-q_0}\left(\frac{\partial\nu}{\partial x_i} - \frac{A^{12}_{x_i}(x_i-\nu)}{t\phi(Q_0)(\nu-q_0)(x_i-q_0)^2}\right).$$

We need to prove that the last equality coincides with (108). Let us rewrite this equality, using the expression (89) for $\partial_{x_i}\nu$, and collecting in the first line the terms that match some terms from (108).

$$\frac{\partial A^{12}_\nu}{\partial x_i} = \frac{1}{2}\frac{A^{12}_{x_i}-A^{12}_\nu}{x_i-\nu} + \frac{2A^{12}_{x_i}A^{12}_\nu}{t\phi(Q_0)(\nu-q_0)(x_i-q_0)^2} - \frac{1}{2}\frac{\partial\nu}{\partial x_i}\sum_{a_j\neq\nu}\frac{A^{12}_\nu - A^{12}_{a_j}}{\nu-a_j}$$

$$-\frac{1}{2}\frac{A^{12}_{x_i}}{x_i-q_0} + \frac{\partial\nu}{\partial x_i}\left(\frac{2A^{12}_\nu}{\nu-q_0}+\frac{1}{2}\sum_{a_j\neq\nu}\frac{A^{12}_{a_j}}{a_j-q_0}\right) - \frac{2A^{12}_{x_i}A^{12}_\nu}{t\phi(Q_0)(\nu-q_0)^2(x_i-q_0)}.$$

Simplifying the factor of $\frac{\partial\nu}{\partial x_i}$ in the second line using (98) and comparing this equality to (108), we see that what remains to show is:

$$A^{12}_{x_i}A^{12}_\nu\frac{X_\nu - X_{x_i}}{x_i-\nu} - \frac{\partial\nu}{\partial x_i}A^{12}_\nu\sum_{a_j\neq\nu}A^{12}_{a_j}\frac{X_{a_j}-X_\nu}{\nu-a_j} - \frac{\partial\nu}{\partial x_i}\frac{A^{12}_\nu}{\phi(Q_0)(\nu-q_0)}\sum_{a_j\neq\nu}\frac{A^{12}_{a_j}}{(a_j-q_0)^2}$$

$$-\frac{\partial\nu}{\partial x_i}\frac{A^{12}_\nu}{\phi(Q_0)(\nu-q_0)^2}\sum_{a_j\neq\nu}\frac{A^{12}_{a_j}}{(a_j-q_0)} = -\frac{t}{4}\frac{A^{12}_{x_i}}{x_i-q_0} + \frac{3t}{4}\frac{\partial\nu}{\partial x_i}\frac{A^{12}_\nu}{\nu-q_0} - \frac{2A^{12}_{x_i}A^{12}_\nu}{\phi(Q_0)(\nu-q_0)^2(x_i-q_0)}.$$

With the help of identities (98) and (101) as well as definition (95) for the X_{a_j}, we rewrite this as

$$\frac{A^{12}_{x_i}A^{12}_\nu\sum_k\frac{1}{a_k-q_0}}{2\phi(Q_0)(x_i-\nu)}\left(\frac{1}{x_i-q_0}-\frac{1}{\nu-q_0}\right) - \frac{\partial\nu}{\partial x_i}\frac{A^{12}_\nu\sum_k\frac{1}{a_k-q_0}}{2\phi(Q_0)}\sum_{a_j\neq\nu}\frac{A^{12}_{a_j}}{\nu-a_j}\left(\frac{1}{\nu-q_0}-\frac{1}{a_j-q_0}\right)$$

$$+\frac{\partial\nu}{\partial x_i}\frac{2(A^{12}_\nu)^2}{\phi(Q_0)(\nu-q_0)^3} = -\frac{t}{4}\frac{A^{12}_{x_i}}{x_i-q_0} - \frac{t}{4}\frac{\partial\nu}{\partial x_i}\frac{A^{12}_\nu}{\nu-q_0} - \frac{2A^{12}_{x_i}A^{12}_\nu}{\phi(Q_0)(\nu-q_0)^2(x_i-q_0)}.$$

Using expression (89) for the derivative of ν in the right hand side, we see that the first two terms on that side cancel each other. Using the same expression in the last term of the left hand side, we see that that term cancels the last term of the right hand side. Calculating the differences in the remaining two terms and applying the identity (98) from Lemma 8, we finish the proof. □

9.2. Proof of Theorem 5 for the (11)**-component.** In this section, we prove equations for $A^{11}_{a_j}$ imposed by the constrained Schlesinger system first in the case of a_j different from ν and x_i and then in the case $a_j=\nu$. These two cases are considered separately in the two proposition of this section.

9.2.1. *The case* $a_j\neq\nu$. The equation in the next proposition is equation (15) of the Schlesinger system written for the matrix (11)-coefficients.

PROPOSITION 6. *Let a_j be different from ν and x_i. Then the functions A^{12} defined by* (87) *and A^{11} defined by* (97) *with β_{a_k} given by* (95) *satisfy for $i = 1, 2$:*

$$\frac{\partial A^{11}_{a_j}}{\partial x_i} = \frac{A^{12}_{x_i}A^{21}_{a_j} - A^{21}_{x_i}A^{12}_{a_j}}{x_i - a_j} + \frac{\partial \nu}{\partial x_i}\frac{A^{12}_{\nu}A^{21}_{a_j} - A^{21}_{\nu}A^{12}_{a_j}}{\nu - a_j}.$$

Proof. Let us rewrite the equation from the proposition in terms of the β_{a_j} and group the terms in a certain way that will be useful later:

$$(109)\quad \frac{\partial \beta_{a_j}}{\partial x_i} = \left(\frac{1}{2}\frac{A^{12}_{x_i}}{x_i - a_j} + \frac{A^{12}_{x_i}A^{12}_{a_j}}{t(x_i - a_j)}\left(\frac{\beta_{a_j}}{A^{12}_{a_j}} - \frac{\beta_{x_i}}{A^{12}_{x_i}}\right)\right)$$
$$\times \frac{\beta_{a_j}}{A^{12}_{a_j}} - \frac{1}{2}\frac{A^{12}_{a_j}}{x_i - a_j}\frac{\beta_{x_i}}{A^{12}_{x_i}} + \frac{A^{12}_{x_i}A^{12}_{a_j}}{t(x_i - a_j)}\left(\frac{\beta_{a_j}}{A^{12}_{a_j}} - \frac{\beta_{x_i}}{A^{12}_{x_i}}\right)\frac{\beta_{x_i}}{A^{12}_{x_i}}$$
$$+ \frac{\partial \nu}{\partial x_i}\left\{\left(\frac{1}{2}\frac{A^{12}_{\nu}}{\nu - a_j} + \frac{A^{12}_{\nu}A^{12}_{a_j}}{t(\nu - a_j)}\left(\frac{\beta_{a_j}}{A^{12}_{a_j}} - \frac{\beta_{\nu}}{A^{12}_{\nu}}\right)\right)\frac{\beta_{a_j}}{A^{12}_{a_j}} - \frac{1}{2}\frac{A^{12}_{a_j}}{\nu - a_j}\frac{\beta_{\nu}}{A^{12}_{\nu}}\right.$$
$$\left. + \frac{A^{12}_{\nu}A^{12}_{a_j}}{t(\nu - a_j)}\left(\frac{\beta_{a_j}}{A^{12}_{a_j}} - \frac{\beta_{\nu}}{A^{12}_{\nu}}\right)\frac{\beta_{\nu}}{A^{12}_{\nu}}\right\}.$$

To prove the proposition, we need to show that derivatives of the quantities β_{a_j} defined by (95) are given by the right hand side of (109) for a_j different from x_i and ν.

Differentiation of β_{a_j}.. Let a_j be different from ν and x_i and let us differentiate the quantities β_{a_j} defined by (95), using (103) from the proof of Proposition 4 for $\partial_{x_i}A^{12}_{a_j}$, as well as (80) for $\frac{\partial \phi(Q_0)}{\partial x_i}$, and (91) for $\frac{\partial q_0}{\partial x_i}$:

$$(110)\quad \partial_{x_i}\beta_{a_j} = \left(\partial_{x_i}A^{12}_{a_j}\right)\frac{\beta_{a_j}}{A^{12}_{a_j}} + A^{12}_{a_j}\partial_{x_i}\left(\frac{1}{\phi(Q_0)(a_j - q_0)^2} + X_{a_j}\right)$$
$$= A^{12}_{a_j}\frac{\partial X_{a_j}}{\partial x_i} + \left(\frac{1}{2}\frac{A^{12}_{x_i} - A^{12}_{a_j}}{x_i - a_j} + \frac{2}{t}\frac{A^{12}_{x_i}A^{12}_{a_j}}{x_i - a_j}\left(\frac{\beta_{a_j}}{A^{12}_{a_j}} - \frac{\beta_{x_i}}{A^{12}_{x_i}}\right)\right)\frac{\beta_{a_j}}{A^{12}_{a_j}}$$
$$+ \frac{A^{12}_{a_j}A^{12}_{x_i}\sum_{a_j}\frac{1}{a_j - q_0}}{2t\phi^2(q_0)(a_j - q_0)^2(x_i - q_0)^2} + \frac{A^{12}_{a_j}}{2\phi(Q_0)(a_j - q_0)^2(x_i - q_0)}$$
$$- \frac{2A^{12}_{a_j}A^{12}_{x_i}}{t\phi^2(q_0)(a_j - q_0)^3(x_i - q_0)^2}$$
$$+ \frac{\partial \nu}{\partial x_i}\left\{\frac{1}{2}\frac{A^{12}_{\nu} - A^{12}_{a_j}}{\nu - a_j}\frac{\beta_{a_j}}{A^{12}_{a_j}} + \frac{2}{t}\frac{A^{12}_{\nu}A^{12}_{a_j}}{\nu - a_j}\frac{\beta_{a_j}}{A^{12}_{a_j}}\left(\frac{\beta_{a_j}}{A^{12}_{a_j}} - \frac{\beta_{\nu}}{A^{12}_{\nu}}\right)\right.$$
$$+ \frac{A^{12}_{a_j}A^{12}_{\nu}\sum_{a_j}\frac{1}{a_j - q_0}}{2t\phi^2(q_0)(a_j - q_0)^2(\nu - q_0)^2} + \frac{A^{12}_{a_j}}{2\phi(Q_0)(a_j - q_0)^2(\nu - q_0)}$$
$$\left. - \frac{2A^{12}_{a_j}A^{12}_{\nu}}{t\phi^2(q_0)(a_j - q_0)^3(\nu - q_0)^2}\right\}.$$

It remains to calculate the derivative of X_{a_j}. We find

$$\frac{\partial X_{a_j}}{\partial x_i} = \partial_{x_i}\left\{-\frac{1}{2}\frac{\sum_k \frac{1}{a_k - q_0}}{(a_j - q_0)\phi(Q_0)} - \Omega(p_\infty)\right\}$$

$$= \frac{\sum_k \frac{1}{a_k-q_0}}{2\phi(Q_0)(a_j-q_0)^2}\left(\frac{A^{12}_{x_i}}{t\phi(Q_0)(x_i-q_0)^2}+\frac{\partial\nu}{\partial x_i}\frac{A^{12}_{\nu}}{t\phi(Q_0)(\nu-q_0)^2}\right)$$

$$-\frac{\sum_k \frac{1}{a_k-q_0}}{4(a_j-q_0)\phi(Q_0)}$$

$$\times\left(\sum_k \frac{1}{a_k-q_0}\left(\frac{A^{12}_{x_i}}{t\phi(Q_0)(x_i-q_0)^2}+\frac{\partial\nu}{\partial x_i}\frac{A^{12}_{\nu}}{t\phi(Q_0)(\nu-q_0)^2}\right)\right.$$

$$\left.+\frac{1}{x_i-q_0}+\frac{\partial\nu}{\partial x_i}\frac{1}{\nu-q_0}\right)$$

$$+\frac{\sum_k \frac{1}{(a_k-q_0)^2}}{2\phi(Q_0)(a_j-q_0)}\left(\frac{A^{12}_{x_i}}{t\phi(Q_0)(x_i-q_0)^2}+\frac{\partial\nu}{\partial x_i}\frac{A^{12}_{\nu}}{t\phi(Q_0)(\nu-q_0)^2}\right)$$

$$+\frac{1}{2\phi(Q_0)(a_j-q_0)(x_i-q_0)^2}+\frac{\partial\nu/\partial x_i}{2\phi(Q_0)(a_j-q_0)(\nu-q_0)^2}-\partial_{x_i}\Omega(p_\infty)\,.$$

Rewriting this to collect the terms proportional to $\partial\nu/\partial x_i$ and using Proposition 3 for $\partial_{x_i}\Omega(p_\infty)$, we have

$$\frac{\partial X_{a_j}}{\partial x_i}=\frac{A^{12}_{x_i}\sum_k\frac{1}{a_k-q_0}}{2t\phi^2(q_0)(a_j-q_0)^2(x_i-q_0)^2}-\frac{A^{12}_{x_i}\left(\sum_k\frac{1}{a_k-q_0}\right)^2}{4t\phi^2(q_0)(a_j-q_0)(x_i-q_0)^2} \tag{111}$$

$$-\frac{\sum_k\frac{1}{a_k-q_0}}{4\phi(Q_0)(a_j-q_0)(x_i-q_0)}+\frac{A^{12}_{x_i}\sum_k\frac{1}{(a_k-q_0)^2}}{2t\phi^2(q_0)(a_j-q_0)(x_i-q_0)^2}+\frac{1}{2\phi(Q_0)(a_j-q_0)(x_i-q_0)^2}$$

$$+\frac{\partial\nu}{\partial x_i}\left\{\frac{A^{12}_{\nu}\sum_k\frac{1}{a_k-q_0}}{2t\phi^2(q_0)(a_j-q_0)^2(\nu-q_0)^2}-\frac{A^{12}_{\nu}\left(\sum_k\frac{1}{a_k-q_0}\right)^2}{4t\phi^2(q_0)(a_j-q_0)(\nu-q_0)^2}\right.$$

$$\left.-\frac{\sum_k\frac{1}{a_k-q_0}}{4\phi(Q_0)(a_j-q_0)(\nu-q_0)}+\frac{A^{12}_{\nu}\sum_k\frac{1}{(a_k-q_0)^2}}{2t\phi^2(q_0)(a_j-q_0)(\nu-q_0)^2}+\frac{1}{2\phi(Q_0)(a_j-q_0)(\nu-q_0)^2}\right\}$$

$$-\frac{A^{12}_{x_i}}{t}\frac{\nu-x_i}{x_i-q_0}\,.$$

Finishing the proof for $A^{11}_{a_j}$.. Calculating the difference of (110) and (109), we find the following expression for $\partial_{x_i}X_{a_j}$ which we will then compare with the derivative (111):

$$\frac{\partial X_{a_j}}{\partial x_i}=-\frac{A^{12}_{x_i}}{t(x_i-a_j)}\left(\frac{\beta_{a_j}}{A^{12}_{a_j}}-\frac{\beta_{x_i}}{A^{12}_{x_i}}\right)^2+\frac{1}{2(x_i-a_j)}\left(\frac{\beta_{a_j}}{A^{12}_{a_j}}-\frac{\beta_{x_i}}{A^{12}_{x_i}}\right)$$

$$-\frac{A^{12}_{x_i}\sum_k\frac{1}{a_k-q_0}}{2t\phi^2(q_0)(a_j-q_0)^2(x_i-q_0)^2}-\frac{1}{2\phi(Q_0)(a_j-q_0)^2(x_i-q_0)}$$

$$+\frac{2A^{12}_{x_i}}{t\phi^2(q_0)(a_j-q_0)^3(x_i-q_0)^2}+\frac{\partial\nu}{\partial x_i}\left\{-\frac{A^{12}_{\nu}}{t(\nu-a_j)}\left(\frac{\beta_{a_j}}{A^{12}_{a_j}}-\frac{\beta_{\nu}}{A^{12}_{\nu}}\right)^2\right.$$

$$+\frac{1}{2(\nu-a_j)}\left(\frac{\beta_{a_j}}{A^{12}_{a_j}}-\frac{\beta_{\nu}}{A^{12}_{\nu}}\right)-\frac{A^{12}_{\nu}\sum_k\frac{1}{a_k-q_0}}{2t\phi^2(q_0)(a_j-q_0)^2(\nu-q_0)^2}$$

$$-\frac{1}{2\phi(Q_0)(a_j-q_0)^2(\nu-q_0)}+\frac{2A_\nu^{12}}{t\phi^2(q_0)(a_j-q_0)^3(\nu-q_0)^2}\Bigg\}.$$

Expressing the differences $\frac{\beta_{a_j}}{A_{a_j}^{12}}-\frac{\beta_{x_i}}{A_{x_i}^{12}}$ and $\frac{\beta_{a_j}}{A_{a_j}^{12}}-\frac{\beta_\nu}{A_\nu^{12}}$ as in (96), after some algebraic simplification, we obtain

$$\begin{aligned}
(112)\quad \frac{\partial X_{a_j}}{\partial x_i} &= -\frac{A_{x_i}^{12}(x_i-a_j)}{t\phi^2(q_0)(a_j-q_0)^4(x_i-q_0)^2}-\frac{A_{x_i}^{12}(x_i-a_j)}{t\phi^2(q_0)(a_j-q_0)^2(x_i-q_0)^4}\\
&+\frac{2A_{x_i}^{12}}{t\phi^2(q_0)(a_j-q_0)^2(x_i-q_0)^3}+\frac{A_{x_i}^{12}(x_i-a_j)\sum_k\frac{1}{a_k-q_0}}{t\phi^2(q_0)(a_j-q_0)^3(x_i-q_0)^2}+\frac{A_{x_i}^{12}(x_i-a_j)\sum_k\frac{1}{a_k-q_0}}{t\phi^2(q_0)(a_j-q_0)^2(x_i-q_0)^3}\\
&-\frac{A_{x_i}^{12}(x_i-a_j)\left(\sum_k\frac{1}{a_k-q_0}\right)^2}{4t\phi^2(q_0)(a_j-q_0)^2(x_i-q_0)^2}+\frac{1}{2\phi(Q_0)(a_j-q_0)(x_i-q_0)^2}-\frac{\sum_k\frac{1}{a_k-q_0}}{4\phi(Q_0)(a_j-q_0)(x_i-q_0)}\\
&-\frac{A_{x_i}^{12}\sum_k\frac{1}{a_k-q_0}}{2t\phi^2(q_0)(a_j-q_0)^2(x_i-q_0)^2}+\frac{\partial\nu}{\partial x_i}\Bigg\{-\frac{A_\nu^{12}(\nu-a_j)}{t\phi^2(q_0)(a_j-q_0)^4(\nu-q_0)^2}-\frac{A_\nu^{12}(\nu-a_j)}{t\phi^2(q_0)(a_j-q_0)^2(\nu-q_0)^4}\\
&+\frac{2A_\nu^{12}}{t\phi^2(q_0)(a_j-q_0)^2(\nu-q_0)^3}+\frac{A_\nu^{12}(\nu-a_j)\sum_k\frac{1}{a_k-q_0}}{t\phi^2(q_0)(a_j-q_0)^3(\nu-q_0)^2}+\frac{A_\nu^{12}(\nu-a_j)\sum_k\frac{1}{a_k-q_0}}{t\phi^2(q_0)(a_j-q_0)^2(\nu-q_0)^3}\\
&-\frac{A_\nu^{12}(\nu-a_j)\left(\sum_k\frac{1}{a_k-q_0}\right)^2}{4t\phi^2(q_0)(a_j-q_0)^2(\nu-q_0)^2}+\frac{1}{2\phi(Q_0)(a_j-q_0)(\nu-q_0)^2}-\frac{\sum_k\frac{1}{a_k-q_0}}{4\phi(Q_0)(a_j-q_0)(\nu-q_0)}\\
&-\frac{A_\nu^{12}\sum_k\frac{1}{a_k-q_0}}{2t\phi^2(q_0)(a_j-q_0)^2(\nu-q_0)^2}\Bigg\}.
\end{aligned}$$

To finish the proof of Proposition 6, we need to show that (112) coincides with (111). Let us subtract expression (111) from (112) and plug in (89) for the derivative $\partial_{x_i}\nu$. After that, we can divide by $A_{x_i}^{12}$, and multiply by $t\phi^2(q_0)$, which, after some careful algebraic simplification yields a rational expression for the difference of (112) and (111):

$$\begin{aligned}
&\frac{1}{(a_j-q_0)^3(x_i-q_0)^2(\nu-q_0)}+\frac{1}{(a_j-q_0)(x_i-q_0)^4(\nu-q_0)}+\frac{1}{(a_j-q_0)(x_i-q_0)^2(\nu-q_0)^3}\\
&+\frac{1}{(a_j-q_0)(x_i-q_0)^3(\nu-q_0)^2}+\frac{1}{(a_j-q_0)^2(x_i-q_0)^3(\nu-q_0)}+\frac{1}{(a_j-q_0)^2(x_i-q_0)^2(\nu-q_0)^2}\\
&-\frac{\sum_k\frac{1}{a_k-q_0}}{(a_j-q_0)^2(x_i-q_0)^2(\nu-q_0)}-\frac{\sum_k\frac{1}{a_k-q_0}}{(a_j-q_0)(x_i-q_0)^3(\nu-q_0)}-\frac{\sum_k\frac{1}{a_k-q_0}}{(a_j-q_0)(x_i-q_0)^2(\nu-q_0)^2}\\
&+\frac{\sum_{k\neq l}\frac{1}{(a_k-q_0)(a_l-q_0)}}{2(a_j-q_0)(x_i-q_0)^2(\nu-q_0)}+\frac{\phi^2(q_0)}{x_i-q_0}.
\end{aligned}$$

This rational expression can be shown to vanish identically by a direct calculation as follows. Denoting the five branch points by $a_j, a_j', \nu, x_i, x_i'$, we can write all the sums and $\phi^2(q_0)$ explicitly. Doing this, we see that all terms cancel out and thus we prove that the difference of (112) and (111) is identically zero. This completes the proof of Proposition 6. □

9.2.2. *The case $a_j=\nu$.* The next proposition shows that quantities defined in Section 8 satisfy equation (16) of the constrained Schlesinger system. The equation in the proposition is equation (16) written for the matrix (11)-component.

PROPOSITION 7. *Then the functions* $A^{12}_{a_j}$ *defined by* (87) *and* $A^{11}_{a_j}$ *defined by* (97) *with* β_{a_j} *given by* (95) *satisfy for* $i=1,2$:

$$\frac{\partial A^{11}_\nu}{\partial x_i} = \frac{A^{12}_{x_i}A^{21}_\nu - A^{21}_{x_i}A^{12}_\nu}{x_i-\nu} - \frac{\partial\nu}{\partial x_i}\sum_{\substack{a_j\in B\\ a_j\neq\nu}} \frac{A^{12}_\nu A^{21}_{a_j} - A^{21}_\nu A^{12}_{a_j}}{\nu-a_j}.$$

Proof. Given that $A^{11}_{a_j} = -1/4 - \beta_{a_j}/t$, the equation of the proposition implies that the derivative of β_ν should satisfy

$$\frac{\partial\beta_\nu}{\partial x_i} = \frac{1}{2(x_i-\nu)}\left(A^{12}_{x_i}\frac{\beta_\nu}{A^{12}_\nu} - A^{12}_\nu\frac{\beta_{x_i}}{A^{12}_{x_i}}\right) + \frac{A^{12}_{x_i}A^{12}_\nu}{t(x_i-\nu)}\left(\left(\frac{\beta_\nu}{A^{12}_\nu}\right)^2 - \left(\frac{\beta_{x_i}}{A^{12}_{x_i}}\right)^2\right) \tag{113}$$
$$-\frac{\partial\nu}{\partial x_i}\sum_{a_j\neq\nu}\left\{\frac{1}{2(\nu-a_j)}\left(A^{12}_\nu\frac{\beta_{a_j}}{A^{12}_{a_j}} - A^{12}_{a_j}\frac{\beta_\nu}{A^{12}_\nu}\right) + \frac{A^{12}_\nu A^{12}_{a_j}}{t(\nu-a_j)}\left(\left(\frac{\beta_{a_j}}{A^{12}_{a_j}}\right)^2 - \left(\frac{\beta_\nu}{A^{12}_\nu}\right)^2\right)\right\}.$$

Let us calculate the derivative in the left hand side of (113) and then show that it coincides with the right hand side of (113).

Differentiation of β_ν.. We calculate the derivative of β_ν, representing β_ν as $A^{12}_\nu\frac{\beta_\nu}{A^{12}_\nu}$ and using (107) from Proposition 5 for the derivative of A^{12}_ν and (90) and (89) for derivatives of q_0 and ν as well as (80) for derivative of $\phi(Q_0)$:

(114)

$$\frac{\partial\beta_\nu}{\partial x_i} = \frac{\partial}{\partial x_i}\left\{A^{12}_\nu\left(\frac{1}{\phi(Q_0)(\nu-q_0)^2}+X_\nu\right)\right\} = A^{12}_\nu\left(-\frac{A^{12}_{x_i}(x_i-\nu)\sum_{a_j}\frac{1}{a_j-q_0}}{2t\phi^2(q_0)(\nu-q_0)^3(x_i-q_0)^2}\right.$$
$$+\frac{1}{2\phi(Q_0)(\nu-q_0)^2(x_i-q_0)} + \frac{2A^{12}_{x_i}(x_i-\nu)}{t\phi^2(q_0)(x_i-q_0)^2(\nu-q_0)^4}$$
$$\left.-\frac{\partial\nu}{\partial x_i}\frac{3}{2\phi(Q_0)(\nu-q_0)^3} + \frac{\partial X_\nu}{\partial x_i}\right) + \frac{\beta_\nu}{A^{12}_\nu}\left\{\frac{1}{2}\frac{A^{12}_{x_i}-A^{12}_\nu}{x_i-\nu} + \frac{2}{t}\frac{A^{12}_{x_i}A^{12}_\nu}{x_i-\nu}\left(\frac{\beta_\nu}{A^{12}_\nu} - \frac{\beta_{x_i}}{A^{12}_{x_i}}\right)\right.$$
$$\left.-\frac{\partial\nu}{\partial x_i}\sum_{a_j\neq\nu}\left(\frac{1}{2}\frac{A^{12}_\nu - A^{12}_{a_j}}{\nu-a_j} + \frac{2}{t}\frac{A^{12}_\nu A^{12}_{a_j}}{\nu-a_j}\left(\frac{\beta_{a_j}}{A^{12}_{a_j}} - \frac{\beta_\nu}{A^{12}_\nu}\right)\right)\right\}.$$

Using definition (95) of X_ν and Lemma 3 for $\partial_{x_i}\Omega(p_\infty)$, the derivative of X_ν is found to be

$$\frac{\partial X_\nu}{\partial x_i} = -\frac{1}{2}\frac{A^{12}_{x_i}(x_i-\nu)\sum_{a_k}\frac{1}{(a_k-q_0)^2}}{t\phi^2(q_0)(\nu-q_0)^2(x_i-q_0)^2} + \frac{1}{2\phi(Q_0)(\nu-q_0)(x_i-q_0)^2} \tag{115}$$
$$-\frac{1}{2}\frac{A^{12}_{x_i}(x_i-\nu)\sum_{a_k}\frac{1}{a_k-q_0}}{t\phi^2(q_0)(\nu-q_0)^3(x_i-q_0)^2} + \frac{A^{12}_{x_i}(x_i-\nu)\left(\sum_{a_k}\frac{1}{a_k-q_0}\right)^2}{4t\phi^2(q_0)(\nu-q_0)^2(x_i-q_0)^2} - \frac{\sum_{a_k}\frac{1}{a_k-q_0}}{4\phi(Q_0)(\nu-q_0)(x_i-q_0)}$$

$$+\frac{\partial\nu}{\partial x_i}\left(\frac{1}{2\phi(Q_0)(\nu-q_0)^3} + \frac{\sum_{a_k}\frac{1}{a_k-q_0}}{2\phi(Q_0)(\nu-q_0)^2} - \frac{\sum_{a_k}\frac{1}{a_k-q_0}}{4\phi(Q_0)(\nu-q_0)^2}\right) - \frac{A^{12}_{x_i}(\nu-x_i)}{t(x_i-q_0)}.$$

Finishing the proof for A^{11}_ν.. Subtracting (113) from (114), we see that we need to prove the following equality:

(116)

$$0=\frac{\frac{\beta_{x_i}}{A^{12}_{x_i}}-\frac{\beta_\nu}{A^{12}_\nu}}{2(x_i-\nu)}+\frac{A^{12}_{x_i}\left(\frac{\beta_{x_i}}{A^{12}_{x_i}}-\frac{\beta_\nu}{A^{12}_\nu}\right)^2}{t(x_i-\nu)}-\frac{\partial\nu}{\partial x_i}\sum_{a_j\neq\nu}\left(\frac{\frac{\beta_{a_j}}{A^{12}_{a_j}}-\frac{\beta_\nu}{A^{12}_\nu}}{2(a_j-\nu)}+\frac{A^{12}_{a_j}\left(\frac{\beta_{a_j}}{A^{12}_{a_j}}-\frac{\beta_\nu}{A^{12}_\nu}\right)^2}{t(a_j-\nu)}\right)$$

$$-\frac{A^{12}_{x_i}(x_i-\nu)\sum_{a_j}\frac{1}{a_j-q_0}}{2t\phi^2(q_0)(\nu-q_0)^3(x_i-q_0)^2}+\frac{1}{2\phi(Q_0)(\nu-q_0)^2(x_i-q_0)}+\frac{2A^{12}_{x_i}(x_i-\nu)}{t\phi^2(q_0)(x_i-q_0)^2(\nu-q_0)^4}$$

$$-\frac{\partial\nu}{\partial x_i}\frac{3}{2\phi(Q_0)(\nu-q_0)^3}+\frac{\partial X_\nu}{\partial x_i}.$$

This equality can be shown to hold by a lengthy calculation applying the identities that we have already established. This will reduce the right hand side of (116) to a rational expression. Let us proceed by performing the following steps. First, we replace the differences of the type $\frac{\beta_{x_i}}{A^{12}_{x_i}}-\frac{\beta_\nu}{A^{12}_\nu}$ from equation (96) and use (115) for $\partial_{x_i}X_\nu$. At this step, the term of the form $\frac{\nu-q_0+x_i-q_0}{2\phi(Q_0)(\nu-q_0)^2(x_i-q_0)^2}$ cancels out. Next, we plug in expressions (95) for X_{a_j} and calculate the squares explicitly keeping $\left(\sum_{a_k}\frac{1}{a_k-q_0}\right)^2$ as is for now. It is then useful to split some fractions into partial fractions by writing $x_i-\nu$ as $x_1-q_0-(\nu-q_0)$ and similarly with $a_j-\nu$. This leads to a long expression in which we will now work on the sums of the form $\sum_{a_j\neq\nu}\frac{A^{12}_{a_j}}{(a_j-q_0)^k}$ by first making the sums ranging over all indices and then applying (98) and (101). Finally, let us use (99) and (100) to replace some sums by the corresponding residues.

It remains to calculate explicitly the residues from (99) and (100). Using representation (62) of Ω with the help of normalization coefficients δ_1 and δ_2, a straightforward calculation yields:

$$\operatorname*{res}_{P=Q_0}\frac{\Omega(P)\phi(P)}{(u-q_0)du}=\phi(Q_0)\left(\delta_2+\sum_{a_k\in B}\frac{1}{a_k-q_0}\right)$$

and

$$\operatorname*{res}_{P=Q_0}\frac{\Omega(P)\phi(P)}{(u-q_0)^2du}=\frac{\phi(Q_0)\left(\sum_{a_k}\frac{1}{a_k-q_0}\right)^2}{2}+\frac{\phi(Q_0)\sum_{a_k}\frac{1}{(a_k-q_0)^2}}{2}+\frac{\phi^2(Q_0)\Omega(P_\nu)}{\phi(P_\nu)(\nu-q_0)}$$
$$-\frac{\phi(Q_0)}{(\nu-q_0)^2}-\frac{\delta_2\phi(Q_0)}{\nu-q_0}+\delta_2\phi(Q_0)\sum_{a_k}\frac{1}{a_k-q_0}.$$

Using this to continue our calculation, we see that all terms containing δ_2 cancel out and we obtain (116) in the form

$$0=-\frac{A^{12}_{x_i}}{t\phi^2(q_0)(\nu-q_0)(x_i-q_0)^4}-\frac{A^{12}_{x_i}}{t\phi^2(q_0)(\nu-q_0)^2(x_i-q_0)^3}-\frac{A^{12}_{x_i}}{t\phi^2(q_0)(\nu-q_0)^3(x_i-q_0)^2}$$
$$+\frac{3A^{12}_{x_i}}{t\phi^2(q_0)(\nu-q_0)^4(x_i-q_0)}+\frac{A^{12}_{x_i}\sum_{a_k}\frac{1}{a_k-q_0}}{t\phi^2(q_0)(\nu-q_0)(x_i-q_0)^3}-\frac{2A^{12}_{x_i}\sum_{a_j}\frac{1}{a_j-q_0}}{t\phi^2(q_0)(\nu-q_0)^3(x_i-q_0)}$$

$$-\frac{A^{12}_{x_i}\left(\sum_{a_k}\frac{1}{a_k-q_0}\right)^2}{2t\phi^2(q_0)(\nu-q_0)(x_i-q_0)^2}+\frac{A^{12}_{x_i}\left(\sum_{a_k}\frac{1}{a_k-q_0}\right)^2}{2t\phi^2(q_0)(\nu-q_0)^2(x_i-q_0)}-\frac{\partial\nu}{\partial x_i}\frac{\Omega(P_\nu)}{\phi(P_\nu)(\nu-q_0)^2}$$
$$+\frac{A^{12}_{x_i}\sum_{a_j}\frac{1}{a_j-q_0}}{t\phi^2(q_0)(\nu-q_0)^2(x_i-q_0)^2}-\frac{A^{12}_{x_i}\sum_{a_k}\frac{1}{(a_k-q_0)^2}}{2t\phi^2(q_0)(\nu-q_0)^2(x_i-q_0)}+\frac{A^{12}_{x_i}\sum_{a_k}\frac{1}{(a_k-q_0)^2}}{2t\phi^2(q_0)(\nu-q_0)(x_i-q_0)^2}-\frac{A^{12}_{x_i}(\nu-x_i)}{t(x_i-q_0)}.$$

Now, plugging in expression (89) for the derivative $\partial_{x_i}\nu$, dividing by $A^{12}_{x_i}$, multiplying by $t\phi^2(q_0)(\nu-q_0)(x_i-q_0)$, and using the definition of A^{12}_ν to absorb $\Omega(P_\nu)$, we arrive at a rational expression in which $\phi^2(q_0)$ is rational by (54):

$$0=-\frac{1}{(x_i-q_0)^3}-\frac{1}{(\nu-q_0)(x_i-q_0)^2}-\frac{1}{(\nu-q_0)^2(x_i-q_0)}+\frac{3}{(\nu-q_0)^3}+\frac{\sum_{a_k}\frac{1}{a_k-q_0}}{(x_i-q_0)^2}$$
$$-\frac{2\sum_{a_j}\frac{1}{a_j-q_0}}{(\nu-q_0)^2}-\frac{\left(\sum_{a_k}\frac{1}{a_k-q_0}\right)^2}{2(x_i-q_0)}+\frac{\left(\sum_{a_k}\frac{1}{a_k-q_0}\right)^2}{2(\nu-q_0)}+\frac{4\phi^2(q_0)}{\phi^2(P_\nu)(\nu-q_0)^2}+\frac{\sum_{a_j}\frac{1}{a_j-q_0}}{(\nu-q_0)(x_i-q_0)}$$
$$-\frac{\sum_{a_k}\frac{1}{(a_k-q_0)^2}}{2(\nu-q_0)}+\frac{\sum_{a_k}\frac{1}{(a_k-q_0)^2}}{2(x_i-q_0)}-\phi^2(q_0)(\nu-q_0)(\nu-x_i).$$

This simplifies to

$$0=\frac{1}{(\nu-q_0)^3}+\frac{\sum_{a_k\notin\{x_i,\nu\}}\frac{1}{a_k-q_0}}{(x_i-q_0)^2}-\frac{2\sum_{a_j\notin\{\nu,x_i\}}\frac{1}{a_j-q_0}}{(\nu-q_0)^2}-\frac{1}{(\nu-q_0)^2(x_i-q_0)}+\frac{4\phi^2(q_0)}{\phi^2(P_\nu)(\nu-q_0)^2}$$
$$-\frac{\sum\limits_{\substack{a_k\neq a_j\\ a_k,a_j\notin\{\nu,x_i\}}}\frac{1}{(a_k-q_0)(a_j-q_0)}}{2(x_i-q_0)}+\frac{\sum\limits_{\substack{a_k\neq a_j\\ a_k,a_j\notin\{\nu,x_i\}}}\frac{1}{(a_k-q_0)(a_j-q_0)}}{2(\nu-q_0)}+\frac{\sum_{a_j\notin\{\nu,x_i\}}\frac{1}{a_j-q_0}}{(\nu-q_0)(x_i-q_0)}-\phi^2(q_0)(\nu-q_0)(\nu-x_i),$$

which can be proven by seeing the right hand side as a rational function of ν and proving that this function has no poles in $\mathbb{C}P^1$ and vanishes, for example, at $\nu=x_i$. This completes the proof of Theorem 5 for A^{11}_ν.

9.3. Proof of Theorem 5 for the 21-component. To show that functions $A^{21}_{a_j}$ from Theorem 5 satisfy equations of the constrained Schlesinger system is now quite easy. Let us first rewrite the constrained Schlesinger equations (15)–(16) for the (21)-component of the matrices:

$$\frac{\partial A^{21}_{a_j}}{\partial x_i}=2\frac{A^{21}_{x_i}A^{11}_{a_j}-A^{11}_{x_i}A^{21}_{a_j}}{x_i-a_j}+2\frac{A^{21}_{\nu}A^{11}_{a_j}-A^{11}_{\nu}A^{21}_{a_j}}{\nu-a_j}\frac{\partial\nu}{\partial x_i}\quad\text{for}\quad a_j\neq\nu;\tag{117}$$

$$\frac{\partial A^{21}_{\nu}}{\partial x_i}=2\frac{A^{21}_{x_i}A^{11}_{\nu}-A^{11}_{x_i}A^{21}_{\nu}}{x_i-\nu}-2\frac{\partial\nu}{\partial x_i}\sum_{\substack{a_j\in B\\ a_j\neq\nu}}\frac{A^{21}_{a_j}A^{11}_{\nu}-A^{11}_{a_j}A^{21}_{\nu}}{a_j-\nu}.\tag{118}$$

Equations (117) and (118) can be derived from the preceding results. Namely, from definition (97) of $A^{21}_{a_j}$ from Theorem 5, we get $A^{21}_{a_j}=(\frac{1}{16}-(A^{11}_{a_j})^2)/A^{12}_{a_j}$ for any $a_j\in B$ and thus

$$\frac{\partial A^{21}_{a_j}}{\partial x_i}=-2\frac{A^{11}_{a_j}}{A^{12}_{a_j}}\frac{\partial A^{11}_{a_j}}{\partial x_i}-\frac{A^{21}_{a_j}}{A^{12}_{a_j}}\frac{\partial A^{12}_{a_j}}{\partial x_i}.$$

The rest is a simple straightforward calculation using the fact that $A^{12}_{a_j}$ and $A^{11}_{a_j}$ satisfy equations of Propositions 4-7.

By a relatively long but rather straightforward calculation, using relations (88) and (98)–(101), we verify that the sum of the residue matrices $-A_\infty=\sum_{a_j\in B}A_{a_j}$ is equal to the diagonal matrix $\operatorname{diag}\left(\frac{1}{4}, -\frac{1}{4}\right)$. This implies remaining equations (17) of the constrained Schlesinger system for the matrices A_{a_j} from Theorem 5. □

Acknowledgements

V.D. acknowledges with gratitude support from the Simons Foundation grant no. 854861 and the Science Fund of Serbia grant Integrability and Extremal Problems in Mechanics, Geometry and Combinatorics, MEGIC, Grant No. 7744592. V.S. gratefully acknowledges support from the Natural Sciences and Engineering Research Council of Canada through a Discovery grant and from the University of Sherbrooke.

References

[1] N. I. Akhiezer, *Über einige Funktionen, welche in zwei gegebenen Interwallen am wenigsten von Null abweichen. I Teil*, Izvestiya Akad. Nauk SSSA, VII ser., Otd. mat. est. nauk, 1932, No. 9, p. 1163–1202.

[2] N. I. Akhiezer, *Über einige Funktionen, welche in zwei gegebenen Interwallen am wenigsten von Null abweichen. II Teil*, Izvestiya Akad. Nauk SSSA, VII ser., Otd. mat. est. nauk, 1933, No. 3, p. 309–344.

[3] N. I. Akhiezer, *Über einige Funktionen, welche in zwei gegebenen Interwallen am wenigsten von Null abweichen. III Teil*, Izvestiya Akad. Nauk SSSA, VII ser., Otd. mat. est. nauk, 1933, No. 4, p. 499–536.

[4] N. I. Ahiezer, *Lekcii po Teorii Approksimacii* (Russian), OGIZ, Moscow-Leningrad, 1947. MR25598

[5] N. I. Akhiezer, *Elements of the theory of elliptic functions*, Translations of Mathematical Monographs, vol. 79, American Mathematical Society, Providence, RI, 1990. Translated from the second Russian edition by H. H. McFaden, DOI 10.1090/mmono/079. MR1054205

[6] A. Bogatyrev, *Extremal polynomials and Riemann surfaces*, Springer Monographs in Mathematics, Springer, Heidelberg, 2012. Translated from the 2005 Russian original by Nikolai Kruzhilin, DOI 10.1007/978-3-642-25634-9. MR2953752

[7] A. B. Bogatyrëv, *Chebyshev representation of rational functions* (Russian, with Russian summary), Mat. Sb. **201** (2010), no. 11, 19–40, DOI 10.1070/SM2010v201n11ABEH004123; English transl., Sb. Math. **201** (2010), no. 11-12, 1579–1598. MR2768552

[8] P. Deift, A. Its, A. Kapaev, and X. Zhou, *On the algebro-geometric integration of the Schlesinger equations*, Comm. Math. Phys. **203** (1999), no. 3, 613–633, DOI 10.1007/s002200050037. MR1700154

[9] V. Dragović and M. Radnović, *Geometry of integrable billiards and pencils of quadrics* (English, with English and French summaries), J. Math. Pures Appl. (9) **85** (2006), no. 6, 758–790, DOI 10.1016/j.matpur.2005.12.002. MR2236243

[10] V. Dragović and M. Radnović, *Poncelet porisms and beyond: Integrable billiards, hyperelliptic Jacobians and pencils of quadrics*, Frontiers in Mathematics, Birkhäuser/Springer Basel AG, Basel, 2011, DOI 10.1007/978-3-0348-0015-0. MR2798784

[11] V. Dragović and M. Radnović, *Periodic ellipsoidal billiard trajectories and extremal polynomials*, Comm. Math. Phys. **372** (2019), no. 1, 183–211, DOI 10.1007/s00220-019-03552-y. MR4031799

[12] V. Dragović, M. Radnović, *Isoperiodic families of Poncelet polygons inscribed in a circle and circumscribed about conics from a confocal pencil*, arXiv: 2103.01215, 2021.

[13] V. Dragović and V. Shramchenko, *Algebro-geometric solutions of the Schlesinger systems and the Poncelet-type polygons in higher dimensions*, Int. Math. Res. Not. IMRN **13** (2018), 4229–4259, DOI 10.1093/imrn/rnx015. MR3829181

[14] V. Dragović and V. Shramchenko, *Algebro-geometric approach to an Okamoto transformation, the Painlevé VI and Schlesinger equations*, Ann. Henri Poincaré **20** (2019), no. 4, 1121–1148, DOI 10.1007/s00023-019-00765-8. MR3928374

[15] V. Dragović and V. Shramchenko, *Deformations of the Zolotarev polynomials and Painlevé VI equations*, Lett. Math. Phys. **111** (2021), no. 3, Paper No. 75, 28 pp., DOI 10.1007/s11005-021-01415-z. MR4266169

[16] V. Dragović, V. Shramchenko, *Isoharmonic deformations and constrained Schlesinger systems*, arXiv: 2112.04110.

[17] J. Fay, *Kernel functions, analytic torsion, and moduli spaces*, Mem. Amer. Math. Soc. **96** (1992), no. 464, vi+123, DOI 10.1090/memo/0464. MR1146600

[18] R. Fuchs, *Über lineare homogene Differentialgleichungen zweiter Ordnung mit drei im Endlichen gelegenen wesentlich singulären Stellen* (German), Math. Ann. **63** (1907), no. 3, 301–321, DOI 10.1007/BF01449199. MR1511408

[19] P. Griffiths and J. Harris, *On Cayley's explicit solution to Poncelet's porism*, Enseign. Math. (2) **24** (1978), no. 1-2, 31–40. MR497281

[20] P. G. Grinevich and M. U. Schmidt, *Period preserving nonisospectral flows and the moduli space of periodic solutions of soliton equations: The nonlinear Schrödinger equation*, Phys. D **87** (1995), no. 1-4, 73–98, DOI 10.1016/0167-2789(95)00139-U. MR1361671

[21] F. Hazama, *Twists and generalized Zolotarev polynomials*, Pacific J. Math. **203** (2002), no. 2, 379–393, DOI 10.2140/pjm.2002.203.379. MR1897905

[22] N. J. Hitchin, *Twistor spaces, Einstein metrics and isomonodromic deformations*, J. Differential Geom. **42** (1995), no. 1, 30–112. MR1350695

[23] N. J. Hitchin, *Poncelet polygons and the Painlevé equations*, Geometry and analysis (Bombay, 1992), Tata Inst. Fund. Res., Bombay, 1995, pp. 151–185. MR1351506

[24] E. L. Ince, *Ordinary Differential Equations*, Dover Publications, New York, 1944. MR10757

[25] K. Iwasaki, H. Kimura, S. Shimomura, and M. Yoshida, *From Gauss to Painlevé: A modern theory of special functions*, Aspects of Mathematics, E16, Friedr. Vieweg & Sohn, Braunschweig, 1991, DOI 10.1007/978-3-322-90163-7. MR1118604

[26] A. V. Kitaev and D. A. Korotkin, *On solutions of the Schlesinger equations in terms of Θ-functions*, Internat. Math. Res. Notices **17** (1998), 877–905, DOI 10.1155/S1073792898000543. MR1646648

[27] A. Kokotov and D. Korotkin, *A new hierarchy of integrable systems associated to Hurwitz spaces*, Philos. Trans. R. Soc. Lond. Ser. A Math. Phys. Eng. Sci. **366** (2008), no. 1867, 1055–1088, DOI 10.1098/rsta.2007.2061. MR2377685

[28] V. V. Kozlov, D. Treshchev, *Billiards*, Amer. Math. Soc., Providence RI, 1991.

[29] H. Lebesgue, *Les coniques*, Gauthier-Villars, Paris, 1942.

[30] F. Loreĭ, *Isomonodromic deformations of Lamé connections, the Painlevé VI equation and Okamoto symmetry* (Russian, with Russian summary), Izv. Ross. Akad. Nauk Ser. Mat. **80** (2016), no. 1, 119–176, DOI 10.4213/im8310; English transl., Izv. Math. **80** (2016), no. 1, 113–166. MR3462678

[31] Yu. I. Manin, *Sixth Painlevé equation, universal elliptic curve, and mirror of* $\mathbf{P}^2$, Geometry of differential equations, Amer. Math. Soc. Transl. Ser. 2, vol. 186, Amer. Math. Soc., Providence, RI, 1998, pp. 131–151, DOI 10.1090/trans2/186/04. MR1732409

[32] K. Okamoto, *Studies on the Painlevé equations. I. Sixth Painlevé equation* $P_{\rm VI}$, Ann. Mat. Pura Appl. (4) **146** (1987), 337–381, DOI 10.1007/BF01762370. MR916698

[33] F. Peherstorfer and K. Schiefermayr, *Description of extremal polynomials on several intervals and their computation. I, II*, Acta Math. Hungar. **83** (1999), no. 1-2, 27–58, 59–83, DOI 10.1023/A:1006607401740. MR1682902

[34] E. Picard, Mémoire sur la théorie des fonctions algébriques de deux variables, Journal de Liouville **5** 1889, 135–319.

[35] R. Ramírez-Ros, *On Cayley conditions for billiards inside ellipsoids*, Nonlinearity **27** (2014), no. 5, 1003–1028, DOI 10.1088/0951-7715/27/5/1003. MR3207163

[36] L. Schlesinger, *Über eine Klasse von Differentialsystemen beliebiger Ordnung mit festen kritischen Punkten* (German), J. Reine Angew. Math. **141** (1912), 96–145, DOI 10.1515/crll.1912.141.96. MR1580847

[37] M. L. Sodin and P. M. Yuditskiĭ, *Functions that deviate least from zero on closed subsets of the real axis* (Russian, with Russian summary), Algebra i Analiz **4** (1992), no. 2, 1–61; English transl., St. Petersburg Math. J. **4** (1993), no. 2, 201–249. MR1182392

[38] S. Tabachnikov, *Geometry and billiards*, Student Mathematical Library, vol. 30, American Mathematical Society, Providence, RI; Mathematics Advanced Study Semesters, University Park, PA, 2005, DOI 10.1090/stml/030. MR2168892

[39] E. I. Zolotarev, A problem concerning least quantities, 1868, Complete works of E. I. Zolotarev, vol. 2, Publishing House of the USSR Academy of Sciences, Leningrad 1932, pp. 130–166. (Russian)

Department of Mathematical Sciences, University of Texas at Dallas, 800 West Campbell Road, Richardson TX 75080, USA. Mathematical Institute SANU, Kneza Mihaila 36, 11000 Belgrade, Serbia

Email address: Vladimir.Dragovic@utdallas.edu--the corresponding author

Department of mathematics, University of Sherbrooke, 2500, boul. de l'Université, J1K 2R1 Sherbrooke, Quebec, Canada

Email address: Vasilisa.Shramchenko@Usherbrooke.ca

Contemporary Mathematics
Volume **798**, 2024
https://doi.org/10.1090/conm/798/15983

The universal von Neumann algebra of smooth four-manifolds revisited

Gábor Etesi

Abstract. Making use of its smooth structure only, out of a connected oriented smooth 4-manifold a von Neumann algebra is constructed. It is geometric in the sense that is generated by local operators and as a special four dimensional phenomenon it contains all algebraic (i.e., formal or coming from a metric) curvature tensors of the underlying 4-manifold. The von Neumann algebra itself is a hyperfinite factor of II_1-type hence is unique up to abstract isomorphisms of von Neumann algebras. Over a fixed 4-manifold this universal von Neumann algebra admits a particular representation on a Hilbert space such that its unitary equivalence class is preserved by orientation-preserving diffeomorphisms consequently the Murray–von Neumann coupling constant of this representation is well-defined and gives rise to a new and computable real-valued smooth 4-manifold invariant. Its link with Jones' subfactor theory is noticed as well as computations in the simply connected closed case are carried out.

Application to the cosmological constant problem is also discussed. Namely, the aforementioned mathematical construction allows to reformulate the classical vacuum Einstein equation with cosmological constant over a 4-manifold as an operator equation over its tracial universal von Neumann algebra such that the trace of a solution is naturally identified with the cosmological constant. This framework permits to use the observed magnitude of the cosmological constant to estimate by topological means the number of primordial black holes about the Planck era. This number turns out to be negligable which is in agreement with known density estimates based on the Press–Schechter mechanism.

1. Introduction and summary

The 20-21st century has been witness to a great expansion of mathematics and physics bringing a genuinely two-sided interaction between them. The 1980-90's culmination of discoveries in low dimensional differential topology driven by Yang–Mills theory of particle physics has dramatically changed our understanding of four dimensional spaces: nowadays we know that the interplay between topology and smoothness is unexpectedly complicated precisely in four dimensions leading to the existence of a superabundance of smooth four dimensional manifolds. While traditional invariants of differential topology loose power in three and four dimensions, the new quantum invariants provided by various Yang–Mills theories work exactly in these dimensions allowing an at least partial enumeration of manifolds. It is

2020 *Mathematics Subject Classification.* Primary: 46L10, 83C45, Secondary: 57R55.

perhaps not just an accidence that quantum invariants are applicable precisely in three and four dimensions, equal to the phenomenological dimensions of physical space and space-time.

It is interesting that unlike Yang–Mills theories, classical general relativity—despite its powerful physical content, too—has not contributed to our understanding of four dimensionality yet. This might follow from the fact that general relativity, unlike Yang–Mills theories with their self-duality phenomena, permits formulations in every dimensions greater than four exhibiting essentially the same properties. This is certainly true when general relativity is considered in its usual fully classical differential-geometric school uniform however four dimensionality enters the game here as well if one tries to link differential geometry with non-commutativity [**8**].

The purpose of these notes is therefore twofold. On the *mathematical side* they intend to replace in a substantionally new way the overcomplicated, superfluous and, to confess, at many technical steps erroneous construction exhibited in our earlier work [**14**, Lemmata 2.1, 2.2 and 2.3]. The aim of that construction was to extract from a connected oriented smooth 4-manifold M a II_1 hypefinite factor von Neumann algebra $\mathfrak{R}$. The independent new construction here has exactly the same purpose however, in sharp contrast to our earlier struggles [**13, 14**], it rests on a standard Clifford algebra construction only hence is both conceptionally and technically very simple; thereby reducing the status of the machinery $M \Rightarrow \mathfrak{R}$ to a sort of so-far-unnoticed triviality. In addition, as a further novelty, a connection with the associated smooth 4-manifold invariant γ introduced and studied in [**14**, Lemmata 2.4 and 2.5] with Jones' subfactor theory is observed and new computations with this invariant in the simply connected closed case are exhibited.

Our mathematical results are summarized as follows:

THEOREM 1.1. *Let M be a connected oriented smooth 4-manifold. Making use of its smooth structure only, a von Neumann algebra $\mathfrak{R}$ can be constructed which is geometric in the sense that it is generated by local operators including all bounded complexified algebraic (i.e., formal or stemming from a metric) curvature tensors of M and $\mathfrak{R}$ itself is a hyperfinite factor of type II_1 hence is unique up to abstract isomorphism of von Neumann algebras.*

Moreover $\mathfrak{R}$ admits a representation on a certain separable Hilbert space over M such that the unitary equivalence class of this representation is invariant under orientation-preserving diffeomorphisms of M. Consequently the Murray–von Neumann coupling constant of this representation gives rise to a smooth invariant $\gamma(M) \in [0,1)$. More precisely it satisfies

$$\gamma(M) = 1 - \frac{1}{x}$$

where $x \in \{4\cos^2\left(\frac{\pi}{n}\right) \mid n \geqq 3\} \cup [4, +\infty)$, the set of Jones' subfactor indices.

This invariant has the following basic properties which are helpful in computations:

THEOREM 1.2. *The invariant behaves like $\gamma(\overline{M}) = \gamma(M)$ under reversing orientation, $\gamma(M \setminus Y) = \gamma(M)$ under excision of homologically trivial submanifolds and*

$$\gamma(M\#N) = \frac{\gamma(M) + \gamma(N)}{1 + \gamma(M)\gamma(N)}$$

under connected sum.

The invariant in the closed simply connected case can be characterized with the following properties. These show that unfortunately γ is not sensitive enough yet i.e., its construction demands further improvements in order to effectively distinguish smooth structures:

THEOREM 1.3. *If M', M'' are connected, closed, simply connected, smooth 4-manifolds which are homeomorphic then $\gamma(M') = \gamma(M'')$.*

More precisely if M is a closed, simply connected, smooth 4-manifold then

$$\gamma(M) = \frac{17^{b_2(M)} - 1}{17^{b_2(M)} + 1}$$

holds.

Thus for example $\gamma(S^4) = 0, \gamma(\mathbb{C}P^2) = \frac{8}{9}$ and $\gamma(\mathbb{C}P^1 \times \mathbb{C}P^1) = \frac{144}{145}$. We also find for instance for the $K3$ surface having $b_2(K3) = 22$ that already $1 - \gamma(K3) \approx 1.70 \times 10^{-27}$ and the general asymptotics of γ in the simply connected case is $0 < 1 - \gamma(M) \approx \mathrm{e}^{-\mathrm{const.}\, b_2(M)}$. This indicates that this invariant maps four dimensional smooth structures into $[0, 1)$ in a logarithmic way in some sense.

Let us add some comments here. The set of Jones indices splits into a discrete and a continuous part. Subfactors belonging to the discrete part $\{4\cos^2\left(\frac{\pi}{n}\right) | n \geqq 3\}$ have been completely classified [**21**] and in turn they follow an ADE pattern (with the odditiy that no subfactors corresponding to D_{2k+1} and E_7 exist) [**20**]. The set of subfactors belonging to the continuous portion $[4, +\infty)$ is however very wild and only partial results are known mainly for the subinterval $[4, 5] \subset [4, +\infty)$ or a bit more (cf. e.g. [**18**] for a survey and results). The connected sum formula exhibited here in Theorem 1.2 shows that all non-prime 4-manifolds (i.e., which admit non-trivial connected sum decompositions) belong via their invariant to the continuous regime i.e., $\frac{1}{1-\gamma(M)} \in [4, +\infty)$ furthermore we also find by Theorem 1.3 that this is also true for the prime manifold complex projective plane because $\frac{1}{1-\gamma(\mathbb{C}P^2)} = 9$. This observation strongly hints that smooth 4-manifolds might provide a rich reservoir of subfactors in the wild index range moreover poses the question whether or not smooth 4-manifolds corresponding to the tame (i.e., discrete) range exist (e.g. $\gamma(S^4) = 0$ corresponds to the Jones index $4\cos^2\left(\frac{\pi}{3}\right) = 1$ hence the trivial case $n = 3$ in the discrete part is realized by some 4-manifolds).

The other purpose of these notes, on the *physical side*, is to provide further clarifications to the material collected in [**13, 14**] regarding how to interpret the merely mathematical considerations above within a physical theory as naturally as possible. The outstanding problem of contemporary theoretical physics is how to unify the obviously successful and mathematically consistent *theory of general relativity* with the obviously successful but yet mathematically problematic *relativistic quantum field theory*. It has been generally believed that these two fundamental pillars of modern theoretical physics are in a clash not only by the different mathematical tools they use but even at a deep conceptional level: classical notions of general relativity such as the space-time event, the light cone or the event horizon of a black hole are too "sharp" objects from a quantum theoretic viewpoint meanwhile relativistic quantum field theory is not background independent from the aspect of general relativity.

Our interpretational efforts naturally fit into this context: they rest on the observation made explicit already in Theorem 1.1 that the constructed operator algebra $\mathfrak{R}$ is unique i.e. independent of the particular manifold we have started with and, as a curiosity of four dimensions, it contains among its members algebraic curvature tensors over M. Since curvature plays a central role in classical general relativity it is quite natural to place this interpretation within the broad realm of quantum gravity. More concretely first by recalling the basic dictionary from [**13**, Section 3] we exhibit a straightforward generalization of the classical vacuum Einstein equation with cosmological constant over a 4-manifold M to an operator equation over its von Neumann algebra $\mathfrak{R}$ (see Definition 3.1 below). One interesting feature of this generalization is that the trace of those operators which satisfy this "quantum vacuum Einstein equation" can be identified with the cosmological constant. Second, we use the previous mathematical results to find solutions to this equation whose traces (hence the cosmological constants they give rise to) are equal to the smooth invariant $1-\gamma$ of Theorem 1.1 hence fall into $(0,1]\subset\mathbb{R}$. Consequently they are always *strictly positive* but *small* numbers (cf. [**2**] too). Third, the observed small positive value of the cosmological constant [**23**] allows in this framework to obtain an estimate on the number of primordial black holes in the very early Universe; the magnitude of this number turns out to be $\sim 10^2$ hence their presence is negligable which is in accord with the current conviction in the cosmologist and perhaps particle physicist community [**6**]. Other recent works treating hyperfinite II factors from a gravitational perspective are [**3, 7**].

The present paper is heavily based on our earlier attempts [**13, 14**] but with substantial clarifications and new extensions. It is organized as follows. Section 2 is a self-contained presentation of all the mathematical constructions involved in Theorems 1.1, 1.2 and 1.3 together with their proofs. The language of this part is therefore rather mathematical. Then Section 3 contains the physics by offering a quantum generalization of the classical vacuum Einstein equation (see Definition 3.1) with an application to the cosmological constant problem. The language of this part is therefore rather physical. The physicist-minded reader may skip Section 2 and read Section 3 first hopefully without problem.

2. Mathematical constructions

In this section we shall first exhibit a simple self-contained two-step construction of a von Neumann algebra attached to any oriented smooth 4-manifold. The structure of this algebra will be also explored in some detail. This is then will be followed by introducing a new smooth 4-manifold invariant whose simplest properties will be examined too.

Construction of an algebra. Take the isomorphism class of a connected oriented smooth 4-manifold (without boundary) and from now on let M be a once and for all fixed representative in it carrying the action of its own orientation-preserving group of diffeomorphisms $\mathrm{Diff}^+(M)$. Among all tensor bundles $T^{(p,q)}M$ over M the 2nd exterior power $\wedge^2T^*M\subset T^{(0,2)}M$ is the only one which can be endowed with a pairing in a natural way i.e., with a pairing extracted from the smooth structure (and the orientation) of M alone. Indeed, consider its associated vector space $\Omega^2_c(M):=C^\infty_c(M;\wedge^2T^*M)$ of compactly supported smooth 2-forms on M. Define a pairing $\langle\,\cdot\,,\,\cdot\,\rangle_{L^2(M)}:\Omega^2_c(M)\times\Omega^2_c(M)\to\mathbb{R}$ via integration:

$$\langle\alpha,\beta\rangle_{L^2(M)}:=\int\limits_M\alpha\wedge\beta \tag{2.1}$$

and observe that this pairing is non-degenerate however is *indefinite* in general thus can be regarded as an indefinite scalar product on $\Omega^2_c(M)$. It therefore induces an indefinite real quadratic form Q on $\Omega^2_c(M)$ given by $Q(\alpha) := \langle\alpha,\alpha\rangle_{L^2(M)}$. Let $C(M)$ denote the complexification of the infinite dimensional real Clifford algebra associated with $(\Omega^2_c(M), Q)$. Because Clifford algebras are usually constructed out of definite quadratic forms, we summarize this construction [**19**, Section I.§3] to make sure that the resulting object $C(M)$ is well-defined i.e. is not sensitive for the indefiniteness of (2.1). To begin with, let $V_m \subset \Omega^2_c(M)$ be an m dimensional real subspace and assume that $Q_{r,s} := Q|_{V_m}$ has signature (r, s) on V_m that is, the maximal positive definite subspace of V_m with respect to $Q_{r,s}$ has dimension r while the dimension of the maximal negative definite subspace is s such that $r+s = m$ by the non-degeneracy of $Q_{r,s}$. Then out of the input data $(V_m, Q_{r,s})$ one constructs in the standard way a finite dimensional real Clifford algebra $C_{r,s}(M)$ with unit $1 \in C_{r,s}(M)$ and an embedding $V_m \subset C_{r,s}(M)$ with the property $\alpha^2 = Q_{r,s}(\alpha)1$ for every element $\alpha \in V_m$. This real algebra depends on the signature (r, s) however fortunately its complexification $C_m(M) := C_{r,s}(M) \otimes \mathbb{C}$ is already independent of it. In fact, if $\mathfrak{M}_k(\mathbb{C})$ denotes the algebra of $k \times k$ complex matrices, then it is well-known [**19**, Section I.§3] that $C_0(M) \cong \mathfrak{M}_1(\mathbb{C})$ while $C_1(M) \cong \mathfrak{M}_1(\mathbb{C}) \oplus \mathfrak{M}_1(\mathbb{C})$ and the higher dimensional cases follow from the complex periodicity $C_{m+2}(M) \cong C_m(M) \otimes \mathfrak{M}_2(\mathbb{C})$. Consequently depending on the parity $C_m(M)$ is isomorphic to either $\mathfrak{M}_{2^{\frac{m}{2}}}(\mathbb{C})$ or $\mathfrak{M}_{2^{\frac{m-1}{2}}}(\mathbb{C}) \oplus \mathfrak{M}_{2^{\frac{m-1}{2}}}(\mathbb{C})$. These imply that 2-step-chains of successive embeddings of real subspaces $V_m \subset V_{m+1} \subset V_{m+2} \subset \Omega^2_c(M)$ starting with $V_0 = \{0\}$ and given by iterating $\omega \mapsto \binom{\omega}{0}$ provide us with injective algebra homomorphisms $\mathfrak{M}_{2^{\frac{m}{2}}}(\mathbb{C}) \hookrightarrow \mathfrak{M}_{2^{\frac{m}{2}+1}}(\mathbb{C})$ having the shape $A \mapsto \left(\begin{smallmatrix} A & 0 \\ 0 & A\end{smallmatrix}\right)$. Therefore $C(M)$ is isomorphic to the injective limit of this directed system, that is there exists a linear-algebraic isomorphism

$$C(M) \cong \bigcup_{n=0}^{+\infty} \mathfrak{M}_{2^n}(\mathbb{C}) \tag{2.2}$$

or equivalently

$$C(M) \cong \mathfrak{M}_2(\mathbb{C}) \otimes \mathfrak{M}_2(\mathbb{C}) \otimes \dots$$

because this injective limit is also isomorphic to the infinite tensor product of $\mathfrak{M}_2(\mathbb{C})$'s. For clarity note that being (2.1) a non-local operation, $C(M)$ is a genuine global infinite dimensional object.

It is well-known (cf. [**8**, Section I.3]) that *any* complexified infinite Clifford algebra like $C(M)$ above generates *the* II_1-type hyperfinite factor von Neumann algebra. Let us summarize this procedure too (cf. [**1**, Section 1.1.6]). It readily follows that $C(M)$ possesses a unit $1 \in C(M)$ and its center comprises the scalar multiples of the unit only. Moreover $C(M)$ continues to admit a canonical embedding $\Omega^2_c(M; \mathbb{C}) \subset C(M)$ satisfying $\omega^2 = Q(\omega)1$ where now Q denotes the quadratic form induced by the complex-bilinear extension of (2.1). We also see via (2.2) already that $C(M)$ is a complex $*$-algebra whose $*$-operation (provided by taking Hermitian matrix transpose, a non-local operation) is written as $A \mapsto A^*$. The isomorphism (2.2) also shows that if $A \in C(M)$ then one can pick the smallest $n \in \mathbb{N}$ such that $A \in \mathfrak{M}_{2^n}(\mathbb{C})$ consequently A has a finite trace defined by $\tau(A) := 2^{-n}\mathrm{Trace}(A)$ i.e., taking the usual normalized trace of the corresponding $2^n \times 2^n$ complex matrix. It is straightforward that $\tau(A) \in \mathbb{C}$ does not depend on n. We can then define a sesquilinear inner product on $C(M)$ by $(A, B) := \tau(AB^*)$

which is non-degenerate thus the completion of $C(M)$ with respect to the norm $\|\cdot\|$ induced by $(\cdot\,,\cdot)$ renders $C(M)$ a complex Hilbert space what we shall write as $\mathscr{H}$ and its Banach algebra of all bounded linear operators as $\mathfrak{B}(\mathscr{H})$. Multiplication in $C(M)$ from the left on itself is continuous hence gives rise to a representation $\pi : C(M) \to \mathfrak{B}(\mathscr{H})$. Finally our central object effortlessly emerges as the weak closure of the image of $C(M)$ under π within $\mathfrak{B}(\mathscr{H})$ or equivalently, by referring to von Neumann's bicommutant theorem [**1**, Theorem 2.1.3] we put

$$\mathfrak{R} := (\pi(C(M)))'' \subset \mathfrak{B}(\mathscr{H}) \ .$$

This von Neumann algebra of course admits a unit $1 \in \mathfrak{R}$ moreover continues to have trivial center i.e., is a factor. Moreover by construction it is hyperfinite. The trace τ as defined extends from $C(M)$ to $\mathfrak{R}$ and satisfies $\tau(1) = 1$. Moreover [**1**, Proposition 4.1.4] this trace is unique on $\mathfrak{R}$. Likewise we obtain by extension a representation $\pi : \mathfrak{R} \to \mathfrak{B}(\mathscr{H})$. The canonical inclusion $\Omega^2_c(M;\mathbb{C}) \subset C(M)$ recorded above extends to both $\Omega^2_c(M;\mathbb{C}) \subset \mathscr{H}$ and $\Omega^2_c(M;\mathbb{C}) \subset \mathfrak{R}$. Thus in order to carefully distinguish the two different completions $\mathfrak{R}$ and $\mathscr{H}$ of one and the same object $C(M)$ we shall write $A \in \mathfrak{R}$ but $\hat{B} \in \mathscr{H}$ from now on as usual. This is necessary since $\mathfrak{R}$ and $\mathscr{H}$ are very different for example as $\mathrm{U}(\mathscr{H})$-modules: given a unitary operator $V \in \mathrm{U}(\mathscr{H})$ then $A \in \mathfrak{R}$ is acted upon as $A \mapsto VAV^{-1}$ but $\hat{B} \in \mathscr{H}$ transforms as $\hat{B} \mapsto V\hat{B}$. Using this notation and introducing $\hat{A} := \pi(A)\hat{1}$ the trace always can be written as a scalar product with the image of the unit in $\mathscr{H}$ that is, for every $A \in \mathfrak{R}$ we have

$$\tau(A) = (\hat{A}, \hat{1})$$

yielding a general and geometric expression for the trace.

Exploring the algebra $\mathfrak{R}$. Before proceeding further let us make a digression here to gain a better picture. This is desirable because taking the weak limit like $\mathfrak{R}$ of some explicitly known structure like $C(M)$ often involves a sort of loosing control over the latter. Nevertheless we already know promisingly that $\mathfrak{R}$ is a hyperfinite factor von Neumann algebra of II_1-type. Let us now exhibit some of its elements.

1. Our first examples are the 2-forms themselves as it follows from the already mentioned canonical embedding $\Omega^2_c(M;\mathbb{C}) \subset C(M)$ combined with $C(M) \subset \mathfrak{R}$ yielding $\Omega^2_c(M;\mathbb{C}) \subset \mathfrak{R}$. Next, proceeding along the same lines we can construct an interesting natural continuous embedding

$$i_M : M \longrightarrow \mathfrak{R} \tag{2.3}$$

of any connected oriented smooth 4-manifold M into its $\mathfrak{R}$. To every sufficiently nice closed subset $\emptyset \subsetneqq X \subsetneqq M$ there is an associated linear subspace $\Omega^2_c(M, X;\mathbb{C}) \subset \Omega^2_c(M;\mathbb{C})$ consisting of compactly supported smooth 2-forms vanishing at least along X. Furthermore we have seen that as a by-product of the Clifford algebra construction there exists a canonical embedding $\Omega^2_c(M;\mathbb{C}) \subset \mathscr{H}$ too. In this way to every point $x \in M$ one can attach a closed subspace $V_x \subset \mathscr{H}$ defined by taking the closure of $\Omega^2_c(M, x\,;\mathbb{C})$ within $\mathscr{H}$. Let $Q_x : \mathscr{H} \to V_x$ be the corresponding orthogonal projection. *A priori* $Q_x \in \mathfrak{B}(\mathscr{H})$ however in fact $Q_x \in \mathfrak{R}$. This is because V_x arises as the image of an operator $A_x \in \mathfrak{R}$ which for instance looks like extending $\omega \mapsto f_x\omega$ with a bounded smooth function f_x vanishing at $x \in M$ from $\Omega^2_c(M;\mathbb{C})$ to $\overline{\Omega^2_c(M;\mathbb{C})} \subset \mathscr{H}$ and defined to be zero on $\overline{\Omega^2_c(M;\mathbb{C})}^{\perp} \subset \mathscr{H}$. It readily follows that the resulting map $x \mapsto Q_x$ is injective and continuous hence gives rise to a continuous embedding. To make a comparison, in *algebraic geometry* points are characterized by maximal ideals of an abstractly given commutative ring. Here the corresponding objects would therefore be the maximal two-sided (weakly closed) ideals of a von Neumann algebra (regarded as a non-commutative ring). However in sharp contrast to the commutative situation a tracial factor von Neumann algebra is always *simple* (cf. e.g. [**1**, Proposition 4.1.5]) consequently in our case the concept of ideals cannot be used to characterize points hence the reason we used rather special elements of the von Neumann algebra. (2.3) is a counterpart of embedding Riemannian manifolds into Hilbert spaces via heat kernel techniques [**4**].

2. To see more examples, let us return to the Clifford algebra in (2.2) for a moment. We already know that there exists a canonical embedding $\Omega^2_c(M;\mathbb{C})\subset C(M)$. In addition to this let us find a Clifford module for $C(M)$. Consider again any finite *even* dimensional approximation $C_m(M)=C_{r,s}(M)\otimes\mathbb{C}$ constructed from $(V_m,Q_{r,s})$ where now $V_m\subset\Omega^2_c(M)$ is a real even $m=r+s$ dimensional subspace. Choose any $2^{\frac{m}{2}}$ dimensional complex vector subspace S_m within $\Omega^2_c(M;\mathbb{C})$. If $\mathrm{End}(\Omega^2_c(M;\mathbb{C}))$ denotes the associative algebra of *all* $\mathbb{C}$-linear transformations of $\Omega^2_c(M;\mathbb{C})$ then $S_m\subset\Omega^2_c(M;\mathbb{C})$ induces an embedding $\mathrm{End}\, S_m\subset\mathrm{End}(\Omega^2_c(M;\mathbb{C}))$ moreover we know that $\mathrm{End}\, S_m\cong\mathfrak{M}_{2^{\frac{m}{2}}}(\mathbb{C})\cong C_m(M)$. Therefore we obtain a non-canonical inclusion $C_m(M)\subset\mathrm{End}(\Omega^2_c(M;\mathbb{C}))$ for every fixed $m\in 2\mathbb{N}$. Furthermore $S_m\subset S_{m+2}\subset\Omega^2_c(M;\mathbb{C})$ given by $\omega\mapsto\binom{\omega}{0}$ induces a sequence $C_m(M)\subset C_{m+2}(M)\subset\mathrm{End}(\Omega^2_c(M;\mathbb{C}))$ for Clifford algebras which is compatible with the previous ascending chain of their matrix algebra realizations. Consequently taking the limit $m\to+\infty$ we come up with a non-canonical injective linear-algebraic homomorphism

$$C(M)\subset\mathrm{End}(\Omega^2_c(M;\mathbb{C}))\ . \tag{2.4}$$

Of course, unlike $\Omega^2_c(M;\mathbb{C})\subset C(M)$ above, (2.4) does not exist in the finite dimensional case.

Although the algebra $\mathrm{End}(\Omega^2_c(M;\mathbb{C}))$ is yet too huge, we can at least exhibit some of its elements. The simplest ones are the 2-forms themselves because $\Omega^2_c(M;\mathbb{C})\subset C(M)\subset\mathrm{End}(\Omega^2_c(M;\mathbb{C}))$ as we already know.[1] Moreover

$$C^\infty(M;\mathrm{End}(\wedge^2T^*M\otimes\mathbb{C}))\subset\mathrm{End}(\Omega^2_c(M;\mathbb{C}))$$

i.e., bundle morphisms are also included. These are local (algebraic) operators but are important because they allow to make a contact with local *four dimensional* differential geometry.[2] A peculiarity of four dimensions is that $C^\infty(M;\mathrm{End}(\wedge^2T^*M\otimes\mathbb{C}))$ contains the space of curvature tensors on M. If (M,g) is an oriented Riemannian 4-manifold then its Riemannian curvature tensor R_g is a member of this subalgebra: with respect to the splitting of 2-forms into their (anti)self-dual parts

[1] The extension of the Clifford multiplication $(V_m\otimes\mathbb{C})\times S_m\to S_m$ by $m\to+\infty$ thus induces a product structure $\Omega^2_c(M;\mathbb{C})\times\Omega^2_c(M;\mathbb{C})\to\Omega^2_c(M;\mathbb{C})$ rendering the space of 2-forms an infinite dimensional non-unital non-commutative associative algebra carrying a non-degenerate symmetric pairing by (2.1); hence resembling a Frobenius algebra. It is interesting that the orientation and smooth structure on M alone induces this structure.

[2] In fact all the constructions so far work for an arbitrary oriented and smooth $4k$-manifold with $k=0,1,2,\dots$ (note that in $4k+2$ dimensions the indefinite pairing (2.1) gives rise to a symplectic structure on $2k+1$-forms).

it looks like (cf. [**24**])

$$(2.5)\quad R_g = \begin{pmatrix} \frac{1}{12}\mathrm{Scal} + \mathrm{Weyl}^+ & \mathrm{Ric}_0 \\ \mathrm{Ric}_0^* & \frac{1}{12}\mathrm{Scal} + \mathrm{Weyl}^- \end{pmatrix} : \begin{matrix} \Omega^+_c(M;\mathbb{C}) \\ \bigoplus \\ \Omega^-_c(M;\mathbb{C}) \end{matrix} \longrightarrow \begin{matrix} \Omega^+_c(M;\mathbb{C}) \\ \bigoplus \\ \Omega^-_c(M;\mathbb{C}) \end{matrix}$$

and more generally, $\mathrm{End}(\Omega^2_c(M;\mathbb{C}))$ contains all algebraic (i.e. formal only, not stemming from a metric) curvature tensors R over M.

To see non-local examples, let us say that $\omega \in L^1_{loc}(M;\wedge^2T^*M\otimes\mathbb{C})$ if $\langle\omega,\varphi\rangle_{L^2(M)}$ given by (2.1) exists for all $\varphi \in \Omega^2_c(M;\mathbb{C})$. Then a tensor field K over $M \times M$ is called an *admissible double 2-form over M* if the map $x \mapsto K(x,y)$ belongs to $C^\infty_c(M;\wedge^2T^*M \otimes \mathbb{C}) = \Omega^2_c(M;\mathbb{C})$ for every $y \in M$ while $y \mapsto K(x,y)$ gives rise to an element of $L^1_{loc}(M;\wedge^2T^*M \otimes \mathbb{C})$ for every $x \in M$. Picking $R \in C^\infty(M;\mathrm{End}(\wedge^2T^*M \otimes \mathbb{C}))$ the map $\omega \mapsto P\omega$ defined by

$$P\omega(x) := \int\limits_{y\in M} K(x,y) \wedge (R(y)\omega(y))$$

is a non-local pseudo-differential operator acting on 2-forms and belongs to $\mathrm{End}(\Omega^2_c(M;\mathbb{C}))$. Furthermore orientation-preserving diffeomorphisms act on 2-forms via non-local pullback hence we conclude that $\mathrm{Diff}^+(M) \subset \mathrm{End}(\Omega^2_c(M;\mathbb{C}))$. Note that the Lie algebra $\mathrm{Lie}(\mathrm{Diff}^+(M)) \cong C^\infty_c(M;TM)$ consisting of compactly supported real vector fields acts $\mathbb{C}$-linearly on $\Omega^2_c(M;\mathbb{C})$ through Lie derivatives hence we also find that $\mathrm{Lie}(\mathrm{Diff}^+(M)) \subset \mathrm{End}(\Omega^2_c(M;\mathbb{C}))$. However these are local (non-algebraic but differential) operators again.

How to decide whether or not these elements of $\mathrm{End}(\Omega^2_c(M;\mathbb{C}))$ belong to $\mathfrak{R}$? The key concept here is the trace. Compared with the above trace expression generally valid on $\mathfrak{R}$, more specific trace formulata are obtained if M is endowed with a normalized Riemannian metric g i.e., the corresponding volume form $\mu_g = *1$ satisfies $\int_M \mu_g = 1$. The unique sesquilinear extension of g induces a positive definite sesquilinear L^2-scalar product

$$(\varphi,\psi)_{L^2(M,g)} := \int\limits_{x\in M} g(\varphi(x),\psi(x))\mu_g(x) = \int\limits_M \varphi \wedge *\overline{\psi}$$

on $\Omega^2_c(M;\mathbb{C})$. If $\{\varphi_1,\varphi_2,\dots\}$ is a smooth orthonormal frame in $\Omega^2_c(M;\mathbb{C})$ then it readily follows that the trace of any $B \in \mathrm{End}(\Omega^2_c(M;\mathbb{C}))$ formally looks like

$$\tau(B) = \lim_{n\to+\infty} \frac{1}{2^n}\sum_{i=1}^{2^n}(B\varphi_i\,,\,\varphi_i)_{L^2(M,g)}$$

and, if exists, is independent of the frame used. Obviously $B \in \mathrm{End}(\Omega^2_c(M;\mathbb{C})) \cap C(M)$ if and only if the sum on the right hand side is constant after finitely many terms; and an inspection of this trace expression shows that in general $B \in \mathrm{End}(\Omega^2_c(M;\mathbb{C})) \cap \mathfrak{R}$ if and only if $\tau(B)$ exists. As a consequence note that $\mathrm{End}(\Omega^2_c(M;\mathbb{C})) \cap \mathfrak{R}$ is already independent of the particular inclusion (2.4). An example: any $\Phi \in \mathrm{Diff}^+(M)$ acts on $\mathrm{End}(\Omega^2_c(M;\mathbb{C}))$ by conjugation hence using (2.4) extends to a unitary operator on $\mathscr{H}$ yielding $|\tau(\Phi)| = 1$ consequently we obtain $\mathrm{Diff}^+(M) \subset \mathrm{End}(\Omega^2_c(M;\mathbb{C})) \cap \mathfrak{R}$.

When $R \in C^\infty(M;\mathrm{End}(\wedge^2T^*M\otimes\mathbb{C}))$ the previous trace formula can be further specified because one can compare the global trace $\tau(R)$ and the local trace function

$x \mapsto \operatorname{tr}(R(x))$ given by the pointwise traces of the local operators $R(x) : \wedge^2 T_x^* M \otimes \mathbb{C} \to \wedge^2 T_x^* M \otimes \mathbb{C}$ at every $x \in M$. Recall that $\mathfrak{R}$ has been constructed as the weak closure of the Clifford algebra (2.2). In fact [**1**, Section 1.1.6] the universality of $\mathfrak{R}$ permits to obtain it from other matrix algebras too, like for instance from $\bigcup_{n=0}^{+\infty} \mathfrak{M}_{6^n}(\mathbb{C})$ whose weak closure therefore is again $\mathfrak{R}$. By the aid of this altered construction we can formally start with

$$\tau(R) = \lim_{n\to+\infty} \frac{1}{6^n} \sum_{i=1}^{6^n} (R\varphi_i \, , \, \varphi_i)_{L^2(M,g)} \; .$$

Fix $n \in \mathbb{N}$, write $M_n := \bigcap_{i=1}^{6^n} \operatorname{supp}\varphi_i \subseteqq M$ and take a point $x \in M_n$. Since $\dim_{\mathbb{C}}(\wedge^2 T_x^* M \otimes \mathbb{C}) = \binom{4}{2} = 6$ the maximal number of completely disjoint linearly independent sub-6-tuples in $\{\varphi_1(x), \varphi_2(x), \dots, \varphi_{6^n}(x)\}$ is equal to $\frac{6^n}{6} = 6^{n-1}$. Moreover it follows from Sard's lemma that with a generic smooth choice for $\{\varphi_1, \varphi_2, \dots\}$ the subset of those points $y \in M_n$ where this number is less than 6^{n-1} has measure zero in M_n with respect to the measure μ_g. Consequently

$$\sum_{i=1}^{6^n} (R\varphi_i, \varphi_i)_{L^2(M_n,g)} = \int\limits_{x\in M_n} \sum_{i=1}^{6^n} g(R(x)\varphi_i(x), \varphi_i(x))\, \mu_g(x) = 6^{n-1} \int\limits_{x\in M_n} \operatorname{tr}(R(x))\mu_g(x) \; .$$

Since $\{\varphi_1, \varphi_2, \dots\}$ is a basis in $\Omega^2_c(M;\mathbb{C})$ therefore $M \setminus \bigcup_{n=0}^{+\infty} M_n$ has measure zero as well we can let $n \to +\infty$ to end up with

$$\tau(R) = \frac{1}{6} \int\limits_M \operatorname{tr}(R)\mu_g \tag{2.6}$$

and again $R \in C^\infty(M; \operatorname{End}(\wedge^2 T^* M \otimes \mathbb{C})) \cap \mathfrak{R}$ if and only if (2.6) exists. Also note that τ is in fact the generalization of the total scalar curvature of a Riemannian manifold. So taking into account (2.4) we obtain useful criteria for checking whether or not an operator in $\operatorname{End}(\Omega^2_c(M;\mathbb{C}))$ belongs to the completion of $C(M)$ which is $\mathfrak{R}$. For instance if the curvature R_g of (M,g) as an operator in (2.5) is bounded which means that

$$\sup_{\|\omega\|_{L^2(M,g)}=1} \|R_g\omega\|_{L^2(M,g)} \leqq K < +\infty$$

then

$$0 \leqq |\tau(R_g)| \leqq \lim_{n\to+\infty} \frac{1}{2^n} \sum_{i=1}^{2^n} |(R_g\varphi_i, \varphi_i)_{L^2(M,g)}| \leqq \lim_{n\to+\infty} \frac{1}{2^n} \sum_{i=1}^{2^n} \|R_g\varphi_i\|_{L^2(M,g)} \leqq K$$

yielding $R_g \in C^\infty(M; \operatorname{End}(\wedge^2 T^* M \otimes \mathbb{C})) \cap \mathfrak{R}$ and more generally

$$R \in C^\infty(M; \operatorname{End}(\wedge^2 T^* M \otimes \mathbb{C})) \cap \mathfrak{R}$$

if R is bounded. Similar condition stems from (2.6).

3. We close the partial comprehension of $\mathfrak{R}$ with an observation regarding its overall structure. The canonical inclusion $\Omega^2_c(M;\mathbb{C}) \subset \mathfrak{R}$ (and likewise $\Omega^2_c(M;\mathbb{C}) \subset \mathscr{H}$) implies that $\mathfrak{R}$ (and likewise $\mathscr{H}$) is generated by all finite products $\omega_1\omega_2\dots\omega_k$ of 2-forms within $C(M)$ however this is not very informative. Rather

consider the well-known short exact sequence of groups involving the fiberwise automorphism group (the gauge group) and the global automorphism group of the vector bundle $\wedge^2T^*M\otimes\mathbb{C}$ as well as the orientation-presering diffeomorphism group of the underlying M respectively:

$$1\longrightarrow C^\infty(M;\mathrm{Aut}(\wedge^2T^*M\otimes\mathbb{C}))\longrightarrow \mathrm{Aut}(\Omega^2_c(M;\mathbb{C}))\longrightarrow \mathrm{Diff}^+(M)\longrightarrow 1$$

whose shape at the Lie algebra level looks like

$$0\longrightarrow C^\infty(M;\mathrm{End}(\wedge^2T^*M\otimes\mathbb{C}))\longrightarrow \mathrm{End}(\Omega^2_c(M;\mathbb{C}))\longrightarrow \mathrm{Lie}(\mathrm{Diff}^+(M))\longrightarrow 0\ .$$

We already have an embedding (2.4). In addition to this there exists an isomorphism of Lie algebras $L: C^\infty_c(M;TM)\to \mathrm{Lie}(\mathrm{Diff}^+(M))$ such that $X\mapsto L_X$ is nothing but taking Lie derivative with respect to a compactly supported real vector field where the first-order $\mathbb{C}$-linear differential operator L_X is supposed to act on 2-forms hence $\mathrm{Lie}(\mathrm{Diff}^+(M))\subset\mathrm{End}(\Omega^2_c(M;\mathbb{C}))$ as we know already too. Therefore the intersection of the latter sequence with $C(M)$ is meaningful and gives

$$0\longrightarrow C^\infty(M;\mathrm{End}(\wedge^2T^*M\otimes\mathbb{C}))\cap C(M)\longrightarrow C(M)\longrightarrow \mathrm{Lie}(\mathrm{Diff}^+(M))\cap C(M)\longrightarrow 0\ .$$

The second term consists of fiberwise algebraic hence local operators having finite trace (2.6) and likewise the fourth terms consists of finite trace Lie derivatives thus belongs to the class of local operators too. Since the vector spaces underlying $C(M)$ considered either as an associative or a Lie algebra are isomorphic we conclude, as an important structural observation, that the overall construction here is geometric in the sense that *the algebra $\mathfrak{R}$ is generated by (trace class) local operators.*

Summing up all of our findings so far: $\mathfrak{R}$ is a hyperfinite factor von Neumann algebra of II_1-type associated to M such that the solely input in its construction has been the pairing (2.1). Hence $\mathfrak{R}$ depends only on the orientation and the smooth structure of M in a functorial way. It contains, certainly among many other non-local operators, the space M itself via (2.3), its orientation-preseving diffeomorphisms as well as its space of bounded algebraic curvature tensors. Nevertheless $\mathfrak{R}$ is geometric in the sense that it is generated by M's finite trace local operators alone. It is remarkable that despite the plethora of smooth 4-manifolds detected since the 1980's their associated von Neumann algebras here are unique up to isomorphisms of von Neumann algebras (cf. e.g. [**1**, Theorem 11.2.2]) offering a sort of justification terming $\mathfrak{R}$ as "universal". One is therefore tempted to look upon $\mathfrak{R}$ as a natural common non-commutative space generalization of either all the pure oriented smooth 4-manifolds M or rather all the (pseudo-)Riemannian ones (M,g). This universality also justifies the simple notation $\mathfrak{R}$ used throughout the text.

A new smooth 4-manifold invariant. After these preliminary considerations we are in a position to approach four dimensional smoothness from a new operator-algebraic direction.

LEMMA 2.1. *Let M be a connected oriented smooth 4-manifold and $\mathfrak{R}$ its von Neumann algebra with trace τ as before. Then there exists a complex separable Hilbert space $\mathscr{I}(M)^\perp$ and a representation $\rho_M:\mathfrak{R}\to\mathfrak{B}(\mathscr{I}(M)^\perp)$ with the following properties. If $\pi:\mathfrak{R}\to\mathfrak{B}(\mathscr{H})$ is the representation constructed above then $\{0\}\subseteqq\mathscr{I}(M)^\perp\subsetneqq\mathscr{H}$ and $\rho_M=\pi|_{\mathscr{I}(M)^\perp}$ holds; therefore, although ρ_M can be the trivial representation, it is surely not unitary equivalent to the standard representation. Moreover the unitary equivalence class of ρ_M is invariant under orientation-preserving diffeomorphisms of M.*

Thus the Murray–von Neumann coupling constant[3] *of* ρ_M *is invariant under orientation-preserving diffeomorphisms. Writing* $P_M : \mathscr{H} \to \mathscr{I}(M)^\perp$ *for the orthogonal projection the coupling constant is equal to* $\tau(P_M) \in [0,1) \subset \mathbb{R}_+$ *consequently* $\gamma(M) := \tau(P_M)$ *is a smooth 4-manifold invariant.*

PROOF. First let us exhibit a representation of $\mathfrak{R}$; this construction is inspired by the general Gelfand–Naimark–Segal technique however exploits the special features of our construction so far as well. Pick a pair (Σ,ω) consisting of an (immersed) closed orientable surface $\Sigma \looparrowright M$ with induced oriantation and $\omega \in \Omega^2_c(M;\mathbb{C})$ which is also closed i.e., $\mathrm{d}\,\omega = 0$. Consider the differential geometric $\mathbb{C}$-linear functional $F_{\Sigma,\omega} : \mathfrak{R} \to \mathbb{C}$ by continuously extending

$$A \longmapsto \int_\Sigma A\omega$$

from $\mathrm{End}(\Omega^2_c(M;\mathbb{C})) \cap \mathfrak{R}$. This extension is unique because $\mathrm{End}(\Omega^2_c(M;\mathbb{C})) \cap \mathfrak{R}$ is norm dense in $\mathfrak{R}$. In case of $F_{\Sigma,\omega}(1) \neq 0$ let $\{0\} \subseteqq I(M) \subseteqq \mathfrak{R}$ be the subset of elements $A \in \mathfrak{R}$ satisfying $F_{\Sigma,\omega}(A^*A) = 0$. In fact for all pairs (Σ,ω) we obviously find $\{0\} \subsetneqq I(M)$ and $I(M) \cap \mathbb{C}1 = \{0\}$ hence $I(M) \subsetneqq \mathfrak{R}$ too. We assert that $I(M)$ is a multiplicative left-ideal in $\mathfrak{R}$ which is independent of (Σ,ω). In the case of $F_{\Sigma,\omega}(1) = 0$ we put $I(M) = \mathfrak{R}$ hence it is again independent of (Σ,ω) but trivially in this way.

Consider the case when $F_{\Sigma,\omega}(1) \neq 0$. Then we can assume that $F_{\Sigma,\omega}(1) = 1$ hence $F_{\Sigma,\omega}$ is a positive functional; applications of the standard inequality

$$|F_{\Sigma,\omega}(A^*B)|^2 \leqq F_{\Sigma,\omega}(A^*A)F_{\Sigma,\omega}(B^*B)$$

show that $I_{\Sigma,\omega}(M)$ defined by the elements satisfying $F_{\Sigma,\omega}(A^*A) = 0$ is a multiplicative left-ideal in $\mathfrak{R}$. Concerning its ω-dependence, let $\omega' \in \Omega^2_c(M;\mathbb{C})$ be another closed 2-form having the property $F_{\Sigma,\omega'}(1) = 1$; since neither ω nor ω' are identically zero, we can pick an *invertible* operator $T \in \mathrm{Aut}(\Omega^2_c(M;\mathbb{C}))$ satisfying $\omega'|_\Sigma = T\omega|_\Sigma$. Then by $F_{\Sigma,\omega'}(A^*A) = F_{\Sigma,\omega}(A^*AT)$ and applying the above inequality in the form

$$|F_{\Sigma,\omega'}(A^*AT)|^2 \leqq F_{\Sigma,\omega}(A^*A)F_{\Sigma,\omega}((AT)^*(AT))$$

we find $I_{\Sigma,\omega'}(M) \subseteqq I_{\Sigma,\omega}(M)$. Likewise, making use of $F_{\Sigma,\omega}(A^*A) = F_{\Sigma,\omega'}(A^*AT^{-1})$ and $|F_{\Sigma,\omega}(A^*AT^{-1})|^2 \leqq F_{\Sigma,\omega'}(A^*A)F_{\Sigma,\omega'}((AT^{-1})^*(AT^{-1}))$ imply $I_{\Sigma,\omega'}(M) \supseteqq I_{\Sigma,\omega}(M)$ too. Consequently $I_{\Sigma,\omega'}(M) = I_{\Sigma,\omega}(M)$.

Concerning the general (Σ,ω)-dependence of $I_{\Sigma,\omega}$ we argue as follows. Let $\eta_\Sigma \in \Omega^2(M;\mathbb{R})$ be a closed real 2-form representing the Poincaré-dual $[\eta_\Sigma] \in H^2(M;\mathbb{R})$ of $\Sigma \looparrowright M$. Equip M with an arbitrary Riemannian metric g; since the corresponding Hodge operator $*$ is an isomorphism on $\wedge^2 T^*M$ we can take a 2-form φ_Σ such that $\eta_\Sigma = *\varphi_\Sigma$. Then via $\int_\Sigma \omega = \int_M \omega \wedge \overline{\eta}_\Sigma = \int_M \omega \wedge *\overline{\varphi}_\Sigma = (\omega, \varphi_\Sigma)_{L^2(M,g)}$ the functional can be re-expressed as $F_{\Sigma,\omega}(A^*A) = (A^*A\omega, \varphi_\Sigma)_{L^2(M,g)}$ in terms of the corresponding definite sesquilinear L^2-scalar product on (M,g). Let $\Sigma' \looparrowright M$ be another closed surface and ω' another closed 2-form such that $F_{\Sigma',\omega'}(1) = 1$. Altering ω and ω' along Σ and Σ' respectively if necessary (which has no effect on $I(M)$ as we have seen) we can pick some compactly supported 2-form Ω on M such that $\Omega|_\Sigma = \omega$ and $\Omega|_{\Sigma'} = \omega'$. Moreover as before take a representative

[3] Also called the $\mathfrak{R}$-*dimension* of a left $\mathfrak{R}$-module hence denoted $\dim_{\mathfrak{R}}$, cf. [**1**, Chapter 8].

$*\varphi_{\Sigma'} \in \Omega^2(M;\mathbb{R})$ for the Poincaré-dual of Σ'. We can use again $T \in \mathrm{Aut}(\Omega^2_c(M;\mathbb{C}))$ satisfying $\varphi_{\Sigma'} = T\varphi_\Sigma$. Thus

$$\begin{aligned} F_{\Sigma',\omega'}(A^*A) =& (A^*A\Omega, \varphi_{\Sigma'})_{L^2(M,g)} = (A^*A\Omega, T\varphi_\Sigma)_{L^2(M,g)} \\ =& (T^*A^*A\Omega, \varphi_\Sigma)_{L^2(M,g)} = F_{\Sigma,\omega}(T^*A^*A) \end{aligned}$$

together with $|F_{\Sigma,\omega}(T^*A^*A)|^2 \leqq F_{\Sigma,\omega}\big((AT)^*AT\big)F_{\Sigma,\omega}(A^*A)$ demonstrate that $I_{\Sigma',\omega'}(M) \subseteqq I_{\Sigma,\omega}(M)$. In the same fashion $F_{\Sigma,\omega}(A^*A) = F_{\Sigma',\omega'}((T^{-1})^*A^*A)$ together with the corresponding inequality convinces us that $I_{\Sigma',\omega'}(M) \supseteqq I_{\Sigma,\omega}(M)$. Thus $I_{\Sigma',\omega'}(M) = I_{\Sigma,\omega}(M)$.

Secondly if (Σ,ω) is such that $F_{\Sigma,\omega}(1) = 0$ then by definition $I_{\Sigma,\omega}(M) = \mathfrak{R}$. Therefore if (Σ',ω') is another pair with $F_{\Sigma',\omega'}(1) = 0$ then $I_{\Sigma,\omega}(M) = I_{\Sigma',\omega'}(M)$ (and equal to $\mathfrak{R}$). We are now convinced that it is correct to write $I_{\Sigma,\omega}(M)$ as $I(M)$. In summary it satisfies $\{0\} \subsetneqq I(M) \subseteqq \mathfrak{R}$.

Let us proceed further by exploiting now the observation made during the construction of $\mathfrak{R}$ that it acts on a Hilbert space $\mathscr{H}$ with scalar product $(\,\cdot\,,\,\cdot\,)$ by the representation π i.e., multiplication from the left. In fact, since $(\hat{A},\hat{B}) = \tau(AB^*)$ we see that π is the standard representation. Consider the space $\{0\} \subsetneqq I(M)I(M)^* \subseteqq \mathfrak{R}$ consisting of all finite sums $A_1B_1 + \cdots + A_kB_k \in \mathfrak{R}$ where $A_i \in I(M)$ and similarly $B_j \in I(M)^*$. It gives rise to a closed linear subspace $\{0\} \subsetneqq \mathscr{I}(M) \subseteqq \mathscr{H}$ by taking the closure of $C(M) \cap I(M)I(M)^*$ within $\mathscr{H} \supset C(M)$. Therefore $\mathscr{I}(M)$ is a well-defined closed subspace of $\mathscr{H}$ which is non-trivial if $F_{\Sigma,\omega}(1) \neq 0$ and coincides with $\mathscr{H}$ whenever $F_{\Sigma,\omega}(1) = 0$. Take its orthogonal complementum $\{0\} \subseteqq \mathscr{I}(M)^\perp \subsetneqq \mathscr{H}$. Note that $\mathscr{I}(M)^\perp$ is isomorphic to $\mathscr{H}/\mathscr{I}(M)$. Taking into account that the subset $I(M)$ is a multiplicative left-ideal $\pi : \mathfrak{R} \to \mathfrak{B}(\mathscr{H})$ given by left-multiplication restricts to $\mathscr{I}(M)$ but even more, since the scalar product on $\mathscr{H}$ satisfies the identity $(\widehat{AB}, \widehat{C}\,) = (\hat{B}, \widehat{A^*C})$ the standard representation restricts to $\mathscr{I}(M)^\perp$ as well. This latter representation is either a unique non-trivial representation if $\mathscr{I}(M)^\perp \neq \{0\}$ (provided by a functional over M with $F_{\Sigma,\omega}(1) \neq 0$ if exists), or the trivial one if $\mathscr{I}(M)^\perp = \{0\}$ (provided by a functional with $F_{\Sigma,\omega}(1) = 0$ which always exists). Keeping these in mind, for a given M we define

$$\rho_M : \mathfrak{R} \to \mathfrak{B}(\mathscr{I}(M)^\perp) \ \text{ to be } \left\{ \begin{array}{l} \pi|_{\mathscr{I}(M)^\perp} \text{ on } \mathscr{I}(M)^\perp \neq \{0\} \text{ if possible,} \\ \pi|_{\mathscr{I}(M)^\perp} \text{ on } \mathscr{I}(M)^\perp = \{0\} \text{ otherwise.} \end{array} \right.$$

The choice is unambigously determined by the topology of M (see the *Remark* below).

From the general theory [**1**, Chapter 8] we know that if $P_M : \mathscr{H} \to \mathscr{I}(M)^\perp$ is the orthogonal projection then $P_M \in \mathfrak{R}$ because $\mathscr{H}$ is the standard $\mathfrak{R}$-module. The Murray–von Neumann coupling constant of ρ_M depends only on the unitary equivalence class of ρ_M and is equal to $\tau(P_M) \in [0,1]$. However observing that ρ_M is surely not isomorphic to π since $\mathscr{I}(M)$ is never trivial the case $\tau(P_M) = 1$ is excluded i.e., $\tau(P_M) \in [0,1)$. Let $\Phi : M \to M$ be an orientation-preserving diffeomorphism. It induces an inner automorphism $A \mapsto \Phi^*A(\Phi^*)^{-1}$ of $\mathfrak{R}$. Taking into account that the scalar product on $\mathscr{H}$ is induced by the trace which is invariant under cyclic permutations this inner automorphism is unitary. Moreover it transforms $I_{\Sigma,\omega}(M)$ into $I_{\Sigma',\omega'}(M) = I_{\Phi(\Sigma),\Phi^*\omega}(M)$ hence $F_{\Sigma,\omega}(1) = 0$ if and only if $F_{\Sigma',\omega'}(1) = 0$ consequently the Hilbert space $\mathscr{I}(M)^\perp$ is invariant under Φ. Thus Φ

transforms ρ_M into a new representation $\Phi^*\rho_M(\Phi^*)^{-1}$ on $\mathscr{I}(M)^\perp$ which is unitary equivalent to ρ_M.

We conclude that $\gamma(M) := \tau(P_M) \in [0,1)$ is a smooth invariant of M as stated. □

REMARK 2.2. Note that $\gamma(M) = 0$ corresponds to the situation when ρ_M is the trivial representation on $\mathscr{I}(M)^\perp = \{0\}$. To avoid this we have to demand $F_{\Sigma,\omega}(1) \neq 0$ which by the closedness assumptions on $\Sigma \looparrowright M$ and $\omega \in \Omega^2_c(M;\mathbb{C})$ is in fact a *topological condition*: it is equivalent that

$$\frac{1}{2\pi\sqrt{-1}}F_{\Sigma,\omega}(1) = \frac{1}{2\pi\sqrt{-1}}\int\limits_\Sigma \omega = \langle[\Sigma],[\omega]\rangle \in \mathbb{C}$$

as a pairing of $[\Sigma] \in H_2(M;\mathbb{Z})$ and $[\omega] \in H^2(M;\mathbb{C})$ in homology is not trivial. Hence $\gamma(M) = 0$ iff $H_2(M;\mathbb{C}) = H_2(M;\mathbb{Z})\otimes\mathbb{C} = \{0\}$ (or equivalently, $H^2(M;\mathbb{C}) = H^2(M;\mathbb{Z})\otimes\mathbb{C} = \{0\}$). Thus unfortunately $\gamma(M) = 0$ for all acyclic or aspherical manifolds (including homology 4-spheres). Examples in the simply connected case are $M = S^4, \mathbb{R}^4, R^4$ (this latter is any exotic or fake $\mathbb{R}^4$) while $M = S^3\times S^1$ is a non-simply connected example.

Jones' subfactor theory (for a summary cf. e.g. [**1**, Section 9.4] or [**8**, Chapter V.10]) imposes an interesting restriction on the possible spectrum of γ just constructed.

LEMMA 2.3. *Let M be a connected oriented smooth 4-manifold and $\gamma(M) \in [0,1)$ its smooth invariant. Then $\gamma(M) = 1 - \frac{1}{x}$ where $x \in \{4\cos^2\left(\frac{\pi}{n}\right) \mid n \geqq 3\}\cup[4,+\infty)$ that is, an element from the set of all possible Jones' subfactor indices.*

PROOF. Taking into account that $\{0\} \subsetneqq I(M)I(M)^* \subseteqq \mathfrak{R}$ is self-adjoint it generates a von Neumann subalgebra; more precisely put

$$\mathfrak{I}(M) := (\pi(C(M)\cap I(M)I(M)^*))'' \subset \mathfrak{B}(\mathscr{H})\ .$$

Since the set $C(M)\cap I(M)I(M)^*$ is already weakly dense in $I(M)I(M)^*$ which is a non-trivial left-ideal of $\mathfrak{R}$ its only defect to be a von Neumann algebra is that $C(M)\cap I(M)I(M)^*$ is not weakly closed and possesses no unit; therefore its weak closure $\mathfrak{I}(M)$ is a factor. Thus it is a subfactor of $\mathfrak{R}$ therefore it possesses a corresponding Jones index denoted as usual by $[\mathfrak{R} : \mathfrak{I}(M)] \in \{4\cos^2\left(\frac{\pi}{n}\right) \mid n \geqq 3\}\cup[4,+\infty)$.

The left-ideal $\{0\} \subsetneqq C(M)\cap I(M)I(M)^* \subseteqq C(M)$ acts on itself by letf-multiplications. Since both the von Neumann algebra $\mathbb{C}\cong Z(\mathfrak{R}) \subsetneqq \mathfrak{I}(M) \subseteqq \mathfrak{R}$ and the Hilbert space $\{0\} \subsetneqq \mathscr{I}(M) \subseteqq \mathscr{H}$ arise as appropriate closures of this same left-ideal its left action on itself extends to a representation of $\mathfrak{I}(M)$ on $\mathscr{I}(M)$ which, taking into account that $\mathscr{I}(M) \neq \{0\}$, is equivalent to the standard representation of $\mathfrak{I}(M)$. We also know already that $\mathscr{H} = \mathscr{I}(M)\oplus\mathscr{I}(M)^\perp$ as $\mathfrak{R}$-modules by Lemma 2.1. Recalling the basic properties of the dimension function of a left von Neumann algebra module over the von Neumann algebra itself (cf. e.g. [**1**, Chapter

8]) we collect:

$$\begin{aligned}
\dim_{\mathfrak{R}} \mathscr{H} &= \dim_{\mathfrak{R}} \mathscr{I}(M) + \dim_{\mathfrak{R}} \mathscr{I}(M)^{\perp} \quad \text{(additivity)}\\
\dim_{\mathfrak{R}} \mathscr{H} &= 1 \quad \text{(the standard representation of} \mathfrak{R})\\
\dim_{\mathfrak{R}} \mathscr{I}(M)^{\perp} &= \gamma(M) \quad \text{(by Lemma 2.1 and Footnote 3)}\\
\dim_{\mathfrak{J}(M)} \mathscr{I}(M) &= [\mathfrak{R} : \mathfrak{J}(M)] \dim_{\mathfrak{R}} \mathscr{I}(M) \quad \text{(dimension comparison)}\\
\dim_{\mathfrak{J}(M)} \mathscr{I}(M) &= 1 \quad \text{(the standard representation of } \mathfrak{J}(M))
\end{aligned}$$

from which it follows that $\gamma(M) = 1 - \frac{1}{[\mathfrak{R} : \mathfrak{J}(M)]}$ yielding the desired result. □

PROOF OF THEOREM 1.1. This theorem follows from the general construction of $\mathfrak{R}$ as we outlined here and in particular from Lemmata 2.1 and 2.3. □

Next we collect some basic useful properties of the invariant.

LEMMA 2.4. *(Reversing orientation.) If M is a connected oriented smooth 4-manifold and $\overline{M}$ is its orientation-reversed form then $\gamma(\overline{M}) = \gamma(M)$.*

(Excision principle.) Let M be a connected oriented smooth 4-manifold and $\emptyset \subsetneqq Y \subset M$ a submanifold so that $M \setminus Y \subsetneqq M$ is connected and the embedding $i : M \setminus Y \to M$ induces an isomorphism $i_* : H_2(M \setminus Y; \mathbb{Z}) \to H_2(M; \mathbb{Z})$ on the 2nd homology. Then $M \setminus Y$ with induced orientation and smooth structure is a connected oriented smooth 4-manifold satisfying $\gamma(M \setminus Y) = \gamma(M)$.

(Gluing principle.) Let M and N be two connected, oriented smooth 4-manifolds and write $M\#N$ for their connected sum. With induced orientation $M\#N$ is a connected, oriented smooth 4-manifold. Its smooth invariant satisfies

$$\gamma(M\#N) = \frac{\gamma(M) + \gamma(N)}{1 + \gamma(M)\gamma(N)} \quad .$$

PROOF. The first assertion is obvious from γ's construction carried out in the proof of Lemma 2.1.

Regarding the second assertion $M \setminus Y$ is a connected oriented smooth 4-manifold by assumption consequently admits an associated von Neumann algebra $\mathfrak{R}(M \setminus Y)$ which is also a hyperfinite II_1 factor. The space of compactly supported smooth forms on $M \setminus Y$ can be identified with the linear subspace consisting of compactly supported smooth forms on M which vanish along $Y \subset M$ that is, $\Omega^2_c(M \setminus Y; \mathbb{C}) \cong \Omega^2_c(M, Y; \mathbb{C}) \subset \Omega^2_c(M; \mathbb{C})$. By the aid of this embedding it is easy to see that in the L^2-topology induced by any Riemannian metric $\Omega^2_c(M \setminus Y; \mathbb{C})$ is dense in $\Omega^2_c(M; \mathbb{C})$ which immediately implies that $C(M \setminus Y) = C(M)$ for the corresponding complexified Clifford algebras; thus taking closures we come up with $\mathscr{H}(M \setminus Y) = \mathscr{H}(M)$ and then $\mathfrak{R}(M \setminus Y) = \mathfrak{R}(M)$. By definition $\mathfrak{R}(M \setminus Y)$ acts via $\rho_{M\setminus Y}$ on $\mathscr{I}(M \setminus Y)^{\perp}$ and $\mathfrak{R}(M)$ acts via ρ_M on $\mathscr{I}(M)^{\perp}$. By assumption $M \setminus Y \subsetneqq M$ induces an isomorphism on the 2nd homology hence $\mathscr{I}(M \setminus Y)^{\perp}$ and $\mathscr{I}(M)^{\perp}$ are simultaneously trivial or not. In fact $\mathscr{I}(M \setminus Y)^{\perp} = \mathscr{I}(M)^{\perp}$. This is because recall that the condition $A \in I(M \setminus Y)$ can be written as $F_{\Sigma,\varphi_\Sigma}(A^*A) = (A^*A\varphi_\Sigma, \varphi_\Sigma)_{L^2(M\setminus Y, g)} = 0$ where φ_Σ is the Hodge-dual of the harmonic representative $\eta_\Sigma$of the Poincaré-dual of $\Sigma \looparrowright M \setminus Y$; since Y has zero 4 dimensional Lebesgue measure $A \in I(M \setminus Y)$ if and only if $A \in I(M)$; therefore $\mathscr{I}(M \setminus Y) = \mathscr{I}(M)$ yielding the same for the complementums in $\mathscr{H}(M \setminus Y) = \mathscr{H}(M)$. Consequently both $\rho_{M\setminus Y}$ and ρ_M

are representations of $\mathfrak{R}(M \setminus Y) = \mathfrak{R}(M)$ on $\mathscr{I}(M \setminus Y)^{\perp} = \mathscr{I}(M)^{\perp}$. By definition $\gamma(M \setminus Y) = \dim_{\mathfrak{R}(M\setminus Y)} \mathscr{I}(M \setminus Y)^{\perp}$ and likewise $\gamma(M) = \dim_{\mathfrak{R}(M)} \mathscr{I}(M)^{\perp}$. Therefore $\gamma(M \setminus Y) = \gamma(M)$ as stated.

Concerning the third assertion note that the γ-invariant is a well-defined map from (the category) Man^4 of all orientation-preserving diffeomorphism classes of connected, oriented smooth 4-manifolds into the real interval $[0, 1) \subset \mathbb{R}$. But Man^4 forms a commutative semigroup with unit S^4 under the connected sum operation $\#$. That is, if $X, Y, Z \in \mathsf{Man}^4$ and $S^4 \in \mathsf{Man}^4$ is the 4-sphere then $X\#Y \cong Y\#X$ and $(X\#Y)\#Z \cong X\#(Y\#Z)$ and $X\#S^4 \cong X$. Pick $M, N \in \mathsf{Man}^4$ with their connected sum $M\#N \in \mathsf{Man}^4$ and consider the corresponding invariants $\gamma(M), \gamma(N), \gamma(M\#N) \in [0, 1)$. Define $\bullet : [0, 1) \times [0, 1) \to [0, 1)$ by setting $\gamma(M) \bullet \gamma(N) := \gamma(M\#N)$. The $\bullet$-operation is therefore well-defined and has the properties $\gamma(X) \bullet \gamma(Y) = \gamma(Y) \bullet \gamma(X)$ and $(\gamma(X) \bullet \gamma(Y)) \bullet \gamma(Z) = \gamma(X) \bullet (\gamma(Y) \bullet \gamma(Z))$ and $\gamma(X) \bullet \gamma(S^4) = \gamma(X)$. These ensure us that $([0, 1), \bullet)$ is a unital commutative semigroup and $\gamma : (\mathsf{Man}^4, \#) \to ([0, 1), \bullet)$ is a unital semigroup homomorphism. Mapping the unique semigroup structure of $[0, +\infty) \subset \mathbb{R}$ to $[0, 1)$ by $\alpha \mapsto \tanh \alpha$ gives $x \bullet y = \frac{x+y}{1+xy}$ with 0 being the unit. Thus $\gamma(S^4) = 0$ (as we already know) but otherwise γ is not trivial because $\gamma(M) > 0$ if $b_2(M) > 0$. This yields the shape for $\gamma(M) \bullet \gamma(N)$ as stated. □

PROOF OF THEOREM 1.2. This theorem follows from Lemma 2.4. □

More generally one can demonstrate by techniques from 4-manifold theory and the gluing principle the following non-trivial but quite insensitive behaviour of γ in the closed simply connected case.

LEMMA 2.5. *If M' and M'' are connected, simply connected, closed smooth 4-manifolds which are homeomorphic then $\gamma(M') = \gamma(M'')$.*

Take $y \in [0, 1)$ and the sequence

$$R_0(y) := 0\,,\ R_1(y) := y\,,\ \dots\,,\ R_k(y) := \frac{y + R_{k-1}(y)}{1 + yR_{k-1}(y)}\,,\ \dots$$

for all $k = 0, 1, 2, \dots$ representing the semigroup $\{0\} \cup \mathbb{N}$ inside $[0, 1)$. Then for any connected, simply connected, closed smooth 4-manifold

$$\gamma(M) = R_{b_2(M)}\left(\tfrac{8}{9}\right) = \frac{17^{b_2(M)} - 1}{17^{b_2(M)} + 1}\,.$$

PROOF. Concerning the first assertion if M' and M'' are as required then there exists an integer $k \geqq 0$ such that $M'\#k(S^2 \times S^2) \cong M''\#k(S^2 \times S^2)$ (cf. e.g. [**15**, Theorem 9.1.12]). Thus we know that we have the equality $\gamma(M'\#k(S^2 \times S^2)) = \gamma(M''\#k(S^2 \times S^2))$. Then applying the gluing principle

$$\frac{\gamma(M') + \gamma(k(S^2 \times S^2))}{1 + \gamma(M')\gamma(k(S^2 \times S^2))} = \frac{\gamma(M'') + \gamma(k(S^2 \times S^2))}{1 + \gamma(M'')\gamma(k(S^2 \times S^2))}$$

and observing that the map $x \mapsto \frac{x+x_0}{1+xx_0}$ is invertible we obtain that $\gamma(M') = \gamma(M'')$.

Concerning the second assertion, if M_1 and M_2 are connected, closed, simply connected, smooth then there exist integers $k_1, l_1 \geqq 0$ and $k_2, l_2 \geqq 0$ such that

$$M_1\#k_1\mathbb{C}P^2\#l_1\overline{\mathbb{C}P^2} \cong M_2\#k_2\mathbb{C}P^2\#l_2\overline{\mathbb{C}P^2}$$

(cf. e.g. [**15**, Theorem 9.1.14]) which shows that

$$\gamma(M_1\#k_1\mathbb{C}P^2\#l_1\overline{\mathbb{C}P^2}) = \gamma(M_2\#k_2\mathbb{C}P^2\#l_2\overline{\mathbb{C}P^2})\ .$$

Now put $y := \gamma(\mathbb{C}P^2) = \gamma(\overline{\mathbb{C}P}^2)$. Then by the gluing principle

$$\frac{\gamma(M_1) + R_{k_1+l_1}(y)}{1+\gamma(M_1)R_{k_1+l_1}(y)} = \frac{\gamma(M_2) + R_{k_2+l_2}(y)}{1+\gamma(M_2)R_{k_2+l_2}(y)}\ .$$

Let $M_1 := M$ be arbitrary and $M_2 := S^4$ hence $\gamma(M_2) = 0$. Then we can suppose that $k_1 + l_1 \leqq k_2 + l_2$ therefore

$$\frac{\gamma(M) + R_{k_1+l_1}(y)}{1+\gamma(M)R_{k_1+l_1}(y)} = R_{k_2+l_2}(y)$$

from which again by invertability we find $\gamma(M) = R_{k_2+l_2-k_1-l_1}(y)$ hence setting $n := k_2 + l_2 - k_1 - l_1 \geqq 0$ we obtain $\gamma(M) = R_n(y)$. Moreover it is clear from the proof that in fact $n = b_2(M)$.

A closed formula arises if we write $R_k(y) = \frac{a_k(y)}{b_k(y)}$ and observe that

$$\begin{pmatrix} a_k(y) \\ b_k(y) \end{pmatrix} = \begin{pmatrix} 1 & y \\ y & 1 \end{pmatrix}^k \begin{pmatrix} 0 \\ 1 \end{pmatrix}\ .$$

The eigenvalues of $\left(\begin{smallmatrix} 1 & y \\ y & 1 \end{smallmatrix}\right)$ are equal to $1 \pm y$ with eigenvectors $\left(\begin{smallmatrix} 1 \\ \pm 1 \end{smallmatrix}\right)$ respectively. Thus inserting the decomposition $\left(\begin{smallmatrix} 0 \\ 1 \end{smallmatrix}\right) = \frac{1}{2}\left(\begin{smallmatrix} 1 \\ 1 \end{smallmatrix}\right) - \frac{1}{2}\left(\begin{smallmatrix} 1 \\ -1 \end{smallmatrix}\right)$ and $k = b_2(M)$ into the above equation we end up with

$$\gamma(M) = R_{b_2(M)}(y) = \frac{(1+y)^{b_2(M)} - (1-y)^{b_2(M)}}{(1+y)^{b_2(M)} + (1-y)^{b_2(M)}}\ .$$

Our last task is to find the precise value of $y \in [0, 1)$. Endowing $\mathbb{C}P^2$ with the orientation compatible with its complex structure and putting the Fubini–Study metric the group $\mathrm{U}(3)$ acts by isometries on $\mathbb{C}P^2$. Then the underlying orientation-preserving diffeomorphisms induce an action of the complexification $\mathrm{U}(3)^{\mathbb{C}} = \mathrm{GL}(3;\mathbb{C})$ on $\mathfrak{R}$ by unitary inner automorphisms. Consequently, on the one hand, there exists an invariant subfactor $\mathfrak{J}(\mathbb{C}P^2) \subset \mathfrak{R}$ such that $\mathfrak{R} = \mathfrak{M}_3(\mathfrak{J}(\mathbb{C}P^2))$. In the proof of Lemma 2.1 we have seen that by exploiting the flexibility of $F_{\Sigma,\omega}$ the condition $A \in I(\mathbb{C}P^2)$ can be written as $F_{\Sigma,\varphi_\Sigma}(A^*A) = (A^*A\varphi_\Sigma, \varphi_\Sigma)_{L^2(M,g)} = 0$ where φ_Σ is the Hodge-dual of the harmonic representative η_Σ of the Poincaré-dual of $\Sigma \looparrowright M$ i.e., $I(\mathbb{C}P^2)$ can be defined in terms of harmonic 2-forms only. However the isometries of $\mathbb{C}P^2$ leave the space of harmonic 2-forms unchanged consequently, on the other hand, the subfactor $\mathfrak{I}(\mathbb{C}P^2) \subset \mathfrak{R}$ generated by $I(\mathbb{C}P^2)I(\mathbb{C}P^2)^* \subset \mathfrak{R}$ as in Lemma 2.3 is invariant under $\mathrm{GL}(3;\mathbb{C})$, too. Therefore $\mathfrak{J}(\mathbb{C}P^2) = \mathfrak{I}(\mathbb{C}P^2)$ implying for the Jones index in question that

$$[\mathfrak{R} : \mathfrak{I}(\mathbb{C}P^2)] = [\mathfrak{M}_3(\mathfrak{J}(\mathbb{C}P^2)) : \mathfrak{J}(\mathbb{C}P^2) \otimes 1_{\mathrm{GL}(3;\mathbb{C})}] = 3^2 = 9\ .$$

Thus putting together everyting we get via Lemma 2.3 that $y = R_1(y) = \gamma(\mathbb{C}P^2) = 1 - \frac{1}{9} = \frac{8}{9}$ and the proof is now complete. □

REMARK 2.6. It follows that $\log(1 - R_k(y)) \approx k \log \frac{1-y}{1+y}$ for k large enough (here "large" depends on y). This yields an interesting estimate for the 2nd Betti number of a closed simply connected manifold in terms of its γ-invariant. On substituting $\gamma(M) = R_{b_2(M)}(\frac{8}{9})$ we get

$$b_2(M) \approx \frac{\log(1-\gamma(M))}{\log\frac{1}{17}} \quad \text{if } b_2(M) \text{ is large enough .} \tag{2.7}$$

PROOF OF THEOREM 1.3. This theorem follows from Lemma 2.5. □

REMARK 2.7. We have seen that $\mathfrak{R}$ is especially interesting in the four dimensional case since it contains curvature tensors. However actually this has not been used during the construction of γ in the proof of Lemma 2.1. Moreover, as recorded in Lemma 2.5, despite its construction depending on the smooth structure, γ is in fact a topological invariant only at least in the simply connected closed case. Nevertheless γ is computable and non-trivial. In order to appreciate γ as it is we close this section by rapidly introducing another huge class of invariants whose members look very natural and meet the demands above however turn out to be trivial.

By the aid of the canonical inclusions $C(M) \subset \mathscr{H}$ and $C(M) \subset \mathfrak{R} \subset \mathfrak{B}(\mathscr{H})$ and the Riesz representation theorem an immense class of functionals on $\mathfrak{R}$ arises by picking any $T \in \mathfrak{B}(\mathscr{H})$ and continuously extending the map $A \mapsto (\hat{T}, \widehat{A^*}) = \tau(TA)$ from $C(M)$ to $\mathfrak{R}$ giving rise to a continuous $\mathbb{C}$-linear map $F_T : \mathfrak{R} \to \mathbb{C}$ of the form

$$F_T(A) := \tau(TA)$$

which is also positive if $\tau(T) > 0$. Interesting examples emerge if instead of a plain manifold M as in Lemma 2.1 we start now with an oriented Riemannian one (M, g). For instance take the Hodge operator $*$ induced by the metric and orientation and put $T := \frac{1}{2}(1 + *)$ hence $TA = A^+$ is the "self-dual part" of A. Or assume that the curvature is bounded hence $R_g \in \mathfrak{R}$ and satisfies $\tau(R_g) \neq 0$ i.e. the metric itself is sufficiently generic[4] and put $T := \pm R_g$. Let $\{0\} \subseteqq I(M) \subsetneqq \mathfrak{R}$ denote the usual multiplicative left-ideal generated by this functional i.e. $A \in I(M)$ if and only if $F_T(A^*A) = 0$. Now if T' is another element satisfying $\tau(T') \neq 0$ then by assumption both $0 \neq \hat{T}, \hat{T}' \in \mathscr{H}$ hence there exists an invertible element $S \in \mathfrak{B}(\mathscr{H})^\times$ such that $T' = ST$; consequently

$$\begin{aligned} 0 \leqq |F_{T'}(A^*A)|^2 &= |\tau(STA^*A)|^2 = |\tau(TA^*AS)|^2 = |F_T(A^*AS)|^2 \\ &\leqq F_T(A^*A)F_T((AS)^*AS) \end{aligned}$$

and likewise $0 \leqq |F_T(A^*A)|^2 \leqq F_{T'}(A^*A)F_{T'}((AS^{-1})^*AS^{-1})$. This demonstrates the independence of $I(M)$. Take again the self-adjoint subspace $I(M)I(M)^*$ and let $\{0\} \subseteqq \mathscr{I}(M) \subsetneqq \mathscr{H}$ be the closure of $C(M) \cap I(M)I(M)^*$ and consider its orthogonal complementum $\{0\} \subsetneqq \mathscr{I}(M)^\perp \subseteqq \mathscr{H}$ and the corresponding projection $P_M : \mathscr{H} \to \mathscr{I}(M)^\perp$. Also take the subfactor $\mathbb{C} \cong Z(\mathfrak{R}) \subseteqq \mathfrak{I}(M) \subsetneqq \mathfrak{R}$ provided by the weak closure of $C(M) \cap I(M)I(M)^*$. Very similarly to the construction in

[4]Note that whenever M is compact it can accommodate a metric g for which either $\tau(R_g) \neq 0$ or $\tau(R_g) = 0$ i.e. both cases can occur. This is obvious for flat manifolds admitting a metric with $R_g = 0$ hence $\tau(R_g) = 0$. But note that (2.6) says that $\tau(R_g)$ is proportional to the total scalar curvature hence $\tau(R_g) = 0$ for all Ricci-flat manifolds too. But even more if M is any compact oriented manifold then take a smooth function $f : M \to \mathbb{R}$ which is (i) strictly negative in one point and (ii) satisfies $\int_M f\mu_h = 0$ with respect to some auxiliary metric h. Then (i) implies by [**5**, Theorem 4.35] the existence of a Riemannian metric g having f as its scalar curvature hence via (ii) and (2.6) satisfying $\tau(R_g) = 0$.

the proof of Lemma 2.1 we observe that $\mathscr{I}(M)^\perp$ is a left $\mathfrak{R}$-module; hence define a new smooth 4-manifold invariant by its coupling constant more precisely

$$\delta(M) := \tau(P_M) \ .$$

The immediate observation is that $\delta(M) \in (0,1]$. However the analogue of Lemma 2.3 is not valid because $\mathscr{I}(M)$ is not necessarily the standard $\mathfrak{I}(M)$-module anymore as the case of $\mathscr{I}(M) = \{0\}$ and the corresponding $\mathfrak{I}(M) = Z(\mathfrak{R}) \cong \mathbb{C}$ shows. The computationally useful Lemma 2.4 also no longer holds (except the invariance under reversing orientation). In fact since we can suppose that T is self-adjoint and positive definite therefore $T = S^2$ with a same type hence invertible operator $S \in \mathfrak{B}(\mathscr{H})$; thus using the norm $\|\cdot\|$ on $\mathscr{H}$ we find $0 = F_T(A^*A) = \|\widehat{(AS)^*}\|^2$ implying $AS = 0$ i.e. $A = 0$. Thus $I(M) = 0$ yielding $\mathscr{I}(M)^\perp = \mathscr{H}$. Thus the fact is that always $\delta(M) = 1$.

3. Physical interpretation

In this section we cannot resist temptation and introduce the basics of a physical interpretation of the material collected so far. Namely, upon reversing the approach of Section 2, we shall replace the immense class of classical space-times of general relativity with a single universal "quantum space-time" allowing us to lay down the foundations of a manifestly four dimensional, covariant, non-perturbative and genuinely quantum theory of gravity. This construction is natural, simple and self-contained. More precisely here in Section 3 not *one particular 4-manifold*—physically regarded as a particular classical space-time—but *the unique hyperfinte* II_1 *factor von Neumann algebra*—physically viewed as the universal quantum space-time—is declared to be the primarily given object. Let us see how it works.

Observables, fields, states and the gauge group. Let $\mathscr{H}$ be an abstractly given infinite dimensional complex separable Hilbert space and $\mathfrak{R} \subset \mathfrak{B}(\mathscr{H})$ be a type II_1 hyperfinite factor hence tracial von Neumann algebra acting on $\mathscr{H}$ by the standard representation. We call $\mathfrak{R}$ the *algebra of (bounded) observables*, its tangent space $T_1\mathfrak{R} \supset \mathfrak{R}$ consisting of the Fréchet derivatives of 1-parameter families of observables at the unit $1 \in \mathfrak{R}$ the *algebra of fields*, while $\mathscr{H}$ the *state space* in this quantum theory. The subgroup $\mathrm{U}(\mathscr{H}) \cap \mathfrak{R}$ of the unitary group of $\mathscr{H}$ operating as inner automorphisms on $\mathfrak{R}$ is the *gauge group*. Note that the gauge group acts on both $\mathfrak{R}$ and $\mathscr{H}$ but in a different way.

What kind of quantum theory is the one in which $\mathfrak{R}$ plays the role of the algebra of *physical observables*? We have seen in Section 2 that $\mathfrak{R}$, when regarded to operate on a differentiable manifold, contains all bounded (complexified) algebraic curvature tensors along it whenever the real dimension of this manifold is precisely *four*. Moreover $\mathfrak{R}$ is generated by local operators more precisely by trace class algebraic operators R acting like *multiplication* operators and by trace class operators L_X acting like *first order differential* operators. Picking canonically conjugate pairs i.e. necessarily unbounded self-adjoint operators $R, L_X \in T_1\mathfrak{R}$ satisfying $[R, L_X] = 1$ (assuming $\hbar = 1$) shows that $T_1\mathfrak{R}$ contains a CCR algebra hence revealing the *bosonic* character of $\mathfrak{R}$. Recall that in classical general relativity *local* gravitational phenomena are caused by the curvature of space-time. Hence interpreting $\mathfrak{R}$ physically as an operator algebra of local observables the corresponding quantum theory is expected to be a *four dimensional quantum theory of pure gravity* (hence a bosonic theory). In this way we fulfill the *Heisenberg dictum* that a quantum theory should completely and unambigously be formulated and interpreted in terms of its local physical observables (and not the other way round). In modern understanding by a *physical theory* one means a two-level description of a bundle of phenomenologically interrelated natural phenomena: the theory possesses a *syntax* provided by its mathematical core structure and a *semantics* which is the meaning i.e. interpretation of the bare mathematical model in terms of physical concepts. In this context our quantum theory is not merely a mathematical theory anymore but a physical theory. This is because the bare mathematical structure $\mathfrak{R}$ (together with a representation π on $\mathscr{H}$) is dressed up i.e., interpreted by assigning a physical (in fact, gravitational) meaning to the experiments consistently performable by the aid of this structure (i.e., the usual quantum measurements of operators $A \in \mathfrak{R}$ in pure states $v \in \mathscr{H}$ or in more general ones, see below). In our opinion it is of particular interest that the geometrical dimension—equal to four—is fixed at the semantical level only and it matches the known phenomenological dimension of space-time. This is in sharp contrast to e.g. string theory where the geometrical dimension of the theory is fixed already at its syntactical level i.e., by its mathematical structure (namely, demaninding the theory to be free of conformal anomaly) and it turns out to be much higher than the phenomenological dimension of space-time. As a further clarification it is emphasized that the aim here is *not* to quantize general relativity in some way but rather conversely: being a successful theory at the classical level, general relativity should be derived from this abstractly given quantum theory by taking an appropriate "classical limit" as naturally as possible. Unfortunately, this program yet cannot be carried out completely here.

Observables as the universal space of all space-times. Taking into account the embeddings (2.3) and the universality of $\mathfrak{R}$ the *unique* algebra of observables $\mathfrak{R}$ can be considered as the collection of *all* classical space-times and we can interpret the appearance of the gauge group as the manifestation of the diffeomorphism gauge symmetry of classical general relativity in this quantum theory. Indeed, orientation-preserving diffeomorphisms of M interchange its points as well as act on the corresponding operators in $\mathfrak{R}$ (projections) by unitary inner automorphisms i.e. $\mathrm{Diff}^+(M)$ embeds into $\mathrm{Aut}\,\mathfrak{R}$. Reformulating this in a more geometric language we can say that classical space-times appear as special orbits within $\mathfrak{R}$ of its gauge group. The operators in $\mathfrak{R}$ representing geometric points via (2.3) are special operators namely projections. Consequently the full non-commutative algebra $\mathfrak{R}$ is not exhausted by operators representing points of space-time; it certainly contains much more operators—e.g. various projections which are not of geometric origin—therefore this "universal quantum space-time" is more than a bunch of all classical space-times. As a comparison, in *algebraic quantum field theory* [**17**] one starts with a particular smooth 4-manifold M and considers an assignment $U \mapsto \mathfrak{R}(U)$ describing local algebras of observables along all open subsets $\emptyset \subsetneqq U \subseteqq M$. However in our case, quite conversely, space-times are secondary structures only and all of them are injected into the unique observable algebra $\mathfrak{R}$ which is considered to be primary.

More examples of observables, fields and states. Let us take a closer look of the elements of $\mathfrak{R}$ and of $T_1\mathfrak{R} \supset \mathfrak{R}$. Taking into account the items above concerning the interpretation of the mathematical results of Section 2 we have agreed to identify some elements of $\mathfrak{R}$ up to finite accuracy with four dimensional bounded algebraic (i.e., formal) curvature tensors of *all possible* smooth 4-manifolds. In fact one can more accurately reformulate some basic concepts of classical general relativity within this quantum (field-)theoretic framework. The only thing we have to do is to refine or improve the constructions in Section 2 by taking into account not only the non-degeneracy but the indefiniteness of the pairing (2.1) too.

Remember that given a connected oriented Riemannian 4-manifold (M, g) one can introduce the Hodge operator $*$ which satisfies $*^2 = \mathrm{Id}_{\Omega^2_c(M)}$ hence it induces a splitting

$$\Omega^2_c(M) = \Omega^+_c(M) \oplus \Omega^-_c(M)$$

into (anti)self-dual or ±1-eigenspaces. Thus the Hodge star as an operator on $\Omega^2_c(M)$ with respect to this splitting has the form $* = \left(\begin{smallmatrix} 1 & 0 \\ 0 & -1 \end{smallmatrix}\right)$ hence is symmetric; likewise the curvature tensor R_g of (M, g) looks like (2.5) i.e., is also a pointwise symmetric operator. We summarize these facts by writing that $*, R_g \in C^\infty(M; S^2 \wedge^2 T^*M) \subset C^\infty(M; \mathrm{End}(\wedge^2 T^*M))$. In addition R_g obeys the algebraic Bianchi identity $b(R_g) = 0$. The vacuum Einstein equation with cosmological constant $\Lambda \in \mathbb{R}$ reads as $\mathrm{Ric} = \Lambda g$ therefore it is equivalent to the vanishing of the traceless Ricci part Ric_0 of R_g in (2.5). From these it readily follows, as noticed in [**24**], that (M, g) is Einstein i.e., satisfies the vacuum Einstein equation with cosmological constant if and only if $*R_g = R_g*$ or equivalently $*R_g* = R_g$.

Now if R_g is a bounded operator then we have seen in Section 2 that its complex-linear extension gives rise to an element $R_g \in \mathfrak{R}$ which is moreover self-adjoint i.e., $R_g^* = R_g$ because R_g is symmetric. Therefore its trace is always real. Likewise for $*$ satisfying $\tau(*) = 0$. The algebraic Bianchi identity has the form $b(R_g) = 0$ where b is a certain fiberwise averaging operator on symmetric bundle morphisms satisfying $b^2 = b$ and $b^* = b$ with respect to the metric. This gives rise to a g-orthogonal decomposition $S^2 \wedge^2 T^*_xM \otimes \mathbb{C} = \ker b_x \oplus \mathrm{im}\, b_x$ at every point $x \in M$. A further subtlety of *four dimensions* is that $\mathrm{im}\, b_x = \mathbb{C} * \mathrm{Id}_{\wedge^2 T^*_x M \otimes \mathbb{C}} \subset S^2 \wedge^2 T^*_xM \otimes \mathbb{C}$ holds [**24**]. Consequently $b(R_g) = 0$ is equivalent to saying that $g(R_g, *) = 0$ i.e., within $S^2 \wedge^2 T^*M \otimes \mathbb{C}$ the operator R_g is g-orthogonal to $*$ at every $x \in M$. By exploiting that $*$ is a pointwise symmetric operator we can use the pointwise equality $g(R_g, *) = \mathrm{tr}(R_g*)$ and then (2.6) to obtain

$$\tau(R_g*) = \frac{1}{6}\int_M \mathrm{tr}(R_g*)\mu_g = 0$$

which geometrically means that $(\hat{R}_g, \hat{*}) = 0$ i.e., $\hat{R}_g$ is perpendicular to $\hat{*}$ within $\mathscr{H}$. Hence the Bianchi identity. Concerning the Einstein condition if R_g comes from an Einstein metric with cosmological constant $\Lambda \in \mathbb{R}$ then we already know that $*R_g* = R_g$ which means that $\mathrm{Ric}_0 = 0$ in (2.5). Moreover taking into account that $\mathrm{Weyl}^\pm \in C^\infty(M; S^2 \wedge^2 T^*M \otimes \mathbb{C})$ in (2.5) are traceless their pointwise scalar products with the identity of $C^\infty(M; S^2 \wedge^2 T^*M \otimes \mathbb{C})$ look like $g(\mathrm{Weyl}^\pm, 1) = \mathrm{tr}(\mathrm{Weyl}^\pm) = 0$. Using (2.6) this shows again that the corresponding vectors $\widehat{\mathrm{Weyl}}^\pm \in \mathscr{H}$ are perpendicular to $\hat{1} \in \mathscr{H}$ i.e., $(\widehat{\mathrm{Weyl}}^\pm, \hat{1}) = 0$. Moreover $\mathrm{Scal} = 4\Lambda$ consequently via (2.5) we get

$$\tau(R_g) = (\hat{R}_g\, , \hat{1}) = \frac{1}{12}(\widehat{\mathrm{Scal}}, \hat{1}) = \frac{4\Lambda}{12}(\hat{1}, \hat{1}) = \frac{\Lambda}{3}\ . \tag{3.1}$$

Finally note that in the previous analysis the signature of the metric plays no significant role because of an involved complexification (we have used Riemannian signature just for convenience).

All of these serve as a motivation for the following operator-algebraic reformulation or generalization of the classical vacuum Einstein equation with cosmological constant:

DEFINITION 3.1. Let M be a connected oriented smooth 4-manifold and $\mathfrak{R}$ its II_1-type hyperfinite factor von Neumann algebra as before.

(i) A refinement of $\mathfrak{R}$ is a pair $(\mathfrak{R}, *)$ where $1 \neq * \in \mathfrak{R}$ and satisfies $*^2 = 1$;

(ii) An operator $Q \in \mathfrak{R}$ solves the quantum vacuum Einstein equation with respect to $(\mathfrak{R}, *)$ if

$$\begin{cases} Q^* & = Q \quad \text{(self-adjointness)} \\ \tau(Q*) & = 0 \quad \text{(algebraic Bianchi identity)} \\ *Q* & = Q \quad \text{(Einstein condition);} \end{cases}$$

(iii) The trace $\tau(Q) =: \frac{\Lambda}{3} \in \mathbb{R}$ is called the corresponding quantum cosmological constant;

(iv) $Q \in \mathfrak{R}$ as above is called a vacuum state if it is positive semi-definite too (hence its corresponding quantum cosmological constant is non-negative).

Note that the curvature tensor R_g of an Einstein manifold (M, g) if bounded as an operator always solves the quantum vacuum Einstein equation with respect to the canonical refinement provided by metric (anti)self-duality. However in sharp contrast to the classical Einstein equation whose solution is a metric therefore is non-linear, its quantum generalization is linear hence easily solvable. Of course this is beacuse in the generalization we do not demand the solution (which can be any non-local operator) to originate from a metric. Indeed, given $B \in \mathfrak{R}$ then $S = \frac{1}{2}(B + B^*)$ is self-adjoint and then picking an arbitrary refinement $(\mathfrak{R}, *)$ and taking into account that $*$ is always self-adjoint, the averaged operator $Q := \frac{1}{2}(S + *S*) - \tau(S*)* \in \mathfrak{R}$ is automatically a solution of the quantum vacuum Einstein equation; moreover using $\tau(*) = 0$ its trace $\frac{\Lambda}{3}$ is equal to $\tau(B)$ hence is independent of the particular refinement. Observe that this linearity permits a BRST-like reformulation of Definition 3.1 too i.e. introducing a linear operator $\Delta : \mathfrak{R} \to \mathfrak{R}$ such that $\ker\Delta$ is precisely equal to the solution space and $\Delta^2 = 0$.

It is important to note that, contrary to smooth solutions, many singular solutions of classical general relativity theory cannot be interpreted as observables because their curvatures lack being bounded operators hence do not belong to $\mathfrak{R}$. As a result, we expect that the classical Schwarzschild or Kerr black hole solutions and more generally gravitatioal fields of isolated bodies (cf. [**10, 11**]) give rise not to *observables* in $\mathfrak{R}$ but rather *fields* in $T_1\mathfrak{R} \supset \mathfrak{R}$.

Questions and answers. First let us clarify what the answers in this proposed quantum theory are because this is easier. Staying safely within the orthodox framework i.e., the *Copenhagen interpretation* and the standard mathematical formulation of quantum theory (but relaxing this latter somewhat), given an observable represented by $A \in \mathfrak{R}$ and a general (i.e., not necessarily pure) state also represented by an element $B \in \mathfrak{R}$ in the observable algebra (regarded as a "density matrix" operator over the state space $\mathscr{H}$) we declare that an *answer* is formally like

The expectation value of the observable quantity A in the state B is $\tau(AB) \in \mathbb{C}$

where $\tau : \mathfrak{R} \to \mathbb{C}$ is the unique finite trace on the hyperfinite II_1 factor von Neumann algebra $\mathfrak{R}$. Note that in order not to be short sighted, at this level of generality we require neither the observable $A \in \mathfrak{R}$ to be self-adjoint nor the state $B \in \mathfrak{R}$ to be positive semi-definite self-adjoint and normalized (however these can be imposed if they turn out to be necessary, cf. e.g. the characterization of a vacuum state in Definition 3.1 above) hence our answers can be complex numbers in general. Nevertheless $\tau(AB)$ is well-posed i.e., is finite and invariant under the gauge group of this theory namely the unitary automorphisms of $\mathfrak{R}$ thus it is indeed an "answer"—at least syntactically.

Now we come to the most difficult problem namely what are the meaningful questions here? This problem is already fully at the foggy semantical level. The orthodox approach says that a *question* should formally be like

From the collection Spec$A \subset \mathbb{C}$ *of "all possible values the pointer of the experimental instrument designed to measure A in the laboratory can assume", which does occur in B?*

and the answer (as defined before) to this question is obtained through a *measurement.* Thus let us make a short interlude concerning the measurement. After performing the physical experiment designed to answer the question above should we expect that the state $B \in \mathfrak{R}$ will necessarily "collapse" to an eigenstate $B_\lambda \in \mathfrak{R}$ of A? In our opinion *no* and this is an essential difference between gravity and quantum mechanics. Namely, in quantum mechanics an ideal observer compared to the physical object to be observed is *infinitely large* hence the immense physical interaction accompanying the measurement procedure drastically disturbs the entity leading to the collapse of its state. However, in sharp contrast to this, in gravity an ideal observer is *infinitely small* hence it is reasonable to expect that measurements might *not* alter gravitational states. This is in accordance with our old experience concerning measurements in astronomy.

Concerning the problem of its *meaning*, since $A, B \in \mathfrak{R}$ have something to do with the curvature of local portions of space-time, it is not easy to assign a straightforward meaning to the above question. Therefore instead of offering a general solution to this problem at this initial state of the art, let us rather consider some special cases. For example if (M, g) is a classical non-singular space-time and (N, h) is another one, then their *classical geometrical* habitants may find that $0 \neq \tau(R_g R_h) \in \mathbb{C}$ in general. Consequently "physical contacts" between different classical geometries can already occur (whatever it means). It follows from the construction of the trace in (2.6) that $\tau(R_g R_h)$ arises by integrating the local trace function $\mathrm{tr}(R_g R_h)$ along $X = M \cap N$. Note that making use of the embeddings of M, N into $\mathfrak{R}$ given by (2.3) taking intersection (maybe empty) is meaningful. Thus given a nearly flat space-time (M, g) "we live in" i.e., an observer satisfying $|R_g(x)|_g \approx 0$ for all $x \in M$ and a different geometry (N, h) "we observe" i.e., a state R_h then we expect $|\tau(R_g R_h)| \approx 0$ in accord with our physical intuition that frequent encounters with different geometries in quantum gravity should occur rather in the strong gravity regime of space-times (hence the reason we do not experience such strange things).

As another example for measurement, consider the energy. In this universal quantum theory the only distinguished non-trivial self-adjoint gauge invariant operator is the identity $1 \in \mathfrak{R}$. Therefore the only natural candidate for playing the role of a *Hamiltonian* responsible for dynamics in this theory is $1 \in \mathfrak{R}$ (in natural units $c = \hbar = G = 1$). Therefore this dynamics is trivial. Nevertheless, quite interestingly, it coincides with the *modular dynamics* introduced by Connes and Rovelli [**9**] because it is associated with the *tracial* state τ on $\mathfrak{R}$ having the identity as its modular operator. Therefore the odd-looking equality (3.1) can be interpreted by saying that the expectation value (in a sequence of measurements) of the energy in the state represented by a vacuum curvature tensor R_g is the number $\tau(R_g)$ and this energy is equal to the cosmological constant constituent $\frac{\Lambda}{3}$ of the corresponding vacuum space-time (M, g). Note that R_g is at least self-adjoint (but perhaps not positive definite neither normalized) hence the energy is a real number but *a priori* can assume negative values too, cf. item (iv) of Definition 3.1.

An application to the cosmological constant problem. The last worry concerning the non-positivity of the energy strongly motivates to test this formalism on the so-called cosmological constant problem. By this we mean the complex problematics introduced into current standard cosmology (based on the cosmological principle and the corresponding Friedmann–Lemaître–Robertson–Walker or FLRW model) and standard particle physics (based on the Standard Model) by the experimental verification [**23**] of the existence of a strictly *positive* but *small* cosmological constant in 1998. Recall that in cosmology the cosmological constant is defined as $\Lambda = 3\left(\frac{H_0}{c}\right)^2\Omega_\Lambda$ where H_0 is the Hubble "constant" (it actually varies with time) whose dimenion is s^{-1}, c is the speed of light having dimension ms^{-1} and the dimensionless number $\Omega_\Lambda \approx 0.69$ is the ratio of the dark energy density and the critical material energy density (Ω_Λ also changes in time). Thus the dimension of the physical cosmological constant is m^{-2}. Based on astronomical measurements $\Lambda \approx 2.89 \times 10^{-122}\,\ell_{\mathrm{Planck}}^{-2}$ where $\ell_{\mathrm{Planck}} \approx 1.62 \times 10^{-35}$ m. In order to gain compatibility with the mathematical expressions obtained in Section 2, from now on we shall use the dimensionless number

$$\frac{\Lambda}{3}\ell^2_{\mathrm{Planck}} \approx 0.96 \times 10^{-122}$$

expressing the magnitude of the cosmological constant in natural Planck units. We claim that this sort of (i.e., a strictly positive but small) cosmological constant naturally appears if classical general relativity is replaced with its formal quantum generalization as introduced above. The cosmological constant problem has been investigated from a closely related angle by other authors too, cf. [**2**].

Before seeing this however let us briefly and roughly clarify what do we mean by this "replacement" from a physical viewpoint. We also want to see how a cosmological context has popped up so suddenly. The modern physical and mathematical basis for thinking about space, time, gravity and the structure and history of the Universe (cosmology) has been provided by classical general relativity in the form of a pseud-Riemannian 4-manifold (M, g) such that g is subject to Einstein's field equation. We have seen in Section 2 especially in Lemma 2.1 that out of this input data (in fact from M alone) an operator algebra $\mathfrak{R}$ and a Hilbert space $\mathscr{I}(M)^\perp$ carrying a representation ρ_M of $\mathfrak{R}$ can be constructed. Moreover the representation ρ_M on $\mathscr{I}(M)^\perp$ is the restriction of the unique standard representation π of $\mathfrak{R}$ on a Hilbert space $\mathscr{H}$ containing $\mathscr{I}(M)^\perp$. The latter objects namely $\mathfrak{R}, \mathscr{H}, \pi$ are all *unique* up to corresponding isomorphisms. The two structures (namely the "classical" (M, g) on the one side and the "quantum" $(\mathfrak{R}, \mathscr{H}, \pi)$ on the other) are naturally connected mathematically by the observation made in Section 2 that $\mathfrak{R}$ contains bounded curvature tensors and by Definition 3.1 offering a re-interpretation of the vacuum Einstein equation in an operator-theoretic language. Remember the plain technicality that a curvature tensor must be *bounded* in order to belong to the operator algebra. However it has been known for a long time that in general relativity the gravitational field of an isolated massive configuration cannot be regular everywhere [**10, 11**]. Thus we are forced to move towards a global non-singular cosmological context when trying to work out a physical connection between the two structures. Therefore hereby we make a *working hypothesis*: the unique abstract triple $(\mathfrak{R}, \mathscr{H}, \pi)$ as a quantum theory describes the Universe from the Big Bang till the Planck time $t_{\text{Planck}} \approx 5.39 \times 10^{-44}$ s while a highly non-unique pair (M, g) as a space-time in classical general relativity describes its evolution from t_{Planck} onwards. A usual choice for (M, g) is the FLRW solution with or without cosmological constant. The "moment" t_{Planck} would therefore formally or symbolically label the emergence or onset of a space-time structure in the course of the history of the Universe[5] establishing a physical correspondence between the two aforementioned mathematical structures. We symbolically write this correspondence as

$$\begin{array}{cccc} (\mathfrak{R}, \mathscr{H}, \pi) & \Longrightarrow & \left(Q_M \in \mathfrak{R}, \mathscr{I}(M)^\perp \subset \mathscr{H}, \rho_M = \pi|_{\mathscr{I}(M)^\perp}\right) & \Longleftrightarrow & M \\ & \Longrightarrow & \left(Q_M, R_g \in \mathfrak{R}, \mathscr{I}(M)^\perp \subset \mathscr{H}, \rho_M = \pi|_{\mathscr{I}(M)^\perp}\right) & \Longleftrightarrow & (M, g) \end{array}$$

and having a sort of spontaneous symmetry breaking or phase transition in mind occuring at t_{Planck}. Actually we broke this process into two parts: "first" the emergence of the plain manifold structure M provided by a choice of an (in our convention dimensionless) operator $Q_M \in \mathfrak{R}$ satisfying the quantum vacuum Einstein equation as in Definition 3.1 and "then" the emergence of the geometry g hence the causal structure over M too provided by a further (in our convention dimensionless) curvature operator $R_g = Q_M + 8\pi T \in \mathfrak{R}$ satisfying the classical Einstein equation with already fixed cosmological constant $\frac{\Lambda}{3}\ell^2_{\text{Planck}} = \tau(Q_M)$ and some matter content described by a further (in our convention dimensionless) stress-energy tensor T. This passage from the unique quantum regime $(\mathfrak{R}, \mathscr{H}, \pi)$ to a particular classical regime (M, g) could probably be more rigorously captured in the framework of algebraic quantum field theory [**17**] as switching from the unique representation π to a different particular representation ρ_M of $\mathfrak{R}$. In this framework one can also formally label the transition with $T_{\text{Planck}} \approx 1.42 \times 10^{32}$ K; this high temperature is reasonable if we make a further technical observation that π is induced by the unique tracial state τ on $\mathfrak{R}$ hence one can imagine that $(\mathfrak{R}, \mathscr{H}, \pi)$ describes a quantum statistical mechanical system being in an infinitely high temperature Kubo–Martin–Schwinger or KMS equilibrium state τ (hence having trivial modular dynamics in the sense of [**9**]).

After this admittedly yet incomplete trial to set up some physical picture we can embark upon the treatment of the cosmological constant problem in this framework. Let M be a connected oriented smooth 4-manifold and consider its associated projector $P_M \in \mathfrak{R}$ from Lemma 2.1. Recall that an operator like P_M has been used to encode in the simplest way a particular connected oriented smooth 4-manifold M within its operator algebra $\mathfrak{R}$ which is however universal. Thus P_M is a distinguished operator in this sense. But even more, it is self-adjoint hence taking an

---

[5]We are aware of, but cannot do better, how nonsense it is, strictly speaking, to talk about the emergence of space and time at the moment t_{Planck}; this assertion requires the contradictory pre-supposition of time before t_{Planck}.

arbitrary refinement $(\mathfrak{R}, *)$ its average

$$Q_M := \frac{1}{2}(1-P_M+*(1-P_M)*)-\tau((1-P_M)*)* = 1-\frac{1}{2}(P_M+*P_M*)+\tau(P_M*)*$$

is an operator which solves the quantum vacuum Einstein equation too in the sense of Definition 3.1 with quantum cosmological constant $\frac{\Lambda}{3}\ell^2_{\text{Planck}} = \tau(Q_M)$. The particular choice for Q_M is verified by Lemma 2.1 making it sure that $\tau(Q_M) = \tau(1-P_M) = 1-\gamma(M)$ thus

$$\frac{\Lambda}{3}\ell^2_{\text{Planck}} = 1-\gamma(M)$$

in terms of the dimensionless γ-invariant of M. The further restriction imposed upon the spectrum of $\gamma(M)$ in Lemma 2.3 guarantees that actually

$$\frac{\Lambda}{3}\ell^2_{\text{Planck}} \in \left(0, \frac{1}{4}\right] \bigcup \left\{ \frac{1}{4}\cos^{-2}\left(\frac{\pi}{n}\right) \,\middle|\, n = \dots 5, 4, 3 \right\} \subset (0, 1]$$

thus it is permitted to be arbitrary close but never equal to zero. That is, whatever M is, in our model *the (dimensionless) cosmological constant is always a small positive number* (thus, using again the terminology of Definition 3.1, in fact Q_M is a vacuum state). This is consistent with the aforementioned emprical value $\approx 0.96 \times 10^{-122}$ of the cosmological constant.

To be more specific we can impose some conditions on M still compatible with cosmological experience. We can suppose that M has the structure $X \setminus \{\text{point}\}$ where X is a connected simply connected closed smooth 4-manifold; this is the case for the FLRW model in the relevant cases $k = 0, -1$. Moreover we can suppose that the 2nd Betti number of X is non-trivial: $b_2(X) > 0$ implying $b_2(M) > 0$ too. This assumption does not hold in FLRW cosmology but it can be physically interpreted as the necessary topological condition for M i.e., the space-time to contain black holes at t_{Planck}. Indeed, taking e.g. a survey on known solutions [**25**], we can make an interesting observation that apparently all explicitly known 4 dimensional black hole solutions in vacuum general relativity have the property that because of some general reason their black hole content is even topologically recognizable as a two dimensional "hole" in space-time. This means that their instantaneous event horizons represented by immersed connected oriented surfaces in M give rise to non-trivial elements in $H_2(M;\mathbb{Z})^{\text{free}} \cong \mathbb{Z}^{b_2(M)}$. One can indeed prove this at least for stationary black holes [**12**]. Using this picture the number $b_2(M)$ can therefore be physically interpreted as the number of primordial black holes in the space-time at its onset. The assumption $0 < b_2(M)$ implies that strictly $0 < \frac{\Lambda}{3}\ell^2_{\text{Planck}} \lneqq 1$ such that letting $b_2(M) \to +\infty$ we find $0 \leftarrow \frac{\Lambda}{3}\ell^2_{\text{Planck}}$. Consequently, the observed magnitude of the cosmological constant allows us to estimate in this model the number $N \approx b_2(M)$ of primordial black holes just around the Planck era. Plugging the empirical value of the cosmological constant into (2.7) we find

$$N \approx \frac{\log(\frac{\Lambda}{3}\ell^2_{\text{Planck}})}{\log\frac{1}{17}} \sim 10^2$$

yielding a number which is negligable. This qualitative result obtained by topological means is compatible with the agreement in the physicist literature that primordial black holes are essentially absent from the very early Universe, cf. e.g. [**6, 16**] (and the hundreds of references therein). Note that the usual considerations leading to this conclusion are based on the application of the Press–Schechter mechanism **[22]** for primordial black hole formation; hence our topological result is an independent confirmation of these considerations.

Acknowledgements

The author is grateful to T. Asselmeyer-Maluga, B. Gyenti and I. Nikolaev for the stimulating discussions. There are no conflicts of interest to declare that are relevant to the content of this article. The work meets all ethical standards applicable here. All the not-referenced results in this work are fully the author's own contribution. No funds, grants, or other financial supports were received. Data sharing is not applicable to this article as no datasets were generated or analysed during the underlying study.

References

[1] C. Anantharaman and S. T. Popa, *An introduction to* II_1 *factors*, preprint, 331 pp. (no year), available at `http://www.math.ucla.edu/~popa/books.html`

[2] T. Asselmeyer-Maluga and J. Król, *How to obtain a cosmological constant from small exotic* $\mathbb{R}^4$, Phys. Dark Universe **19**, 66-77 (2018)

[3] T. Asselmeyer-Maluga, J. Król, and A. Wilms, *Big Bang and topology*, preprint, 17 pp. (2022)

[4] P. Bérard, G. Besson, and S. Gallot, *Embedding Riemannian manifolds by their heat kernel*, Geom. Funct. Anal. **4** (1994), no. 4, 373–398, DOI 10.1007/BF01896401. MR1280119

[5] A. L. Besse, *Einstein manifolds*, Springer-Verlag Berlin (1987);

[6] B. Carr and F. Kühnel, *Primordial black holes as dark matter: recent developments*, Ann. Rev. Nucl. Part. Sci. **70**, 355-394 (2020)

[7] V. Chandrasekaran, R. Longo, G. Penington, and E. Witten, *An algebra of observables for de Sitter space*, J. High Energy Phys. **2** (2023), Paper No. 82, 55 pp., DOI 10.1007/jhep02(2023)082. MR4547343

[8] A. Connes, *Noncommutative geometry*, Academic Press, New York (1994)

[9] A. Connes and C. Rovelli, *von Neumann algebra automorphisms and time-thermodynamics relation in generally covariant quantum theories*, Classical Quantum Gravity **11** (1994), no. 12, 2899–2917. MR1307019

[10] A. Einstein, *Demonstración de la no existencia de campos gravitacionales sin singularidades de masa total no nula*, Revista de la Universidad Nacional de Tucumán **A2**, 11-15 (1942)

[11] A. Einstein and W. Pauli, *On the non-existence of regular stationary solutions of relativistic field equations*, Ann. of Math. (2) **44** (1943), 131–137, DOI 10.2307/1968759. MR7992

[12] G. Etesi, *A global uniqueness theorem for stationary black holes*, Comm. Math. Phys. **195** (1998), no. 3, 691–697, DOI 10.1007/s002200050408. MR1641027

[13] G. Etesi, *Gravity as a four dimensional algebraic quantum field theory*, Adv. Theor. Math. Phys. **20** (2016), no. 5, 1049–1082, DOI 10.4310/ATMP.2016.v20.n5.a3. MR3604552

[14] G. Etesi, *The universal von Neumann algebra of smooth four-manifolds*, Adv. Theor. Math. Phys. **25** (2021), no. 4, 995–1024, DOI 10.4310/ATMP.2021.v25.n4.a3. MR4407970

[15] R. E. Gompf and A. I. Stipsicz, *4-manifolds and Kirby calculus*, Graduate Studies in Mathematics, vol. 20, American Mathematical Society, Providence, RI, 1999, DOI 10.1090/gsm/020. MR1707327

[16] A. M. Green, *Primordial black holes: sirens of the early universe*, in: *Quantum aspects of black holes* (ed.: X. Calmet) FTPH **178**, 129-149, Springer (2014)

[17] R. Haag, *Local quantum physics*, Springer–Verlag, Berlin (1993)

[18] V. F. R. Jones, S. Morrison, and N. Snyder, *The classification of subfactors of index at most 5*, Bull. Amer. Math. Soc. (N.S.) **51** (2014), no. 2, 277–327, DOI 10.1090/S0273-0979-2013-01442-3. MR3166042

[19] H. B. Lawson Jr. and M-L. Michelsohn, *Spin geometry*, Princeton Univ. Press, Princeton, New Jersey (1989);

[20] A. Ocneanu, *Quantized groups, string algebras and Galois theory for algebras*, Operator algebras and applications, Vol. 2, London Math. Soc. Lecture Note Ser., vol. 136, Cambridge Univ. Press, Cambridge, 1988, pp. 119–172. MR996454

[21] S. Popa, *Classification of subfactors: the reduction to commuting squares*, Invent. Math. **101** (1990), no. 1, 19–43, DOI 10.1007/BF01231494. MR1055708

[22] W. H. Press and P. Schechter, *Formation of galaxies and clusters of galaxies by self similar gravitational condensation*, Astrophys. Journ. **187**, 425-438 (1974);

[23] A. Riess et al., *Observational evidence from supernovae for an accelerating universe and a cosmological constant*, Astronomical Journ. **116**, 1009-1038 (1998);

[24] I. M. Singer and J. A. Thorpe, *The curvature of 4-dimensional Einstein spaces*, Global Analysis (Papers in Honor of K. Kodaira), Univ. Tokyo Press, Tokyo, 1969, pp. 355–365. MR256303

[25] H. Stephani, D. Kramer, M. MacCallum, C. Hoenselaers, and E. Herlt, Exact solutions of Einstein's field equations, Cambridge Univ. Press, Cambridge (2003).

Department of Algebra and Geometry, Institute of Mathematics, Budapest University of Technology and Economics, Műegyetem rkp. 3., H-1111 Budapest, Hungary

Email address: etesi@math.bme.hu, etesigabor@gmail.com

Contemporary Mathematics
Volume **798**, 2024
https://doi.org/10.1090/conm/798/15984

Advances in quantum permutation groups

Amaury Freslon

Abstract. We survey several results, ranging over the past fifteen years, concerning the quantum permutation groups and their quantum subgroups around three themes: quantum information, dynamics, and probability theory. In each case, we try to be as self-contained as possible and convey some intuition concerning the role and importance of quantum permutations, as well as key ideas of the proofs.

1. Introduction

Permutation groups are central objects in many areas of mathematics, and ones that any mathematician has to know about. This is obvious in algebra of course, where they provide fundamental examples of finite non-commutative groups with the nice property that many of their features, for instance their representation theory, are both rich and amenable to a complete and explicit description. Moreover, they are in a sense unavoidable among finite groups since by Cauchy's embedding theorem, any finite group embeds into a permutation group. But they also have deep connections to infinite groups at the level of representation theory, as Schur-Weyl duality illustrates.

Permutation groups also naturally appear in combinatorics in a dynamical way, since when studying finite structures one is naturally led to consider the corresponding symmetry group, which is a group of permutations. This leads to a vast field of study of which one of the most important examples is graph theory, which itself has connections to many other areas of mathematics.

Even though this may be less well-known, permutation groups have connections as well with probability theory. In fact, as the book [**24**] shows with a wealth of examples, many situations from both probability theory and statistics lead to models having the structure of a homogeneous space for some permutation group. This can be fruitfully exploited to turn knowledge about the representation theory of S_N into information on the random processes under study. Moreover, random walks on permutation groups, which mix algebraic properties, geometric intuition and probabilistic reasoning, are a lively topic with connections to many concrete problems.

2020 *Mathematics Subject Classification.* Primary 20G42, 05E10, 05C25, 05C10, 46L53, 46L54.

Permutations are finite, hence compact, groups. The theory of compact groups was extended by S.L. Woronowicz to that of compact quantum groups in the eighties. The purpose of that extension was to generalize tools from harmonic analysis to objects like deformations of Lie groups, but since it uses the setting of operator algebras, it also provides a useful notion of quantum symmetry in the context of non-commutative geometry. As a consequence, once the foundations of the theory were laid, it was important to find examples. One line of thought, pursued by Sh. Wang, was to construct compact quantum groups as "quantum symmetries" of spaces (see Section 3 for more on this). In particular, he discovered in [**72**] that a space with N points has a compact quantum group of automorphisms, called the *quantum permutation group* on N points and denoted by S_N^+.

As we have explained in the beginning of this introduction, permutation groups are ubiquitous in mathematics. Therefore, it is natural to try to see whether the quantum permutation groups enjoy a similar status. The purpose of the present survey is to try to convince the reader that S_N^+ is indeed a central object in mathematics, or at least in several areas of it.

Fifteen years ago, a survey [**5**] on quantum permutation groups was written by T. Banica, J. Bichon and B. Collins. Taking over where they stopped, this leaves us with fifteen years of works to describe, which is of course too much. We therefore have to make choices, and we will focus on a few topics which seem especially relevant to us and adapted to our goal of showing how quantum permutation groups have a role parallel to that of classical permutations. Let us also mention that recently, M. Weber wrote a survey [**73**] on quantum permutation matrices. Though very close in spirit, quantum permutation matrices are more general objects than quantum permutation groups (see Section 2.7 for explanations) and their study can lead to many questions of different nature, which are explained in [**73**]. Therefore, despite some overlap, we highly recommend it as a companion reading.

We have decided to cover three main subjects. The first one is the connection between quantum permutation groups and quantum information theory. This is based on recent results which made a remarkable contact between quantum groups and genuine quantum physics, which by the way justifies in a sense the use of the adjective "quantum". Besides giving physical motivation for the study of compact quantum groups, the results of [**51**] which we will describe below also triggered many new question concerning quantum automorphism groups of graphs, some of which we will survey. This also raises many interesting problems and the subject is evolving rapidly.

The second subject is quantum symmetry. We will first explain the result of Sh. Wang motivating the original definition of S_N^+ and then move to a more general problem: what are the classical spaces on which quantum permutations can act non-trivially ? For classical permutation, these are just finite spaces. As we will see, the situation in the quantum world is both more complicated and more rigid in a sense.

Eventually, the third subject that we treat is non-commutative probability theory. There, one very striking result is the analogue of the de Finetti theorem for free probability proven by C. Köstler and R. Speicher in [**45**], which once again gives convincing evidence that S_N^+ plays the role of S_N in non-commutative mathematics. After explaining this in some detail, we will consider the problem of random walks and for the sake of clarity, we will focus on one specific example involving random transpositions.

As already mentioned, we had to make choices, and consequently some important works on quantum permutations will not be explained below. Among them is the connection to the theory of partitions. This was in fact already known since the founding work of T. Banica [**3**] and explained in the survey [**5**], but was only made systematic in the seminal paper [**8**] by T. Banica and R. Speicher. That connection proved useful for instance for the study of the algebraic structure of quantum permutation groups, like in [**21**] where it is proven that these are residually finite. Another topic which is absent is that of cohomology. There are several cohomology theories that one may study for compact quantum groups. One of them is Hochschild cohomology with trivial coefficients, which was computed by J. Bichon, U. Franz and M. Gerhold in [**14**] (see also [**12**] for the computation of the cohomological dimension). Another one is ℓ^2-cohomology, and the corresponding ℓ^2-Betti numbers were computed by D. Kyed, S. Raum, S. Vaes and M. Valvekens in [**47**]. One may also compute the cohomology of a given representation in terms of Hilbert space valued cocycles. This is particularly important in connection to the classification of non-commutative Lévy processes. Several results in that direction were obtained by U. Franz, A. Kula and A. Skalski in [**29**].

2. On graphs and games

One of the most important results concerning quantum permutation groups in the recent years is a surprising connection with quantum information theory discovered in [**51**]. A nice feature of that connection is that on the one hand, it enabled to solve some problem from quantum information theory (see Sections 2.1 and 2.4 below), and on the other hand it solved several open questions concerning quantum groups (see Section 2.3 for some examples). On top of that, these results provide a very nice and rather natural way to motivate the very definition of quantum permutation groups. We will therefore follow that path hereafter.

2.1. The graph isomorphism game. The connection alluded to above is based on a "game". As always in quantum information theory, the game is played by two players named *Alice* (denoted by A) and *Bob* (denoted by B). In this so-called *graph isomorphism game*, they cooperate to win against the *Referee* (denoted by R) leading the game. The rules are given by two finite graphs[1] X and Y with vertex sets having the same cardinality N, and both are known to A and B. At each round of the game, R sends a vertex $v_A \in X$ to A and a vertex $v_B \in X$ to B. Each of them answers with a vertex $w_A \in Y$, $w_B \in Y$ of the other graph and they win the round if the following conditions are matched:

- If $v_A = v_B$ then $w_A = w_B$;
- If v_A and v_B are distinct and connected by an edge, then w_A and w_B are distinct and connected by an edge;
- If v_A and v_B are distinct and not connected by an edge, then w_A and w_B are distinct and not connected by an edge.

[1]For simplicity, by "graph" we will always mean a non-oriented graph without loops and multiple edges, even though these restrictions are not always necessary for the results presented below to hold true.

REMARK 2.1. This is not the most general version of the graph isomorphism game, but will be sufficient for our purpose. We refer the reader to [**1**] for a more comprehensive exposition. Note that we assume here that X and Y have the same number of vertices, so that the game is what is called *bisynchronous.*

Now, the crucial point is that once the game starts, A and B **cannot communicate in any way** (this is usually referred to as the *non-signalling condition*). The situation can be summarized thanks to the following picture:

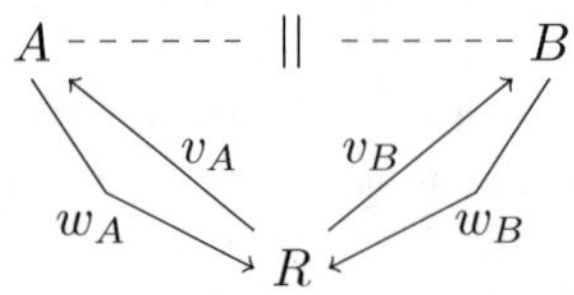

The question one asks is then: under which condition on the graphs X and Y can the players devise a strategy which wins whatever the given vertices are ? It is not very difficult to see that the answer is the following,

PROPOSITION 2.2. *There exists a perfect strategy if and only if X and Y are isomorphic. Moreover, perfect strategies are then in one-to-one correspondence with isomorphisms between X and Y.*

PROOF. A strategy is given by two functions

$$f_A, f_B : V(X) \to V(Y)$$

giving, for a vertex x, the answers of A and B respectively. Because of the first winning condition, the answers must be the same if the questions are the same, hence $f_A(x) = f_B(x)$ for all $x \in V(X)$, i.e. $f_A = f_B$. Moreover, if x and x' are distinct vertices connected by an edge, then so must be $f_A(x)$ and $f_B(x') = f_A(x')$. Distinctness means that f_A is injective, hence bijective, and connection by an edge means that f_A is a graph homomorphism, concluding the proof. □

REMARK 2.3. One may try to improve the classical setting by including some (shared) randomness. Even though this may have an effect on the performance of the strategy when the graphs X and Y are not isomorphic, it has no influence on the existence of a perfect classical strategy.

This settles the problem in classical information theory. In the quantum world however, A and B can refine their strategy without communicating through the use of *entanglement.* This means that they can set up a quantum mechanical system and then split it into two parts, such that manipulating one part instantly modifies the other one. We will not go into the details, but it turns out that this gives more strategies, which are said to be *quantum*[2]. Using these quantum strategies, the previous proposition can be improved. Before giving a precise statement, let us fix some notations:

- Given a Hilbert space H, we denote by $\mathcal{B}(H)$ the bounded (i.e. continuous) linear maps on H;

[2]The concept of quantum strategy turns out to be subtle, depending on the type of operators allowed. We here use the term in a purposely vague sense.

- Given a graph X on N vertices which are numbered from 1 to N, we denote by A_X the adjacency matrix of A, that is, the matrix with (i,j)-th coefficient equal to 1 if the vertices i and j are connected and to 0 otherwise.

The first characterization of perfect quantum strategies was given in [**1**, Thm 5.8]. To state it, let us first recall that a *C*-algebra* is a closed subalgebra (for the operator norm) of $\mathcal{B}(H)$ which is stable under taking adjoints. If A is a unital C*-algebra, a linear form $\tau : A \to \mathbf{C}$ is said to be

- A state if $\tau(x) \geqslant 0$ for all positive operators $x \in A \subset \mathcal{B}(H)$ and $\tau(1) = 1$;
- Tracial if $\tau(xy) = \tau(yx)$ for all $x, y \in A$.

We can now state the first important result of this text, which is due to Atserias-Mančinska-Roberson-Šamal-Severini-Varvitsiotis:

THEOREM 2.4. *Perfect quantum strategy are in one-to-one correspondence with matrices $P = (p_{ij})_{1\leqslant i,j\leqslant N}$ with coefficients in $\mathcal{B}(H)$ for some Hilbert space H, such that:*

(1) *p_{ij} is an orthogonal projection for all $1 \leqslant i,j \leqslant N$;*
(2) $\displaystyle\sum_{k=1}^{N} p_{kj} = \mathrm{Id}_H = \sum_{k=1}^{N} p_{ik}$ *for all $1 \leqslant i,j \leqslant N$;*
(3) $A_X P = P A_Y$;
(4) *The C*-algebra generated by the operators p_{ij} for $1 \leqslant i,j \leqslant N$ in $\mathcal{B}(H)$ has a tracial state.*

REMARK 2.5. From the perspective of quantum physics, the characterization above is reasonable. Indeed, a family of orthogonal projections summing up to one is given by a *measurement system*, i.e. a complete system of pairwise commuting observables. In this context, the projections come from the eigenspaces of the observables. We are therefore considering a collection of measurement systems with compatibility conditions coming from the graphs.

REMARK 2.6. It is not straightforward to produce a pair of graphs for which there is a perfect quantum strategy but no classical one. The first example, given for instance in [**1**, Fig 1 and 2], has 24 vertices and may be the smallest possible one. Even though explicit examples are hard to find (see [**65**] for other examples), it is quite natural that they exist. Indeed, graph isomorphism is a decidable problem, while it can be shown that quantum graph isomorphism is an undecidable problem (see [**1**]).

An intriguing point of Theorem 2.4 is the operator-valued matrices which appear in the statement. To understand them, let us consider the case $H = \mathbf{C}$. Then, the coefficients are scalars and since they are projections, they all equal either 0 or 1. Moreover, the sum on any row is 1, hence there is exactly one non-zero coefficient on each row. The same being true for columns, we have a permutation matrix! We should therefore think of the operator-valued matrices as quantum versions of permutations and this leads to the following definition:

DEFINITION 2.7. Let H be a Hilbert space. A *quantum permutation matrix* in H is a matrix $P = (p_{ij})_{1\leqslant i,j\leqslant N}$ with coefficients in $\mathcal{B}(H)$ such that:

- p_{ij} is an orthogonal projection for all $1 \leqslant i,j \leqslant N$;

- $\displaystyle\sum_{k=1}^{N} p_{kj} = \mathrm{Id}_H = \sum_{k=1}^{N} p_{ik}$ for all $1 \leqslant i, j \leqslant N$.

REMARK 2.8. Even though we will focus here on a particular point of view on quantum permutation matrices, which is that of quantum groups, these objects are more general and have connections to many interesting problems. We refer the reader to [**73**] for a survey of the subject.

Moreover, with this point of view, the last point of Theorem 2.4 has a nice interpretation. Indeed, if σ is a permutation and the corresponding permutation matrix P_σ satisfies

$$A_X P_\sigma = P_\sigma A_Y,$$

then this means that the bijection induced by σ between the vertices of X and those of Y respects the edges. In other words, it is a graph isomorphism. Therefore, if the conditions of Theorem 2.4 are matched, one says that the graphs are *quantum isomorphic.*

The last condition in Theorem 2.4 is of a slightly different nature. The tracial state alluded to can be thought as a map of the form

$$x \mapsto \langle x(\psi), \psi\rangle,$$

where $\psi \in H$ is the state prepared in advance by Alice and Bob. However, the traciality condition means that ψ must have a very specific form with respect to the operators p_{ij}. In that sense, the condition seems genuinely restrictive. However, one of the most remarkable results of [**51**] is that it is in fact superfluous!

THEOREM 2.9 (Lupini-Mančinska-Roberson). *There exists a perfect quantum strategy if and only if there exists a quantum permutation matrix P such that*

$$A_X P = P A_Y.$$

REMARK 2.10. The statement of Theorem 2.9 is different from that of Theorem 2.4. Indeed, the former does not yield a one-to-one correspondence, since there exist quantum permutation matrices such that the C*-algebra generated by the entries does not admit any tracial state. What the theorem claims is that if a quantum permutation matrix intertwining the adjacency matrices exist, then there also exist another one with the tracial state property.

One interesting point for us is that the proof of Theorem 2.9 involves compact quantum groups. But since we have not introduced these objects yet, we will postpone the explanations to Section 2.4.

2.2. The quantum permutation group. The brief discussion of Section 2.1 suggests that quantum permutation matrices are interesting objects which require further study. However, their definition lacks several important features of classical permutation matrices. In particular, there is no obvious way to "compose" quantum permutation matrices, especially if they do not act on the same Hilbert space, so that one could recover an analogue of the group structure of permutations. To overcome this problem, it is quite natural from an operator algebraic point of view to introduce a universal object associated to quantum permutation matrices. And because we are considering bounded operators on Hilbert spaces, C*-algebras are a natural choice.

DEFINITION 2.11. Let $C(S_N^+)$ be the universal C*-algebra (see for instance [**15**, Sec II.8.3] for that notion) generated by N^2 elements $(p_{ij})_{1\leqslant i,j\leqslant N}$ such that

(1) $p_{ij}^2 = p_{ij} = p_{ij}^*$;

(2) For all $1 \leqslant i,j \leqslant N$, $\displaystyle\sum_{k=1}^{N} p_{ik} = 1 = \sum_{k=1}^{N} p_{kj}$.

This will be called the *quantum permutation algebra* on N points.

The fancy notation $C(S_N^+)$ should not scare the reader: it will make sense once more theory is developed, see Example 2.16. We now have a nice object to study, but the link to the classical permutation group is somewhat blurred. To clear it, let us consider the function $c_{ij} : S_N \to \mathbf{C}$ defined by

$$c_{ij}(\sigma) = \delta_{\sigma(i)j}.$$

This is nothing but the function sending the permutation matrix of σ to its (i,j)-th coefficient. By universality, there is a $*$-homomorphism

$$\pi_{\mathrm{ab}} : \left\{ \begin{array}{ccc} C(S_N^+) & \to & \mathcal{F}(S_N) \\ p_{ij} & \mapsto & c_{ij} \end{array} \right.$$

where $\mathcal{F}(S_N)$ is the algebra of all functions from S_N to $\mathbf{C}$. Moreover, since the functions c_{ij} obviously generate the whole algebra $\mathcal{F}(S_N)$, π_{ab} is onto.

REMARK 2.12. It turns out that for $N \leqslant 3$, the map π_{ab} is injective. This results from elementary calculations showing that the defining relations of $C(S_N^+)$ force the generators to pairwise commute. For $N \geqslant 4$ however, $C(S_N^+)$ is non-commutative. Indeed, consider two non-commuting projections p and q in a Hilbert space H. Then, the matrix

$$\begin{pmatrix} p & 1-p & 0 & 0 \\ 1-p & p & 0 & 0 \\ 0 & 0 & q & 1-q \\ 0 & 0 & 1-q & q \end{pmatrix}$$

provides a quantum permutation matrix (which can be extended to larger values of N by adding 0's). One can even refine this example to show that $C(S_N^+)$ is infinite-dimensional as soon as $N \geqslant 4$.

We will now use this link to investigate a possible "group-like" structure on $C(S_N^+)$. At the level of the coefficient functions, the group product satisfies the following equation:

$$c_{ij}(\sigma_1\sigma_2) = \sum_{k=1}^{N} c_{ik}(\sigma_1)c_{kj}(\sigma_2) = \left(\sum_{k=1}^{N} c_{ik} \otimes c_{kj}\right)(\sigma_1, \sigma_2).$$

Here, we have used the fact that the injective map (the symbol $\otimes$ below denotes the tensor products of unital algebras over $\mathbf{C}$)

$$\imath : \left\{ \begin{array}{ccc} \mathcal{F}(S_N) \otimes \mathcal{F}(S_N) & \to & \mathcal{F}(S_N \times S_N) \\ f \otimes g & \mapsto & ((\sigma_1, \sigma_2) \mapsto f(\sigma_1)g(\sigma_2)) \end{array} \right.$$

is onto for dimension reasons. Considering the elements p_{ij} as "coefficient functions", this suggests to encode a kind of "group law" through the map

$$\Delta : p_{ij} \to \sum_{k=1}^{N} p_{ik} \otimes p_{kj}. \tag{1}$$

The previous equation raises two questions. The first one is the meaning of the $\otimes$ symbol since $C(S_N^+)$ is not finite-dimensional. Indeed, there exist several ways of completing the algebraic tensor product of two C*-algebras into a C*-algebra. In this text, we will consider the so-called *minimal* tensor product without specifying it and refer the reader to [**15**, Chap II.9] for details. The second question is whether such a map Δ indeed exists ...

PROPOSITION 2.13. *There exists a unique $*$-homomorphism*

$$\Delta : C(S_N^+) \to C(S_N^+) \otimes C(S_N^+)$$

satisfying Equation (1).

PROOF. Let us set, for $1 \leqslant i, j \leqslant N$,

$$q_{ij} = \sum_{k=1}^{N} p_{ik} \otimes p_{kj}.$$

It is easily checked by direct computation that the q_{ij}'s satisfy Conditions (1) and (2) of Definition 2.11. The existence of Δ then follows from the universal property. $\square$

The map Δ is called the *coproduct* and is a reasonable substitute for the group law. In particular, it satisfies an analogue of the associativity property, called *coassociativity* (which it is enough to prove on the generators, where it is obvious):

$$(\Delta \otimes \mathrm{id}) \circ \Delta = (\mathrm{id} \otimes \Delta) \circ \Delta.$$

The coproduct certainly indicates that we are on the right track to product a "group-like" structure on the quantum permutation algebra. However, we still need a neutral element and an inverse, but instead of trying to translate each of them, we will take advantage of the fact that we are considering a matrix group. Indeed, for any permutation, the corresponding matrix is orthogonal:

$$\sum_{k=1}^{N} c_{ik}(\sigma)c_{jk}(\sigma) = \delta_{ij} = \sum_{k=1}^{N} c_{ki}(\sigma)c_{kj}(\sigma).$$

Since this holds for any σ, it can be written as an equality of functions in $\mathcal{F}(S_N)$, and it turns out that the same equality holds in $C(S_N^+)$:

PROPOSITION 2.14. *For any $1 \leqslant i, j \leqslant N$,*

$$\sum_{k=1}^{N} p_{ik}p_{jk} = \delta_{ij} = \sum_{k=1}^{N} p_{ki}p_{kj}. \tag{2}$$

PROOF. This is a direct consequence of Conditions (1) and (2) once one notices that in a C*-algebra, if a family of projections adds up to a projection, then they must be pairwise orthogonal. Indeed, this implies that $p_{ij}p_{ik} = \delta_{ik}p_{ij}$ and $p_{ij}p_{kj} = \delta_{ik}p_{ij}$ for all $1 \leqslant i, j, k \leqslant N$. $\square$

This means that the quantum permutation algebra is somehow "made of orthogonal matrices" and that property should contain all the information about the unit and the inverse. The work of S.L. Woronowicz has shown in [**74**] and [**76**] that this is indeed enough to recover all the main features of the theory of compact groups of matrices, hence the name "compact matrix quantum group" given to these objects.

DEFINITION 2.15. An *orthogonal compact matrix quantum group* is a pair $(C(\mathbb{G}), U)$ where $C(\mathbb{G})$ is a unital C*-algebra and $U \in M_N(C(\mathbb{G}))$ has coefficients $(u_{ij})_{1\leqslant i,j\leqslant N}$ satisfying

- The elements $(u_{ij})_{1\leqslant i,j\leqslant N}$generate $C(\mathbb{G})$ as a C*-algebra;
- There exists a $*$-homomorphism $\Delta : C(\mathbb{G}) \to C(\mathbb{G}) \otimes C(\mathbb{G})$ such that $\Delta(u_{ij}) = \sum_{k=1}^{N} u_{ik} \otimes u_{kj}$ for all $1 \leqslant i, j \leqslant N$;
- U is orthogonal in the sense that $u_{ij}^* = u_{ij}$ and
$$U^tU = \mathrm{id}_{M_N(C(\mathbb{G}))} = {}^tUU.$$

For the sake of simplicity, we will from now on simply write "compact quantum group" instead "orthogonal compact matrix quantum group". We will not recall the whole theory of these objects but only explain the features which are directly relevant to our exposition. The reader can find comprehensive treatments in [**69**] and [**55**].

EXAMPLE 2.16. Let O_N denote the group of all orthogonal matrices in $M_N(\mathbf{C})$ and let $G \subset O_N$ be a closed subgroup. It is then compact, and as a consequence the algebra $C(G)$ of all continuous complex-valued functions on G is a C*-algebra for the uniform norm. Denote by $c_{ij} : G \to \mathbf{C}$ the function sending an element of G to its (i, j)-th coefficient. Then, the matrix

$$U = (c_{ij})_{1\leqslant i,j\leqslant N} \in M_N(C(G))$$

is orthogonal because all the matrices in G are. Moreover, recall that for a compact Hausdorff space X, the map sending $f \otimes g \in C(X) \otimes C(X)$ to the map $(x, y) \mapsto f(x)g(y)$ is a $*$-isomorphism onto $C(X \times X)$. Thus, seeing the matrix product as a continuous map $m : G \times G \to G$ and denoting by $\Delta : C(G) \to C(G \times G)$ the $*$-homomorphism given by $f \mapsto f \circ m$, we have

$$\begin{aligned}\Delta(c_{ij}) : (g, h) &\mapsto c_{ij}(gh)\\ &= \sum_{k=1}^{N} c_{ik}(g)c_{kj}(h)\\ &= \sum_{k=1}^{N} (c_{ik} \otimes c_{kj})(g, h).\end{aligned}$$

In other words, $(C(G), \Delta)$ is a compact quantum group. Note that this example covers in particular all finite groups and all compact Lie groups.

Conversely, if $\mathbb{G} = (C(\mathbb{G}), U)$ is a compact quantum group and $C(\mathbb{G})$ is a commutative C*-algebra, then there exists a compact group $G \subset O_N$ and an isomorphism $C(\mathbb{G}) \simeq C(G)$ sending U to the matrix of coefficient functions (this follows from the Gelfand-Naimark theorem, see for instance [**15**, Thm II.2.2.4]).

EXAMPLE 2.17. As we have discussed before, if U denotes the matrix with coefficients p_{ij} for $1 \leqslant i, j \leqslant N$, then $(C(S_N^+), U)$ is a compact quantum group called the *quantum permutation group* on N points and denoted by S_N^+. Recall the surjective $*$-homomorphism $\pi_{\mathrm{ab}} : C(S_N^+) \to \mathcal{F}(S_N) = C(S_N)$ introduced earlier. It is straightforward on the generators that

$$\Delta \circ \pi_{\mathrm{ab}} = (\pi_{\mathrm{ab}} \otimes \pi_{\mathrm{ab}}) \circ \Delta,$$

so that S_N is in a sense a "quantum subgroup" of S_N^+.

2.3. Quantum automorphism groups of graphs. If one wants to investigate further the connection between S_N^+ and the graph isomorphism game, it is natural to try to incorporate the adjacency matrix of a graph in Definition 2.11 in order to build an object which somehow reflects "quantum" properties of the graph. That idea leads to the next definition.

DEFINITION 2.18. Let X be a finite graph and let $C(\mathrm{Aut}^+(X))$ be the quotient of $C(S_N^+)$ by the closed $*$-ideal generated by the relations $PA_X = A_XP$. If q_{ij} denotes the image of p_{ij} in this quotient, then $(C(\mathrm{Aut}^+(X)), (q_{ij})_{1\leqslant i,j\leqslant N})$ is again a compact quantum group, called the *quantum automorphism group* of X and denoted by $\mathrm{Aut}^+(X)$.

REMARK 2.19. Definition 2.18 implicitly asserts the existence of a coproduct Δ satisfying the properties of Definition 2.15. Proving this amounts to proving that the ideal

$$I = \ker\left(C(S_N^+) \to C(\mathrm{Aut}^+(X))\right)$$

satisfies the equation

$$\Delta(I) \subset I \otimes C(S_N^+) + C(S_N^+) \otimes I$$

(here Δ is the coproduct of S_N^+). This in turn can be checked on a suitable set of generators of I, namely the coefficients of the matrix $A_XU - UA_X$.

These objects were originally defined by T. Banica in [**4**] (modifying a construction by J. Bichon from [**11**]) to produce new examples of compact quantum groups. However, it turns out that deciding whether $\mathrm{Aut}^+(X)$ is classical or not (that is to say, whether the C*-algebra $C(\mathrm{Aut}^+(X))$ is commutative or not, according to Example 2.16) is difficult in general. We will illustrate this here with a few recent results which stress the complexity of the problem.

One approach is through probabilistic arguments. Indeed, consider for each integer N the set $\mathcal{G}_N$ of all graphs on N vertices. This is a finite set with cardinality $|\mathcal{G}_N|$. Given a property $\mathcal{P}$ which makes sense for a graph, we will say that $\mathcal{P}$ is satisfied *almost surely* if

$$\frac{|\{X \in \mathcal{G}_N \mid X \text{ satisfies } \mathcal{P}\}|}{|\mathcal{G}_N|} \underset{N\to\infty}{\longrightarrow} 1.$$

To give an example of how this concept can be used to investigate the problem, let us introduce a tool from algebraic graph theory.

DEFINITION 2.20. A *coherent algebra* is a subalgebra of $M_N(\mathbf{C})$ which is stable under entry-wise product, taking adjoints, and contains the all-one matrix J_N. If X is a graph on N vertices, its coherent algebra is the intersection (observe that an intersection of coherent algebras is again a coherent algebra) $\mathrm{Coh}(X)$ of all the coherent algebras containing the adjacency matrix A_X.

In the following proof, we will denote by $A \odot B$ the entry-wise product of the matrices A and B.

THEOREM 2.21 (Lupini-Mančinska-Roberson). *The property of having a classical quantum automorphism group is satisfied almost surely.*

PROOF. The key idea is to prove that if $Q = (q_{ij})_{1\leqslant i,j\leqslant N}$ is the matrix of generators of $C(\mathrm{Aut}^+(X))$ as in Definition 2.18, then the set

$$\mathrm{Mor}(Q,Q) = \{T \in M_N(\mathbf{C}) \mid TQ = QT\},$$

which is clearly a subalgebra stable under taking adjoints, is also stable under entry-wise product. Indeed, if $T_1, T_2 \in \mathrm{Mor}(Q,Q)$, then for any $1 \leqslant i,j \leqslant N$,

$$\begin{aligned}
[(T_1 \odot T_2)Q]_{ij} &= \sum_{k=1}^{N} T_{1,ik}T_{2,ik}Q_{kj} \\
&= \sum_{k,k'=1}^{N} T_{1,ik}T_{2,ik'}\delta_{kk'}Q_{kj} \\
&= \sum_{k,k'=1}^{N} T_{1,ik}T_{2,ik'}Q_{kj}Q_{k'j} \\
&= [T_1Q]_{ij}\,[T_2Q]_{ij} \\
&= [QT_1)_{ij}\,[QT_2]_{ij} \\
&= [Q(T_1 \odot T_2)]_{ij}\,.
\end{aligned}$$

Besides, computing the coefficients of the products QJ_N and J_NQ yields, for $1 \leqslant i,j \leqslant N$,

$$\begin{aligned}
[QJ_N]_{ij} &= \sum_{k=1}^{N} Q_{ik} \\
&= 1 \\
&= \sum_{k=1}^{N} Q_{kj} \\
&= [J_NQ]_{ij}.
\end{aligned}$$

It follows that $\mathrm{Mor}(Q,Q)$ is a coherent algebra, and since by definition $A_X \in \mathrm{Mor}(Q,Q)$, we have

$$\mathrm{Coh}(X) \subset \mathrm{Mor}(Q,Q).$$

One may then use a classical result of in [**2**], stating that the property Coh(X) $= M_N(\mathbf{C})$ is almost surely satisfied, to conclude that $\mathrm{Mor}(Q,Q) = M_N(\mathbf{C})$ is almost surely satisfied. We claim that this is enough to finish the proof.

Indeed, let E_{ii} be the matrix with coefficient 1 in position (i,i) and 0 elsewhere. Then, $E_{ii}Q$ is the matrix whose i-th row is that of Q and all other rows are 0, while QE_{ii} is the matrix with i-th column being that of Q and all other ones being 0. As a consequence, if $E_{ii}Q = QE_{ii}$, then Q_{ii} is the only non-zero coefficient in the i-th row and the i-th column. Applying this for all $1 \leqslant i \leqslant N$, we conclude that if $\mathrm{Mor}(Q,Q) = M_N(\mathbf{C})$, then Q is diagonal. But Condition (2) in Definition 2.11 then implies that the diagonal coefficients are equal to 1 and as a corollary, $\dim(C(\mathrm{Aut}^+(X))) = 1$. □

REMARK 2.22. The reader may be surprised that we only used the fact that the diagonal matrices E_{ii} are in $\mathrm{Mor}(Q,Q)$ to get the result, but it is an easy exercise to check that a coherent algebra containing E_{ii} for all $1 \leqslant i \leqslant N$ contains all the matrices E_{ij} for $1 \leqslant E_{ij} \leqslant N$.

Even though this seems discouraging, it is possible to find graphs with interesting quantum automorphism groups. For instance, the complete graph K_N on N vertices satisfies $\mathrm{Aut}^+(K_N) = S_N^+$, since $A_{K_N} = J_N - I_N$ and we have shown in the proof of Theorem 2.21 that any quantum permutation matrix commutes with J_N, hence with A_{K_N}. More generally, one of the only general criteria to guarentee the existence of quantum automorphisms is the following one given by S. Schmidt in [**64**, Thm 2.2].

THEOREM 2.23 (Schmidt). *Let X be a graph and assume that there exist two non-trivial automorphisms σ and τ which are* disjoint *in the sense that $\sigma(x) \neq x$ implies $\tau(x) = x$ and $\tau(x) \neq x$ implies $\sigma(x) = x$. Then, $C(\mathrm{Aut}^+(X))$ is non-commutative.*

PROOF. Let n be the order of σ and let m be the order of τ, so that we can see these automorphisms as defining actions of $\mathbf{Z}_n$ and $\mathbf{Z}_m$ respectively. The disjointness assumption implies that σ and τ commute, hence we also have an action of $\mathbf{Z}_n \times \mathbf{Z}_m$. The idea of the proof is to see that we have more: there is an "action" (in the sense of compact quantum groups, see Definition 3.1 below) of the free product $\mathbf{Z}_n * \mathbf{Z}_m$ on the graph[3]. Concretely, this amounts to building a surjection

$$C(\mathrm{Aut}^+(X)) \to C^*(\mathbf{Z}_n * \mathbf{Z}_m)$$

a task which can be performed through explicit formulæ for the images of the generators, namely

$$v_{ij} \mapsto \begin{cases} \displaystyle\sum_{k=1}^{n} \delta_{\sigma^k(i)j}\delta_{s^k} & \text{if} \qquad \sigma(i) \neq i \\ \displaystyle\sum_{l=1}^{m} \delta_{\tau^k(i)j}\delta_{t^l} & \text{if} \qquad \tau(i) \neq i \\ \delta_{ij} & \text{otherwise} \end{cases},$$

where s and t are the self-adjoint unitaries generating $\mathbf{Z}_n$ and $\mathbf{Z}_m$ respectively. We then conclude by the non-commutativity of the full C*-algebra $C^*(\mathbf{Z}_n * \mathbf{Z}_m)$. □

With this in hand, it is possible to revisit Theorem 2.21 in a positive way by restricting to the set $\mathcal{T}_N$ of *trees*, that is to say connected graphs without cycles, on N vertices.

THEOREM 2.24 (Junk-Schmidt-Weber). *The property of having a non-classical quantum automorphism group is satisfied almost surely for trees, i.e.*

$$\frac{|\{X \in \mathcal{T}_N \mid C(\mathrm{Aut}^+(X)) \text{ is noncommutative}\}|}{|\mathcal{T}_N|} \underset{N\to\infty}{\longrightarrow} 1.$$

PROOF. The proof is an elegant one based on the notion of a *cherry*, which is a triple (x, y, y') of vertices of a tree such that

- Both y and y' have x as their only neighbour;
- x has exactly one neihgbour besides y and y'.

[3]Strictly speaking it is an action of the quantum Pontryagin dual.

Here is a picture of a cherry,

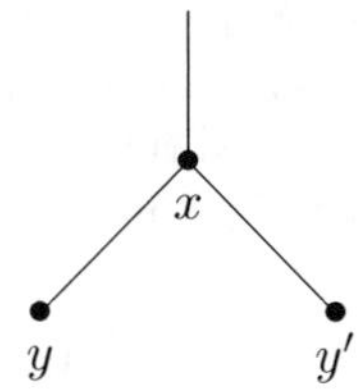

It follows from the definition that the map exchanging y and y' and fixing all the other vertices is an automorphism of the graph. Cherries were first introduced in [**28**], where it is proven that trees almost surely have at least one cherry. As a consequence, having a non-trivial automorphism groups is a property almost surely satisfied among trees. The authors of [**43**] refined this by proving that trees almost surely have *two* cherries. Since two cherries yield two automorphisms satisfying the assumptions of Theorem 2.23, we can then conclude. $\square$

REMARK 2.25. As explained in the proof of Theorem 2.23, having two cherries in fact tells us that $C(\mathrm{Aut}^+(X))$ surjects onto $C^*(\mathbf{Z}_2 * \mathbf{Z}_2)$, so that the proof yields a slightly stronger result: trees almost surely have an infinite-dimensional quantum automorphism group. Improving the result to the existence of three cherries would yield something even stronger since trees would almost surely have a *non-coamenable* quantum automorphism group, a notion which we will not define (see for instance [**10**]).

As already mentioned, there has been much progress recently on the topic of quantum automorphism groups of graphs. For instance, D. Roberson and S. Schmidt gave in [**59**] the first example of a graph whose quantum automorphism group is finite yet not classical. Besides answering an important open question, this shows that the criterion above is not necessary. Let us also mention the work of D. Gromada [**39**] investigating the problem for Cayley graphs of finite abelian groups and several works of S. Schmidt [**63**], [**64**] and [**62**] investigating various families of graphs.

We will end this section with an instance, the first to our knowledge, where compact quantum groups lead to the solution of a problem from graph theory. L. Lovàsz proved in [**50**] the following characterization of graph isomorphism:

THEOREM 2.26 (Lovàsz). *Two graphs X and Y are isomorphic if and only if for any graph Z,*

$$|\mathrm{Hom}(Z, X)| = |\mathrm{Hom}(Z, Y)|.$$

This suggests to investigate the relation of having the same number of homomorphism from a fixed family of graphs. For instance, what if we further impose that Z is *planar* ? This was an open question for several decades, before the answer was given in [**52**].

THEOREM 2.27 (Mančinska-Roberson). *There exists a perfect quantum strategy for the game with X and Y if and only if for any* planar *graph Z,*

$$|\mathrm{Hom}(Z, X)| = |\mathrm{Hom}(Z, Y)|.$$

Even though quantum groups do not appear in the statement, they are crucial in the proof. The idea is that the existence of a perfect quantum strategy is equivalent to the requirement that the corresponding quantum automorphism groups are *monoidally equivalent* (see [**13**] for details on that notion). Monoidal equivalence can be understood if one has a good description of the category of representations of the quantum group as an abstract C*-tensor category (see Section 3.2 below for more on this), and the core result of [**52**] is such a description through matrices whose coefficients are given by counting homomorphisms from planar graphs.

2.4. Back to the game. Before switching to another aspect of quantum permutation groups, we will explain how they enter the proof of Theorem 2.4. Recall that the goal is, from a quantum permutation matrix intertwining the adjacency matrices of the graphs, to produce one such that the C*-algebra generated by its coefficients admits a tracial state. The idea is that if we can arrange the C*-algebra to be contained in $C(\mathbb{G})$ for some compact quantum group $\mathbb{G}$, then the general theory will provide us with a trace for free.

To start with, we will consider the quantum automorphism groups of the disjoint union $X \sqcup Y$ of X and Y. It comes by definition with a quantum permutation matrix Q, which we can write in block-form (with respect to the disjoint union decomposition) as follows:

$$Q = \begin{pmatrix} Q_{11} & Q_{12} \\ Q_{21} & Q_{22} \end{pmatrix}$$

That matrix commutes with the adjacency matrix $A_{X\sqcup Y}$, which is block-diagonal with blocks A_X and A_Y. Since

$$A_{X\sqcup Y}Q = \begin{pmatrix} A_XQ_{11} & A_XQ_{12} \\ A_YQ_{21} & A_YQ_{22} \end{pmatrix} \text{ and } QA_{X\sqcup Y} = \begin{pmatrix} Q_{11}A_X & Q_{12}A_Y \\ Q_{21}A_X & Q_{22}A_Y \end{pmatrix},$$

we see in particular that $A_XQ_{12} = Q_{12}A_Y$. We therefore have a matrix of projections intertwining the adjacency matrices, whose coefficients generate a subalgebra of $C(\mathrm{Aut}^+(X \sqcup Y))$. The point is now that, by a deep result from quantum group theory, there is a tracial state on $C(\mathrm{Aut}^+(X \sqcup Y))$. More precisely, we have the following:

THEOREM 2.28 (Woronowicz). *Let $(\mathbb{G}, U)$ be an orthogonal compact matrix quantum group. Then, there exists a tracial state φ_{Haar} on $C(\mathbb{G})$, which moreover satisfies*

$$(\varphi_{Haar} \otimes \mathrm{id}) \circ \Delta(x) = \varphi_{Haar}(x)1 = (\mathrm{id} \otimes \varphi_{Haar}) \circ \Delta(x)$$

for all $x \in C(\mathbb{G})$.

REMARK 2.29. The equation in Theorem 2.28 is a translation of the invariance property of the Haar measure, hence the name *Haar state* usually given to it. More precisely, if μ_{Haar} denotes the Haar measure of a compact group G, then

$$\int_G f(gh)\mathrm{d}\mu_{\mathrm{Haar}}(h) = \int f(h)\mathrm{d}\mu_{\mathrm{Haar}}(h) = \int f(hg)\mathrm{d}\mu_{\mathrm{Haar}}(h)$$

where $g \in G$ is arbitrary and $f \in C(G)$. Observing that under the isomorphism $C(G) \otimes C(G) \simeq C(G \times G)$, $\Delta(f)$ is the function sending (g, h) to $f(gh)$, we can write the equation above as

$$\left(\mathrm{id} \otimes \int\right) \circ \Delta(f) = \int f = \left(\int \otimes \mathrm{id}\right) \circ \Delta(f).$$

Therefore, a state $\varphi : C(\mathbb{G}) \to \mathbf{C}$ can be called a *Haar state* if it satisfies

$$(\mathrm{id} \otimes \varphi) \circ \Delta(x) = \varphi(x).1 = (\varphi \otimes \mathrm{id}) \circ \Delta(x)$$

for all $x \in C(\mathbb{G})$. It is easy to see that such a state is unique, and its existence is precisely the contents of Theorem 2.28.

But all that we need for our purpose is traciality. Indeed, simply restricting h to the C*-subalgebra of $C(\mathrm{Aut}^+(X \sqcup Y))$ generated by the coefficients of Q_{12} does the job. However, to conclude we still have to prove that Q_{12} is a quantum permutation matrix.

But this is of course hopeless, since one must sum up a line of Q_{11} with a line of Q_{12} to get the identity. There would nevertheless be a way out if at least the sum over each row of Q_{12}, which is a projection, was constant. And the trick is that this is indeed the case! To prove it, we first need to show some elementary relations between the coefficients of Q.

LEMMA 2.30. *Let X and Y be connected graphs, let Q be the fundamental representation of* $\mathrm{Aut}^+(X \sqcup Y)$ *and fix $x, x', x'' \in V(X)$ and $y \in V(Y)$. Then* $q_{xx''}q_{x'y} = 0$.

PROOF. Let us write $a \sim b$ to denote the fact that two vertices of a graph are connected by an edge. The crucial observation is that if $x \sim x'$ but $x'' \nsim y$, then $q_{xx'}q_{x''y} = 0$. To see why this holds, let us get back to the defining relations of $C(\mathrm{Aut}^+(X))$ for a graph X. The (x, y)-th coefficient of the commutation relation $A_X Q = QA_X$ reads

$$\sum_{x \sim z} q_{zy} = \sum_{z \sim y} q_{xz}.$$

Multiplying both sides of the equation by $q_{xx''}$ on the left yields by assumption

$$\sum_{x \sim z} q_{xx''}q_{zy} = 0.$$

Multiplying by $q_{xx''}$ on the right yields a sum of positive operators which equals 0, forcing each operator to be 0. In particular, $q_{xx''}q_{x'y}q_{xx''} = 0$ and since these are projections, it follows that $q_{xx''}q_{x'y} = 0$.

The proof now relies on some elementary algebraic facts. Recall that the k-*distance graph* of a graph X is the graph $X^{(k)}$ where two vertices are connected if and only if their distance in X equals k. It is a fact that the corresponding adjacency matrix $A_{X^{(k)}}$ always belongs to $\mathrm{Coh}(X) \subset \mathrm{Mor}(Q, Q)$. Thus, the first part of the proof gives the relation $q_{xx''}q_{x'y}$ as soon as $d(x, x') \neq d(x'', y)$. Observing that in the disjoint union $X \sqcup Y$, the distance between a vertex of X and a vertex of Y is infinite, while the distance between two vertices of the same graph is finite by connectedness, concludes the proof. □

With this in hand, we can prove our claim that rows and columns of Q_{12} all sum up to the same projection.

PROPOSITION 2.31. *Set, for $x \in V(X)$ and $y \in V(Y)$,*

$$Q_x = \sum_{z \in Y} q_{xz} \text{ and } Q_y = \sum_{z \in X} q_{zy}.$$

Then, $Q_x = Q_y$.

PROOF. This is an elementary computation:

$$Q_x(1 - Q_y) = \sum_{z\in X} \sum_{z'\in X} q_{xz}q_{z'y} = 0,$$

hence Q_x is dominated by Q_y. In the same way, $(1 - Q_x)Q_y = 0$ so that Q_y dominates Q_x and the proof is complete. □

Denoting by H' the range of Q_x, we can see Q_{12} as a quantum permutation matrix with coefficients in $\mathcal{B}(H')$, and the end of the proof of Theorem 2.4 is not very difficult.

2.5. To infinity and beyond. Everything we have done in this section is in a finite setting: we defined quantum permutations of a set of N points, and quantum automorphisms of finite graphs. The reader may want to ask what happens for infinite sets or graphs. The answer is that one has to be cautious, as the classical case shows. Indeed, to build a compact quantum group analogous to the group of permutations of an infinite set, say $\mathbf{N}$, one would first need to choose a topology on the classical group of permutations S_∞. One natural topology is the compact-open one, but unfortunately, this does not yield a compact space . . . and not even a locally compact one!

One way around the problem could be to see S_∞ as the inductive limit of the groups S_N where each one is embedded into the next one as the stabilizer of the $N+1$-th point. In quantum terms, this would mean that we consider maps $\pi_N : C(S_{N+1}) \to C(S_N)$ forming a projective system. An inductive limit of compact groups need not be (locally compact) in general, so that working with C*-algebras does not seem to be a good idea. If one wants to keep some analysis in the background, then we may try using von Neumann algebra. This is what D. Goswami and A. Skalski did in [**37**], where they construct a Hopf-von Neumann algebra as a projective limit, thereby giving a candidate for the set of permutations of a countable infinite set. Note however that, even though this constructions can reasonably be called a quantum semi-group of permutations, it lacks some features that one would expect from a quantum group.

Nevertheless, we lack topology in the previous construction. Very recently, C. Voigt gave in [**71**] another contruction which has the merit of providing a compact quantum group. The caveat is that its topology is not as natural as one may expect. The idea is that we can endow S_∞ with the discrete topology. Then, using a categorical construction, one may quantize that object to produce a discrete quantum group $\widehat{S}_\infty^+$. Taking the Pontryagin dual[4] provides a compact quantum group S_∞^+ which is in a sense the quantum permutation group of an infinite countable set.

The same idea can be applied to automorphism groups of infinite graphs, but the resulting objects are in general "very big" (i.e. not separable), which contrasts with the classical case. Indeed, if X is a graph which is locally finite (in the sense that each vertex as finitely many neighbours), then its automorphism group $\mathrm{Aut}(X)$ is locally compact when endowed with the compact-open topology. Therefore, one would like to be able to build in that case a locally compact quantum group in the sense of [**46**] which generalizes $\mathrm{Aut}(X)$. Very recently, L. Rollier and S. Vaes gave in [**60**] a construction doing this.

[4]There is a Pontryagin duality in the world of quantum groups relating compact and discrete objects, but elaborating on it would take us too far away from our subject.

3. Taking action

Quantum groups, like groups, act on spaces and the quantum permutation groups are no exception. In fact, the original definition by Sh. Wang in **[72]** was motivated by the notion of quantum symmetry, thereby suggesting new dynamical problems which recently led to interesting "quantum rigidity" results. We will now explain this, thereby giving another illustration of the importance of quantum permutation groups in several areas of operator algebras, as well as how it acts as a quantum version of S_N. But to start with, we need to introduce a bit of terminology.

3.1. Actions of compact quantum groups. Quantum groups act on topological spaces, but more generally on C*-algebras. For the sake of simplicity, we will focus here on actions on unital C*-algebras, since this is the only kind of example that we will study in detail.

DEFINITION 3.1. A *continuous right action* (hereafter abbreviated into "action") of a compact quantum group $\mathbb{G}$ on a unital C*-algebra A is a $*$-homomorphism

$$\alpha : A \to A \otimes C(\mathbb{G})$$

satisfying the equations

$$(\alpha \otimes \mathrm{id}) \circ \alpha = (\mathrm{id} \otimes \Delta) \circ \alpha \tag{3}$$

and

$$\overline{\mathrm{Span}}\, \{(1 \otimes C(\mathbb{G}))\alpha(A)\} = A \otimes C(\mathbb{G}). \tag{4}$$

Let us briefly explain the meaning of these two equations. The first one translates the fact that we have a "quantum semigroup" action, that is to say compatibility of the action map with the group operation (here embodied by the coproduct). As for the second one, it can in some cases be replaced by a simpler one, namely

$$(\mathrm{id} \otimes \varepsilon) \circ \alpha = \mathrm{id}_A,$$

where $\varepsilon : C(\mathbb{G}) \to \mathbf{C}$ is the unique $*$-homomorphism such that $\varepsilon(u_{ij}) = \delta_{ij}$ for all $1 \leqslant i, j \leqslant N$. It is then analogous to the fact that the neutral element acts trivially. However, ε need not be well-defined on $C(\mathbb{G})$ in general, which is why we need this slightly different condition. Example 3.3 below will give an idea of the connection of that property with the invertibility of the matrix U.

REMARK 3.2. The continuity of the action is not given by a specific axiom since it is contained in the existence of the $*$-homomorphism α itself. The reader is invited to check that if X is a compact space and $A = C(X)$, then the notion above coincides for classical groups with that of a continuous right action. Moreover, there is nothing specific about the action being on the right, the definition of left actions is similar.

EXAMPLE 3.3. If $(\mathbb{G}, U)$ is a compact quantum group, then $\Delta : C(\mathbb{G}) \to C(\mathbb{G}) \otimes C(\mathbb{G})$ is an action: indeed, Equation (3) is just the coassociativity property of the coproduct, while Condition (4) comes from the orthogonality of U. Let us explain that last claim. We have for any $1 \leqslant i, j \leqslant N$, we have

$$\begin{aligned}\sum_{k=1}^{N}(1 \otimes u_{jk})\Delta(u_{ik}) &= \sum_{k=1}^{N}\sum_{l=1}^{N} u_{il} \otimes u_{jk}u_{lk} \\ &= \sum_{l=1}^{N} u_{il} \otimes \sum_{k=1}^{N} u_{jk}u_{lk} \\ &= \sum_{l=1}^{N} u_{il} \otimes \delta_{jl}1 \\ &= u_{ij} \otimes 1.\end{aligned}$$

This means that $u_{ij} \otimes 1 \subset (1 \otimes C(\mathbb{G}))\Delta(C(\mathbb{G}))$. Let us set

$$E = \{x \in C(\mathbb{G}) \mid x \otimes 1 \in \overline{\mathrm{Span}}\{(1 \otimes C(\mathbb{G}))\Delta(C(\mathbb{G}))\}\}.$$

If $x \otimes 1 = \sum(1 \otimes a_i)\Delta(a'_i)$ and $y \otimes 1 = \sum(1 \otimes b_j)\Delta(b'_j)$, then

$$\begin{aligned}xy \otimes 1 &= (x \otimes 1)\sum(1 \otimes b_j)\Delta(b'_j) \\ &= \sum(1 \otimes b_j)(x \otimes 1)\Delta(b'_j) \\ &= \sum(1 \otimes b_j)\sum(1 \otimes a_i)\Delta(a'_i)\Delta(b'_j) \\ &= \sum(1 \otimes b_j a_i)\Delta(a'_i b'_j) \\ &\in E\end{aligned}$$

so that E is an algebra and, because it is closed and stable under the involution by construction, a C*-algebra. Since $u_{ij} \in E$ for all $1 \leqslant i, j \leqslant N$, we have proven that $C(\mathbb{G}) \otimes 1 \subset \overline{\mathrm{Span}}\{(1 \otimes C(\mathbb{G}))\Delta(C(\mathbb{G}))\}$. Obviously $1 \otimes C(\mathbb{G})$ is also contained in this, hence the result.

EXAMPLE 3.4. The quantum permutation group naturally acts on $\mathbf{C}^N$. Indeed, denoting by $(e_i)_{1 \leqslant i \leqslant N}$ the canonical basis of $\mathbf{C}^N$, we can set

$$\alpha_N(e_i) = \sum_{i=1}^{N} e_j \otimes p_{ji}.$$

Condition (3) is satisfied by definition of $C(S_N^+)$ so that we just have to check Condition (4). To do this, let us compute, for $1 \leqslant i, j \leqslant N$,

$$(1 \otimes p_{ji})\alpha(e_i) = \sum_{k=1}^{N} e_k \otimes p_{ki}p_{ji} = e_j \otimes p_{ji}.$$

Summing up these equalities over i then yields

$$e_j \otimes 1 = \sum_{i=1}^{N} e_j \otimes p_{ji} = \sum_{i=1}^{N}(1 \otimes p_{ji})\alpha(e_i) \in \mathrm{span}\{(1 \otimes C(S_N^+))\alpha(\mathbf{C}^N)\}$$

and the space on the right therefore contains $\mathbf{C}^N \otimes 1$. Since it also contains by definition $1 \otimes C(S_N^+)$, the result is proven.

The action of S_N^+ on $\mathbf{C}^N$ is very natural, and in fact it is even universal among such actions of compact quantum groups. The next result, which is the original motivation for the definition of S_N^+, makes this precise. However, we must first introduce another definition. Indeed, observe that any compact (quantum) group can act on any C*-algebra through the *trivial action*

$$\alpha_{\text{triv}} : x \mapsto x \otimes 1.$$

Therefore, to consider that a quantum group embodies genuine quantum symmetries of a space, one should rule out this kind of example. Here is a way of doing so:

DEFINITION 3.5. An action α of a compact quantum group $\mathbb{G}$ on a C*-algebra A is said to be *faithful* if there is no C*-subalgebra $C(\mathbb{H}) \subset C(\mathbb{G})$ such that

$$\Delta(C(\mathbb{H})) \subset C(\mathbb{H}) \otimes C(\mathbb{H})$$

and

$$\alpha(A) \subset A \otimes C(\mathbb{H})$$

except for $C(\mathbb{G})$.

REMARK 3.6. Assuming G to be classical in the definition of faithfulness, it follows that $C(\mathbb{H})$ is the algebra of functions on a quotient group of G. The condition therefore means that the action does not factor through any quotient, which is another way of saying that it is faithful.

Here is now [**72**, Thm 3.1]:

THEOREM 3.7 (Wang). *Let* $\mathbb{G}$ *be a compact quantum group and let* α *be a* faithful *action of* $\mathbb{G}$ *on* $\mathbf{C}^N$. *Then, there is a surjective* $*$*-homomorphism* $\pi : C(S_N^+) \to C(\mathbb{G})$ *such that*

$$\alpha = (\mathrm{id} \otimes \pi) \circ \alpha_N.$$

PROOF. By definition, there exists elements $v_{ij} \in C(\mathbb{G})$ such that the action is given by the formula

$$\alpha_N(e_i) = \sum_{j=1}^{N} e_j \otimes v_{ji}.$$

Equation (3) translates into the equality

$$\Delta(u_{ij}) = \sum_{k=1}^{N} v_{ik} \otimes v_{kj}$$

while the relations $\alpha(e_i)\alpha(e_j) = \alpha(e_ie_j) = \delta_{ij}\alpha(e_i)$ easily yields

$$v_{ij}v_{ij'} = \delta_{jj'}v_{ij}.$$

Similarly, from $e_i^* = e_i$ we deduce that $v_{ij}^* = v_{ij}$ and from the fact that the sum of the e_i's is 1 we conclude that all the relations of Definition 2.11 are satisfied. Moreover, let $C(\mathbb{H})$ be the sub-C*-algebra generated by the elements v_{ij} for $1 \leqslant i, j \leqslant N$. By what precedes,

$$\Delta(C(\mathbb{H})) \subset C(\mathbb{H}) \otimes C(\mathbb{H})$$

and $\alpha(\mathbf{C}^N) \subset \mathbf{C}^N \otimes C(\mathbb{H})$, so that by faithfulness we have $C(\mathbb{H}) = C(\mathbb{G})$. We are now in position to use the universal property of $C(S_N^+)$ to conclude. □

REMARK 3.8. One interpretation of Theorem 3.7 is that if $\mathbb{G}$ is a compact quantum group acting faithfully on $\mathbf{C}^N$, then $\mathbb{G}$ embeds into S_N^+. Note however that this does not imply an analogue of the Cauchy embedding theorem. Indeed, for a finite quantum group $\mathbb{G}$ (that is, with $C(\mathbb{G})$ being finite-dimensional), $C(\mathbb{G})$ is not be isomorphic to some $\mathbf{C}^N$ as soon as $\mathbb{G}$ is not classical, so that its action on itself does not yield an embedding into a quantum permutation group. As a matter of fact, explicit examples of finite quantum groups which are not quantum subgroups of S_N^+ are given in [**6**].

In other words, any faithful action on $\mathbf{C}^N$ factors through S_N^+, so that the latter can rightfully be termed the *quantum automorphism group* of $\mathbf{C}^N$. Of course, a similar statement holds for $\mathrm{Aut}^+(X)$ with respect to compact quantum groups acting on a graph X, once that notion of action is properly defined. Despite that remarkable result, it turns out that the general situation is very difficult to understand. For instance, Sh. Wang showed in [**72**] that if A is a non-commutative finite-dimensional C*-algebra, then it has no quantum automorphism group. The reason is that any action of a quantum group automatically fixes a state (that notion was defined just before Theorem 2.4) on the C*-algebra, but no non-trivial action can fix all of them. As we see, the dynamical theory of compact quantum groups has subtleties making it richer than the classical one.

This suggests that one considers an extra structure on the space and restricts attention to actions preserving that structure. For instance, if one fixes a state on a finite-dimensional C*-algebra (thereby considering a "non-commutative finite measure space"), then there exists a quantum automorphism group by [**72**]. Nevertheless, exploring commutative spaces leads to a series of negative results (see for instance [**36**] and [**17**]) showing that connected spaces tend to have no quantum symetries. However, it is extremely difficult to say something about actions of arbitrary compact quantum groups on arbitrary compact spaces. One way around can be to consider actions of a *given* compact quantum group on arbitrary spaces. This is the line of thought triggered by the following question of D. Goswami:

QUESTION. Can S_N^+ act faithully on a connected compact space ?

It turns out that any action of S_N^+ is either trivial or faithful, and that there does exist non-trivial actions on classical connected spaces. This was shown by H. Huang in [**41**] through an explicit construction.

EXAMPLE 3.9 (Huang). Let Y be a compact connected compact space. If Y_N denotes the disjoint unions of N copies of Y, then there is an isomorphism

$$C(Y_N) \simeq C(Y) \otimes \mathbf{C}^N.$$

Letting S_N^+ act trivially on the first tensorand and through α_N on the second one then provides an action on Y_N. Of course, Y_N is not connected, but we can make it so by gluing the various copies together. To do this, let $Z \subset Y$ be a closed subset and consider an element

$$a = \sum_{i=1}^{N} f_i \otimes e_i \in C(Y) \otimes \mathbf{C}^N$$

such that for any $z \in Z$ and any $1 \leqslant i, j \leqslant N$, $f_i(z) = f_j(z)$. Then, if $\mathrm{ev}_z : C(Y) \to \mathbf{C}$ denotes the evaluation map at some $z \in Z$, we have

$$\begin{aligned}(\mathrm{ev}_z \otimes e_k^* \otimes \mathrm{id}) \circ \alpha(a) &= (\mathrm{ev}_z \otimes e_k^* \otimes \mathrm{id}) \left(\sum_{i=1}^{N} \sum_{j=1}^{N} f_i \otimes e_j \otimes u_{ij} \right) \\ &= \sum_{i=1}^{N} f_i(z) u_{ik} \\ &= f_1(z) \sum_{i=1}^{N} u_{ik} \\ &= f_1(z).\end{aligned}$$

This means that if $A \subset C(Y_N)$ denotes the subalgebra of functions which all coincide on Z, then $\alpha(A) \subset A \otimes C(S_N^+)$. In other words, we have produced an action on A. The Gelfand-Naimark theorem already mentionned in the beginning of this work then shows that $A = C(X)$, where X is the quotient of Y_N obtained by identifying all copies of Z pointwise. This is compact and connected, and not a point if $Z \neq Y$.

There remains to check that the action is non-trivial. To see this, let $x \notin Z$ and let $f \in C(Y)$ be such that $f(x) = 1$ and $f(z) = 0$ for all $z \in Z$ (such a function exists by Urysohn's Lemma). Setting $a = f \otimes e_1$, we see that

$$(\mathrm{ev}_x \otimes e_2^* \otimes \mathrm{id}) \circ \alpha(a) = u_{12} \neq 0,$$

so that $\alpha(a) \neq a \otimes 1$.

Even though that example is convincing, it has some drawback. In particular, the closed subset Z is in a sense invariant under that action, so that it is certainly not transitive. This can be formalized thanks to the following notion:

DEFINITION 3.10. An action α of a compact quantum group $\mathbb{G}$ on a C*-algebra A is said to be *ergodic* if

$$\mathrm{Fix}(\alpha) := \{x \in A \mid \alpha(x) = x \otimes 1\} = \mathbf{C}.1_A.$$

REMARK 3.11. Ergodicity can be shown to be equivalent to the existence of a unique state on A which is invariant under the action. Classically, the state corresponds to a measure on the space which is therefore ergodic with respect to the action, hence the name. Moreover, for any closed invariant subset and any probability measure on that subset, we get an invariant measure for the action by extending it by 0, hence an ergodic action must be minimal.

The actions of Example 3.9 are not ergodic, since constant tuples of functions belong to $\mathrm{Fix}(\alpha)$. This is not the case for the canonical action of S_N^+ on $\mathbf{C}^N$ however.

EXAMPLE 3.12. Let us check that the action α_N of Example 3.4 is ergodic: if $x = \sum_{i=1}^{N} \lambda_i e_i$ is fixed, then

$$(e_k^* \otimes \mathrm{id}) \circ \alpha_N(x) = \lambda_k.1$$

for all $1 \leqslant k \leqslant N$. However, direct computation yields

$$(e_k^* \otimes \mathrm{id}) \circ \alpha_N(x) = \sum_{i=1}^{N} \lambda_i p_{ik}$$

so that $\lambda_k = \sum_{i=1}^{N} \lambda_i p_{ik}$. Multiplying that equation by p_{jk} for some $1 \leqslant j \leqslant k$ leads to $\lambda_k p_{jk} = \lambda_j p_{jk}$ so that all the coefficients are equal, hence $x = \lambda.1$.

Based on these observations, H. Huang asked the following strengthened question:

QUESTION. Can S_N^+ act ergodically on a connected compact space ?

The answer was given in [**32**] and can be summarized as follows:

THEOREM 3.13 (Freslon-Taipe-Wang). *The only ergodic action of S_N^+ on a classical connected space is the trivial action on a point.*

The proof is not a direct one and uses tool from a much more algebraic approach to compact quantum groups. We will in the next section try to convey the main ideas of that approach without dwelling too much on the details, so that we can describe some of the main features of the proof.

3.2. A categorical detour. Quantum groups can be considered under several points of view. In particular, the *Tannaka-Krein duality* established by S.L. Woronowicz in [**75**] shows that they can be seen as a C*-tensor categories equipped with a specific functor. Let us explain how this works.

3.2.1. *Tannaka-Krein duality.* We will need the concept of a representation, but after all, an action is not very far from a representation, so this should not be too difficult. And indeed, consider a linear map $\rho : \mathbf{C}^d \to \mathbf{C}^d \otimes C(\mathbb{G})$ such that

$$(\rho \otimes \mathrm{id}) \circ \rho = (\mathrm{id} \otimes \Delta) \circ \rho.$$

This corresponds in a sense to a semi-group representation of $\mathbb{G}$ and we need an additional condition to ensure that this is in fact a genuine group representation. One way of doing this is to consider the *coefficients* of the representation, that is to say the elements $(\rho_{ij})_{1 \leqslant i,j \leqslant d}$ such that, denoting again by $(e_i)_{1 \leqslant i \leqslant N}$ the canonical basis of $\mathbf{C}^N$,

$$\rho(e_i) = \sum_{j=1}^{d} e_j \otimes \rho_{ji}.$$

DEFINITION 3.14. A *representation of dimension d* of a compact quantum groups $\mathbb{G}$ is a linear map

$$\rho : \mathbf{C}^d \to \mathbf{C}^d \otimes C(\mathbb{G})$$

which satisfies

$$(\rho \otimes \mathrm{id}) \circ \rho = (\mathrm{id} \otimes \Delta) \circ \rho$$

and such that the matrix $\widehat{\rho} = (\rho_{ij})_{1 \leqslant i,j \leqslant d} \in M_d(C(\mathbb{G}))$ is invertible.

EXAMPLE 3.15. Let G be a compact group of matrices and let $\rho : G \to GL_d(\mathbf{C})$ be a representation. Then, the linear map

$$\widetilde{\rho} : \mathbf{C}^d \to \mathbf{C}^d \otimes C(G) \simeq C(G, \mathbf{C}^d)$$

given by

$$\widetilde{\rho}(x) : g \mapsto \rho(g).x$$

satisfies the conditions in Definition 3.14. The first condition is clear and for the second one, first note that ρ_{ij} is nothing but the map giving (i,j)-th coefficient function of $\rho(g)$. Because $\rho(g)$ is invertible, it has an inverse which we denote by $\kappa(g)$. Then, the matrix

$$\widehat{\kappa} = (\kappa_{ij})_{1\leqslant i,j\leqslant d} \in M_d(C(G))$$

is inverse to $\widehat{\rho}$. It is an easy exercise to check that conversely, any finite-dimensional representation of G in the sense of compact quantum groups comes from that construction.

EXAMPLE 3.16. The action $\alpha_N : \mathbf{C}^N \to \mathbf{C}^N \otimes C(S_N^+)$ is in particular a linear map. Moreover, the matrix U formed by the coefficients $(p_{ij})_{1\leqslant i,j\leqslant N}$ is orthogonal, as seen in Proposition 2.14. This means that it is invertible with inverse ${}^tU$, so that α_N is a representation.

EXAMPLE 3.17. More generally, let $(\mathbb{G}, U)$ be a compact matrix quantum group and let

$$\rho_U : \mathbf{C}^N \to \mathbf{C}^N \otimes C(\mathbb{G})$$

be the linear map given by

$$\rho_U(e_i) = \sum_{j=1}^{N} e_j \otimes U_{ji}.$$

The condition $(\rho_U \otimes \mathrm{id}) \circ \rho_U = (\mathrm{id} \otimes \Delta) \circ \rho_U$ directly follows from Definition 2.15. As for invertibility, as in the case of S_N^+ this is given by the fact that the matrix U is orthogonal. This is called the *fundamental representation* of $\mathbb{G}$.

We will now recast compact quantum groups in a categorical setting using representations and this requires the notion of an intertwiner. If ρ_1 and ρ_2 are representations of a compact quantum group $\mathbb{G}$ with dimension d_1 and d_2 respectively, then an intertwiner between ρ_1 and ρ_2 is a linear map $T : \mathbf{C}^{d_1} \to \mathbf{C}^{d_2}$ such that

$$(T \otimes \mathrm{id}) \circ \rho_1 = \rho_2 \circ T.$$

We are ready to introduce the main object of this section.

DEFINITION 3.18. Let $\mathbb{G}$ be a compact quantum group. Its *representation category* is the category $\mathrm{Rep}(\mathbb{G})$ whose objects are finite-dimensional representations and whose morphisms are intertwiners.

This is a C*-category (see for instance [**55**, Chap 2] for details on this notion and the ones below) with an additional structure coming from the tensor product of representations, which is defined as follows: given ρ_1 and ρ_2 of dimension d_1 and d_2 respectively, $\rho_1 \otimes \rho_2$ has dimension $d_1 d_2$ and satisfies

$$(\rho_1 \otimes \rho_2)(e_i \otimes f_j) = \sum_{k=1}^{d_1}\sum_{l=1}^{d_2} e_k \otimes f_l \otimes (\widehat{\rho}_1)_{ki}(\widehat{\rho}_2)_{lj}.$$

Moreover, we have a unitary tensor functor to the category of finite-dimensional Hilbert spaces

$$\mathcal{F}_{\mathbb{G}} : \mathrm{Rep}(\mathbb{G}) \to \mathrm{Hilb}_f$$

sending a representation to the underlying space. Here, by tensor functor we mean that for any two representations ρ_1 and ρ_2, there is an isomorphism

$$\varphi_{\rho_1,\rho_2} : \mathcal{F}_{\mathbb{G}}(\rho_1) \otimes \mathcal{F}_{\mathbb{G}}(\rho_2) \to \mathcal{F}_{\mathbb{G}}(\rho_1 \otimes \rho_2)$$

which is compatible with the associativity of tensor products, in the sense that

$$\varphi_{\rho_1,\rho_2\otimes\rho_3} \circ (\mathrm{id} \otimes \varphi_{\rho_2,\rho_3}) = \varphi_{\rho_1\otimes\rho_2,\rho_3} \circ (\varphi_{\rho_1,\rho_2} \otimes \mathrm{id}) \tag{5}$$

for representations ρ_1, ρ_2 and ρ_3. As for the adjective "unitary", it means that $\mathcal{F}_{\mathbb{G}}$ respects adjoints and that φ_{ρ_1,ρ_2} is a unitary map for all ρ_1 and ρ_2. We now have everything in hand to state the celebrated duality theorem proven by S.L. Woronowicz in [**75**].

THEOREM 3.19 (Woronowicz). *Let $\mathfrak{C}$ be a C*-tensor category and let*

$$\mathcal{F} : \mathfrak{C} \to \mathrm{Hilb}_f$$

be a faithful unitary tensor functor. Then, there exists a compact quantum group $\mathbb{G}$ such that $\mathfrak{C}$ is isomorphic to $\mathrm{Rep}(\mathbb{G})$ *and $\mathcal{F}$ is isomorphic to $\mathcal{F}_{\mathbb{G}}$. Moreover, $\mathbb{G}$ is unique up to isomorphism.*

3.2.2. *A duality theorem for ergodic actions.* Theorem 3.19 tells us that $\mathrm{Rep}(\mathbb{G})$ is in a sense "the same thing" as the quantum group $\mathbb{G}$. As a consequence, any notion concerning quantum groups should have an equivalent categorical description, usually in terms of a specific functor on $\mathrm{Rep}(\mathbb{G})$. Let us apply this principle to ergodic actions.

Given an ergodic action α of $\mathbb{G}$ on a unital C*-algebra A, we would like to build a functor $\mathcal{F}_\alpha$ encoding its properties. We first have to find the image of a given representation ρ. To do this, observe that α is in particular a representation[5] of $\mathbb{G}$ on the vector space underlying A so that we may try to "compare" it with ρ. This will be done through an appropriate notion of morphism between a representation and an action.

DEFINITION 3.20. A *morphism* between ρ and α is a linear map

$$T : \mathbf{C}^d \to A$$

such that

$$(T \otimes \mathrm{id}) \circ \rho = \alpha \circ T.$$

The set of such morphisms is denoted by $\mathrm{Mor}(\rho, \alpha)$.

In other words, a morphism is a way of seeing the representation ρ as a "subrepresentation" of A. We can then put all these morphisms together by setting

$$A_\rho = \left\{T(\xi) \mid T \in \mathrm{Mor}(\rho, \alpha) \text{ and } \xi \in \mathbf{C}^d\right\}.$$

By definition, $\alpha(A_\rho) \subset A_\rho \otimes C(\mathbb{G})$ so that this yields a sub-representation, which is the "largest" one in A equivalent to a sum of copies of ρ (two representations are said to be equivalent if there exists an invertible intertwiner between them). One advantage of that point of view is that we can define a natural sesquilinear form on A_ρ: if $T_1(\xi), T_2(\eta) \in A_\rho$, we simply set

$$(T_1(\xi) \mid T_2(\eta))_{A_\rho} = \langle T_2^* \circ T_1(\xi), \eta\rangle.$$

It follows from the general theory developped by F. Boca in [**16**] that

[5]Well, if A is not finite-dimensional, then we have not defined what this means, and some care must be taken to make this statement true.

(1) A_ρ is finite-dimensional for all ρ;
(2) $\bigcup_\rho A_\rho$ is dense in A;
(3) The sesquilinear form $(\bullet \mid \bullet)$ is an inner product on A_ρ.

Moreover, if $T \in \mathrm{Mor}(\rho_1, \alpha)$ and if $S : \mathbf{C}^d \to \mathbf{C}^{d_2}$ is an intertwiner, then it follows from the definition that $ST \in \mathrm{Mor}(\rho_2, \alpha)$, so that $S(A_{\rho_1}) \subset A_{\rho_2}$. This means that we have a functor

$$\mathcal{F}_\alpha : \mathrm{Rep}(\mathbb{G}) \to \mathrm{Hilb}_f$$

sending ρ to A_ρ and, by construction of the inner product on A_ρ, $\mathcal{F}_\alpha$ is unitary. The only issue is with the tensor structure. Of course, there needs to be an issue: unitary tensor functors correspond, by virtue of Theorem 3.19, to compact quantum groups, and ergodic actions are a more general notion (see for instance Example 3.3). To understand what is going on, let us try to relate $\mathcal{F}_\alpha(\rho_1) \otimes \mathcal{F}_\alpha(\rho_2)$ to $\mathcal{F}_\alpha(\rho_1 \otimes \rho_2)$. There is a natural way to do so: given $T(\xi) \in A_{\rho_1}$ and $S(\eta) \in A_{\rho_2}$, we may form the element $(T \otimes S)(\xi \otimes \eta) \in A_{\rho_1 \otimes \rho_2}$. This yields an isometric (with respect to the previously defined inner product) embedding

$$\imath_{\rho_1,\rho_2} : A_{\rho_1} \otimes A_{\rho_2} \hookrightarrow A_{\rho_1 \otimes \rho_2},$$

but there is no reason for it to be surjective in general. We therefore need to weaken our definition of a tensor functor.

DEFINITION 3.21. Let $(\mathbb{G}, U)$ be a compact quantum group. A unitary functor

$$\mathcal{F} : \mathrm{Rep}(\mathbb{G}) \to \mathrm{Hilb}_f$$

is said to be a *weak unitary tensor functor* if there exist isometric embeddings

$$\imath_{\rho_1,\rho_2} : \mathcal{F}(\rho_1) \otimes \mathcal{F}(\rho_2) \hookrightarrow \mathcal{F}(\rho_1 \otimes \rho_2)$$

for any two representations ρ_1 and ρ_2, satisfying Equation (5) as well as

$$P_{\rho_1,\rho_2 \otimes \rho_3} \circ (\mathrm{id} \otimes P_{\rho_1,\rho_2}) = P_{\rho_1 \otimes \rho_2,\rho_3} \circ (P_{\rho_1,\rho_2} \otimes \mathrm{id}),$$

where P_{ρ_1,ρ_2} is the range projection[6] of $\imath_{\rho_1,\rho_2}$.

We can now state an analogue of Theorem 3.19 for ergodic actions, which is due to C. Pinzari and J. Roberts in [**58**].

THEOREM 3.22 (Pinzari-Roberts). *Let $(\mathbb{G}, U)$ be an orthogonal compact matrix quantum group and let $\mathcal{F}$ be a weak unitary tensor functor on* $\mathrm{Rep}(\mathbb{G})$. *Then, there exists an ergodic action α of $\mathbb{G}$ on a unital C^*-algebra A such that $\mathcal{F}$ is isomorphic to $\mathcal{F}_\alpha$.*

We will not explain the proof here because it would take us too far from our subject, and rather turn our attention to Theorem 3.13.

3.3. Quantum rigidity. We will now explain the proof of Theorem 3.13, using the previous setting. This requires a bit more detail on the representation theory of S_N^+. More precisely, let us call a representation *irreducible* if its self-intertwiners form a one-dimensional space. Then, by a fundamental result of S.L. Woronowicz in [**74**], any representation splits into a direct sum of irreducibles, so that understanding the latter is enough to describe the whole representation theory. The irreducible representations of S_N^+ were classified by T. Banica in [**3**].

[6]The range projection of an isometry $\imath : H_1 \to H_2$ is the orthogonal projection in H_2 onto the range of $\imath$, or equivalently the operator $\imath\imath^*$.

THEOREM 3.23 (Banica). *The irreducible representations of S_N^+ can be indexed by the integers in such a way that ρ_0 is the trivial representation (the one underlying the trivial action on* $\mathbf{C}$*), $\rho_U = \rho_0 \oplus \rho_1$ and for any $n \in \mathbf{N}$,*

$$\rho_1 \otimes \rho_n = \rho_{n-1} \oplus \rho_n \oplus \rho_{n+1}.$$

A straightforward induction yields the more general formula

$$\rho_k \otimes \rho_n = \bigoplus_{i=|k-n|}^{k+n} \rho_i. \tag{6}$$

Now, if X is a compact space and α an ergodic action of S_N^+ on $C(X)$, then we have a corresponding weak unitary tensor functor $\mathcal{F}_\alpha$ and the idea is that the commutativity of the C*-algebra $C(X)$ forces some relations on the possible images of the representations ρ_n.

To be more precise, we need to introduce some notations. For $n \in \mathbf{N}$, it follows from Equation (6) that $(\mathbf{C}^N)^{\otimes n}$ contains a unique subrepresentation equivalent to ρ_n. Denoting by H_n the subspace on which it acts and using once again Equation (6), there exists for each $0 \leqslant k \leqslant 2n$ a unique subspace of $H_n \otimes H_n$ such that the corresponding subrepresentation of $\rho_n \otimes \rho_n$ is ρ_k. We will denote by $P_k^{n,n}$ the orthogonal projection onto that subspace. By definition, this is an intertwiner, hence it makes sense to consider the linear map $\mathcal{F}_\alpha(P_k^{n,n})$.

LEMMA 3.24. *With the previous notations, assume that $\mathcal{F}_\alpha(P_k^{n,n}) \neq 0$ for some $k \geqslant 2$ and $n \geqslant k/2$. Then, the range of $P_k^{n,n}$ is contained in an eigenspace of the flip map σ_n on $H_n \otimes H_n$, that is to say the map sending $x \otimes y$ to $y \otimes x$ for all pure tensors $x \otimes y$.*

PROOF. Theorem 3.22 comes with an explicit description of the algebra A, or more precisely of a dense $*$-subalgebra, which we will use. Let us set recall that the *conjugate space* of a Hilbert space H is the Hilbert space $\overline{H}$ with the same real Hilbert space structure and such that for all $\lambda \in \mathbf{C}$, $\lambda.\overline{x} = \overline{\lambda.x}$, and set

$$\mathcal{A} = \bigoplus_{n \in \mathbf{N}} \overline{\mathcal{F}_\alpha(\rho_n)} \otimes H_n$$

and define a product through the formula (with $\imath_{n,m} = \imath_{\rho_n,\rho_m}$)

$$(\overline{\xi}_1 \otimes \eta_1)(\overline{\xi}_2 \otimes \eta_2) = \sum_{k \in \mathbf{N}} \overline{\mathcal{F}_\alpha(P_k^{n,m}) \circ \imath_{n,m}(\xi_1 \otimes \xi_2)} \otimes P_k^{n,m}(\eta_1 \otimes \eta_2).$$

Then, $\mathcal{A}$ embeds as a dense subalgebra of A and the restriction of α coincides on the n-th summand with $\mathrm{id} \otimes \rho_n$.

Applying the product formula for $\xi_1, \xi_2 \in \mathcal{F}_\alpha(\rho_n)$ and $\eta_1, \eta_2 \in H_n$ and using the assumed commutativity of $\mathcal{A}$ then yields that

$$\sum_{n \in \mathbf{N}} \overline{\mathcal{F}_\alpha(P_k^{n,n}) \circ \iota_{n,n}(\xi_1 \otimes \xi_2)} \otimes P_k^{n,n}(\eta_1 \otimes \eta_2)$$

equals

$$\sum_{k=0}^{2n} \overline{\mathcal{F}_\alpha(P_k^{n,n}) \circ \iota_{n,n}(\xi_2 \otimes \xi_1)} \otimes P_k^{n,n}(\eta_2 \otimes \eta_1).$$

Because the projections $P_k^{n,n}$ have pairwise orthogonal ranges, there must in fact be equality for each k. But then, there exists $\lambda \in \mathbf{C}^*$ such that

$$\mathcal{F}_\alpha(P_k^{n,n}) \circ \iota_{n,n}(\xi_1 \otimes \xi_2) = \lambda^{-1}\mathcal{F}_\alpha(P_k^{n,n}) \circ \iota_{n,n}(\xi_2 \otimes \xi_1)$$
$$P_k^{n,n}(\eta_1 \otimes \eta_2) = \lambda P_k^{n,n}(\eta_2 \otimes \eta_1)$$

unless $\mathcal{F}_\alpha(P_k^{n,n}) \circ \iota_{n,n}(\xi_1 \otimes \xi_2) = 0$ and the result follows. □

The main virtue of the necessary condition appearing in Lemma 3.24 is that it can be directly checked by exhibiting vectors which are neither symmetric nor anti-symmetric, hence not in an eigenspace of σ_n.

PROPOSITION 3.25. *Let $N \geqslant 4$ and $n \geqslant 2$. Then, there is a vector in the range of $P_k^{n,n}$ which is not symmetric nor anti-symmetric. Therefore, $\mathcal{F}_\alpha(P_k^{n,n}) = 0$ for any ergodic action α on a classical space.*

PROOF. The proof is done by building an explicit vector and checking that it is not an eigenvector for the flip map σ_n. The crucial point here is that we have a good combinatorial description of the projection $P_k^{n,n}$ using non-crossing partitions, which is given in [**34**]. This is nevertheless quite heavy to write down and we refer to [**32**, Prop 6.4] for details. □

Gathering the two previous facts, one can eventually prove Theorem 3.13.

PROOF OF THEOREM 3.13. Let us write A_n for A_{ρ_n}. By the general theory of [**16**], we have a direct sum decomposition

$$\mathcal{A} = \bigoplus_{n \in \mathbf{N}} A_n.$$

Let now k be the smallest strictly positive integer such that $A_k \neq \{0\}$. Using the explicit description of $\mathcal{A}$ in terms of $\mathcal{F}_\alpha$ given at the beginning of the proof of Lemma 3.24, we see that if $x \in A_n$ and $y \in A_m$, then

$$xy \in \bigoplus_{k=0}^{n+m} A_{n+m}$$

and that the component in a given A_k vanishes if $\mathcal{F}_\alpha(P_k^{n,m}) = 0$. In particular, if $x, y \in A_n$, then the product xy must lie in $A_0 \oplus A_1 \subset A_0 \oplus A_k$. As a consequence,

$$B = A_0 \oplus A_k$$

is a finite-dimensional subalgebra. Moreover, B is stable under the involution and this follows from the definition of the involution in the reconstruction theorem together with the fact that all the irreducible representations of S_N^+ are self-conjugate. Therefore, B is a sub-C*-algebra of A of dimension at least 2. In particular, it contains a non-trivial projection, contradicting connectedness of X since a non-trivial projection in $C(X)$ is the indicator function of a proper clopen subset. □

This result, though satisfying, raises other questions. For instance, we still do not know of any classical space, apart from finite ones, on which S_N^+ can act ergodically. It is therefore tempting to conjecture that such a space must be very far from connected, and the next question pushes that idea to the extreme:

QUESTION. If S_N^+ acts ergodically on a compact space X, then does it follow that X is finite ?

If we drop the commutativity assumption, very little is known. Of course S_N^+ can act non-trivially and even ergodically on non-commutative C*-algebras (starting with $C(S_N^+)$ itself), but as we saw in the proof of Theorem 3.13 the key property of connectedness that we used is the absence of non-trivial projection. The next question is therefore a natural generalization.

QUESTION. Can S_N^+ act ergodically on a projectionless C*-algebra ?

Eventually, going back to H. Huang's example at the beginning, we may even wonder whether it is possible to classify all actions of S_N^+ on a compact topological space. Indeed, given an action on X, we have by the Gelfand-Naimark theorem that $\mathrm{Fix}(\alpha) = C(Y)$ for some quotient space $X \twoheadrightarrow Y$. The fibres are the orbits in the sense of [**42**, Def 4.1], on which the action is ergodic by [**42**, Thm 4.15]. Therefore, a positive answer to the first question above would give us more information on X. It is quite unclear however how far this can be pushed. Let us just summarize this as a general question,

QUESTION. What can be said of a general non-trivial action of S_N^+ on a compact space X ?

There has been a lot of progress in past years on the topic of quantum rigidity. In particular, D. Goswami completed in [**36**] a program to show that compact connected manifold have no quantum symmetries. More generally for metric spaces, the problem is difficult since we do not know in general under which condition there exists a maximal compact quantum group acting (see for instance [**35**] or [**19**] for discussions and examples concerning this problem). Nevertheless, A. Chirvasitu was able to prove in [**18**] that metric spaces generically have no quantum symmetry in the Baire sense, which is somehow an analogue of Theorem 2.21. The two previously mentioned authors also exhibited quantum rigidity for specific metric spaces either coming from Riemaniann manifolds in [**20**] or satisfying some "negative curvature" property in [**17**].

4. Probabilistic, but not random, topics

In this last section, we will explain some interactions between the quantum permutation group S_N^+ and probability theory in a broad sense. More precisely, we will use a setting which may be called *non-commutative probability theory* (the term "quantum" is also sometimes used instead of "non-commutative"). This is a theory which was initiated to develop probabilistic tools in the setting of quantum mechanics, i.e. where random variables are replaced with operators which need not commute (see for instance [**54**] or [**57**] for a detailed introduction). The subject had rapid developments and M. Schürmann showed in particular how $*$-bialgebras where the natural objects to define and study non-commutative stochastic processes (see for instance [**66**]). Since compact quantum groups yield particularly nice examples of $*$-bialgebras, it is very natural to try to investigate random processes on them. We will explain this in Section 4.2.

Another direction taken by non-commutative probability theory is *free probability theory*, introduced by D. Voiculescu (see for instance [**70**] and also [**56**] for a different approach). Once again, the starting point is to consider operators on a Hilbert space as random variables. But the goal is different: one wants to understand the structure of certain algebras generated by these non-commutative random variables. Opposite to the case where the operators commute, which yields classically independent random variables, D. Voiculescu focused on the case where the variables are *free* in the sense that they generate a free product of von Neumann algebras. He then discovered that this freeness can be interpreted as a kind of independence, leading to a very rich theory. Moreover, a deep connection between classical and free independence arises through the study of large random matrices. In that setting, S_N^+ turns out to play a very important role and we will now explain this.

4.1. (Quantum) symetry in (free) probability theory. What an abstract group-theoretical object like S_N or S_N^+ has to do with probability theory mays seem unclear at first sight. However, as P. Diaconis illustrates in depth in his monograph [**24**], group representations are in fact everywhere in probability theory, as well as in statistics. The de Finetti theorem is a very interesting instance of that phenomenon, which moreover has the virtue of admitting a non-commutative counterpart involving the quantum permutation group.

4.1.1. *The de Finetti theorem.* The de Finetti theorem is a deep result which in a sense relates probability theory and statistics, or more precisely describes the relationship between the properties that one can observe on a family of random variables and the independence of the variables. To explain this, let us consider an infinite family $(X_i)_{i\in\mathbf{N}}$ of random variables, all defined on a common probability space $(\Omega, \mathbb{P})$. We want to translate the fact that these variables are somewhat not distinguishable from one another by empirical means. Assuming that we empirically have access at least to the (joint) moments, the next definition is very natural.

DEFINITION 4.1. The family of random variables $(X_i)_{i\in\mathbf{N}}$ is said to be *exchangeable* if its joint distribution is invariant under permutation. Concretely, this means that for any integer $N \in \mathbf{N}$ and any permutation $\sigma \in S_N$, the tuples $(X_1, \cdots, X_N)$ and $(X_{\sigma(1)}, \cdots, X_{\sigma(N)})$ have the same joint distribution. Otherwise said, for any $1 \leqslant n \leqslant N$ and $1 \leqslant i_1, \cdots, i_n \leqslant N$,

$$\mathbb{E}\left(X_{i_1} \cdots X_{i_n}\right) = \mathbb{E}\left(X_{\sigma(i_1)} \cdots X_{\sigma(i_n)}\right)$$

This is really a group-theoretical property: it means that the natural action of the family of permutation groups $(S_N)_{N\in\mathbf{N}}$ given, for $\sigma \in S_N$, by

$$\sigma.X_i = \begin{cases} X_{\sigma^{-1}(i)} & \text{if} \quad i \leqslant N \\ X_i & \text{if} \quad i > N \end{cases}$$

preserves the probability distribution. One therefore sometimes terms this kind of property a *distributional symetry.*

EXAMPLE 4.2. Assume that the variables are all independent and identically distributed. Then, there exist integers $n_1, \cdots, n_N$ such that

$$\begin{aligned} \mathbb{E}\left(X_{i_1} \cdots X_{i_n}\right) &= \mathbb{E}\left(X_1^{n_1} \cdots X_N^{n_N}\right) \\ &= \mathbb{E}\left(X_1^{n_1}\right) \cdots \mathbb{E}\left(X_N^{n_N}\right) \\ &= \mathbb{E}\left(X_{\sigma(1)}^{n_{\sigma(1)}}\right) \cdots \mathbb{E}\left(X_{\sigma(N)}^{n_{\sigma(N)}}\right) \\ &= \mathbb{E}\left(X_{\sigma(i_1)} \cdots X_{\sigma(i_n)}\right) \end{aligned}$$

where independence was used in the second and fourth lines and identical distribution in the third one. Thus, $(X_i)_{i\in\mathbf{N}}$ is exchangeable.

EXAMPLE 4.3. Independence is far from necessary, as the following trivial example shows: if $X_i = X_1$ for all $i \in \mathbf{N}$, then the family $(X_i)_{i\in\mathbf{N}}$ is exchangeable, simply because all joint moments are just ordinary moments of X_1.

To get a better understanding of the general situation, observe that both examples above satisfy the following general criterion: if the variables are conditionnally independent and identically distributed, then they are exchangeable. The problem in general is therefore to find a σ-algebra with respect to which these conditions are achieved, and there is a natural candidate for that.

DEFINITION 4.4. Let $(X_i)_{i\in\mathbf{N}}$ be a family of random variables. Their *tail algebra* is the σ-algebra

$$\mathcal{T} = \bigcap_{N\in\mathbf{N}} \sigma(X_i \mid i \geqslant N).$$

Intuitively, an event is measurable with respect to $\mathcal{T}$ if it does not depend upon the outcome of any fixed finite number of the variables $(X_i)_{i\in\mathbf{N}}$. We can now state the motivating theorem of this section, which relates permutational symetry with conditional independence. Even though the name of B. de Finetti stays attached to it, the latter only proved it for Bernoulli random variables in [**23**, Chap III], and the general statement hereafter is due to E. Hewitt and L. Savage in [**40**].

THEOREM 4.5 (de Finetti, Hewitt-Savage). *The family of random variables $(X_i)_{i\in\mathbf{N}}$ is exchangeable if and only if the variables are independent and identically distributed* conditionnally to their tail algebra.

4.1.2. *A very short introduction to free probability.* We will now explain how this result can be generalized to a non-commutative setting. As already mentioned, this basically means that random variables should be replaced by operators on a Hilbert space. It is nevertheless convenient to assume a bit more structure on the set of the variables.

DEFINITION 4.6. A *non-commutative measure space* is given by a *von Neumann algebra* $\mathcal{N}$ acting on a Hilbert space H that is, a sub-C*-algebra $\mathcal{N} \subset B(H)$ which is closed under the *strong operator topology*, that is, the topology for which a sequence $(T_n)_{n\in\mathbf{N}}$ of operators converges to T if and only if $\|T_n(\xi) - \xi\| \to 0$ for all $\xi \in H$.

We will moreover assume that $\mathcal{N}$ admits a tracial state $\tau : \mathcal{N} \to \mathbf{C}$. Then, one may think of τ as the integration functional with respect to a probability measure, and we will accordingly call $(\mathcal{N}, \tau)$ a *non-commutative probability space.* Given an element $X \in \mathcal{N}$, which we may think of as a non-commutative random variable, its *moments* are the numbers

$$m_k(X) = \tau(X^k).$$

Classically, there are many instances where the moments of a random variable enable to recover the whole distribution, for instance if its support is compact. In the non-commutative setting, if we assume X to be self-adjoint (i.e. $X^* = X$), then X has a compact spectrum[7] $\mathrm{Sp}(X)$ and its moments therefore define a unique probability measure on $\mathrm{Sp}(X)$ which is called the *spectral measure of* X. Even though we will not use that measure later on, we will restrict ourselves to the case of self-adjoint random variables.

[7]The spectrum of an element X in a complex unital algebra A is the set of $\lambda \in \mathbf{C}$ such that $X - \lambda.1$ is not invertible.

Exchangeability is, as we have explained above, an invariance property of the joint distribution under the action of the symetric groups. Based on this, it is quite natural to mimic the action α_N of S_N^+ to define a quantum version of exchangeability. This means that the action of S_N^+ on a given X_i for some $1 \leqslant i \leqslant N$ should be

$$X_i \mapsto \sum_{j=1}^{N} X_j \otimes p_{ji}.$$

This can be straightforwardly extended to monomials in the X_i's,

$$X_{i_1} \cdots X_{i_n} \mapsto \sum_{j_1, \cdots, j_n} X_{j_1} \cdots X_{j_n} \otimes p_{j_1 i_1} \cdots p_{j_n i_n}.$$

To obtain a quantum notion of exchangeability we simply have to translate the fact that τ should be invariant under the above map.

DEFINITION 4.7. A family of non-commutative random variables $(X_i)_{i \in \mathbf{N}}$ in a non-commutative probability space $(\mathcal{N}, \tau)$ is said to be *quantum exchangeable* if for any $N \in \mathbf{N}$ and $1 \leqslant i_1, \leqslant, i_n \leqslant N$,

$$\tau(X_{i_1} \cdots X_{i_n}) = \sum_{j_1, \cdots, j_n = 1}^{N} \tau(X_{j_1} \cdots X_{j_n}) p_{j_1 i_1} \cdots p_{j_n i_n}. \tag{7}$$

To convince the reader that this is the correct invariance condition, let us make a sanity check. If Equation (7) is satisfied, then it also holds when p_{ij} is replaced by the coefficients of any quantum permutation matrix. In particular, it should hold for the coefficients of a permutation matrix P_σ for $\sigma \in S_N$. In that case, $p_{ij} = \delta_{i\sigma(j)}$, so that Equation (7) becomes

$$\tau(X_{i_1} \cdots X_{i_n}) = \tau(X_{\sigma^{-1}(i_1)} \cdots X_{\sigma^{-1}(i_n)}).$$

Therefore, quantum exchangeability implies classical exchangeability. In order to make sense of the statement of Theorem 4.5, we now have to define the tail algebra of a family of non-commutative random variables, and this is rather straightforward.

DEFINITION 4.8. Let $(X_i)_{i \in \mathbf{N}}$ be non-commutative random variables in a non-commutative probability space $(\mathcal{N}, \tau)$. Their *tail algebra* is the von Neumann subalgebra

$$\mathcal{N}_{\mathrm{tail}} = \overline{\bigcap_{i \in \mathbf{N}} \langle X_j \mid j \geqslant i \rangle}$$

where the closure is taken in the strong operator topology.

It follows from the general theory of von Neumann algebras that there exists a linear map $\mathbb{E}_{\mathrm{tail}} : \mathcal{N} \to \mathcal{N}_{\mathrm{tail}}$ which is a *conditional expectation* in the sense that it satisfies the following properties:

- $\mathbb{E}_{\mathrm{tail}}(axb) = a\mathbb{E}_{\mathrm{tail}}(x)b$ for all $x \in \mathcal{N}$ and $a, b \in \mathcal{N}_{\mathrm{tail}}$;
- $\mathbb{E}_{\mathrm{tail}}(1) = 1$;
- $\|\mathbb{E}_{\mathrm{tail}}\| = 1$.

REMARK 4.9. If τ is not assumed to be tracial, the the existence of a conditional expectation can still be proven as soon as one assumes the variables to be classically exchangeable, see [**45**, Prop 4.2].

The last ingredient needed is a non-commutative analogue of independence. Note that because X_i and X_j are not assumed to commute, independence is of no use: how can we compute for instance $\mathbb{E}_{\text{tail}}(X_1X_2X_1X_2)$? What we are looking for is a rule which enables us to compute any mixed moment of the variables in terms of their individual moments in a coherent way. It turns out that they are only two ways of doing this (a precise meaning for that sentence is given, and proven, in [**67**]): classical independence and *free independence.*

The problem is that we have to introduce free independence not only in the scalar setting, but also in the conditional one and that entails subtleties. Plainly, we are going to explain what it means for non-commutative random variables to be freely independent with respect to the condition expectation $\mathbb{E}_{\text{tail}}$. This definition should be a statement about the conditional joint moments of the variables and these would classically be random variables of the form $\mathbb{E}_{\text{tail}}(P(X_1)Q(X_2))$ for polynomials $P, Q \in \mathbf{C}[X]$. Here, first note that we should take an alternating product of polynomials since X_1 and X_2 do not commute. But this is not sufficient, since the polynomials should have coefficients in $\mathcal{N}_{\text{tail}}$, and the latter need not commute with X_1 nor with X_2. Hence, the following definition is the correct one:

DEFINITION 4.10. Given a von Neumann algebra $\mathcal{N}$ and a von Neumann subalgebra $\mathcal{M}$, a *non-commutative $\mathcal{M}$-valued polynomial* is a linear combination of elements of the form

$$b_0Xb_1X\cdots b_{n-1}Xb_n$$

for some $b_0, \cdots, b_n \in \mathcal{M}$. The space of such polynomials is denoted by $\mathcal{M}\langle X\rangle$.

We are now ready for the definition.

DEFINITION 4.11. The variables $(X_i)_{i\in\mathbf{N}}$ are said to be *conditionally freely independent* with respect to $\mathcal{N}_{\text{tail}}$ if for any $i_1 \neq i_2 \neq \cdots \neq i_n$ and any polynomials $p_1, \cdots, p_m \in \mathcal{N}_{\text{tail}}\langle X\rangle$ such that

$$\mathbb{E}_{\text{tail}}(p_k(X_{i_k})) = 0$$

for all $1 \leqslant k \leqslant n$, we have

$$\mathbb{E}_{\text{tail}}\left(p_1(X_{i_1})\cdots p_n(X_{i_n})\right) = 0$$

The definition may seem complicated and obscure at first sight, but the main point is that it enables to compute by induction any mixed moment. Let us do this in a simple case. Setting $Y_i = X_i - \mathbb{E}(X_i)$, we have for $i \neq j$

$$\begin{aligned}
\mathbb{E}_{\text{tail}}(X_iX_j) &= \mathbb{E}_{\text{tail}}\left((Y_i + \mathbb{E}_{\text{tail}}(X_i))(Y_j + \mathbb{E}_{\text{tail}}(X_j))\right) \\
&= \mathbb{E}_{\text{tail}}(Y_iY_j) - \mathbb{E}_{\text{tail}}\left(\mathbb{E}_{\text{tail}}(X_i)Y_j\right) - \mathbb{E}_{\text{tail}}\left(Y_i\mathbb{E}_{\text{tail}}(X_j)\right) \\
&+ \mathbb{E}_{\text{tail}}\left(\mathbb{E}_{\text{tail}}(X_i)\mathbb{E}_{\text{tail}}(X_j)\right) \\
&= \mathbb{E}_{\text{tail}}(Y_iY_j) - \mathbb{E}_{\text{tail}}(X_i)\mathbb{E}_{\text{tail}}\left(Y_j\right) - \mathbb{E}_{\text{tail}}\left(Y_i\right)\mathbb{E}_{\text{tail}}(X_j) \\
&+ \mathbb{E}_{\text{tail}}(X_i)\mathbb{E}_{\text{tail}}(X_j) \\
&= 0 - 0 - 0 + \mathbb{E}_{\text{tail}}(X_i)\mathbb{E}(X_j) \\
&= \mathbb{E}_{\text{tail}}(X_i)\mathbb{E}_{\text{tail}}(X_j).
\end{aligned}$$

REMARK 4.12. For products of two variables, the result in the free case in the same as in the classical one. To see the difference, one should use non-commutativity. An indeed, assuming for instance $\mathbb{E}_{\text{tail}}(X_1) = 0 = \mathbb{E}_{\text{tail}}(X_2)$ for simplicity, free independence yields

$$\mathbb{E}_{\text{tail}}(X_1X_2X_1X_2) = 0,$$

while for classically independent random variables,

$$\mathbb{E}_{\text{tail}}(X_1X_2X_1X_2) = \mathbb{E}_{\text{tail}}(X_1^2X_2^2) = \mathbb{E}_{\text{tail}}(X_1^2)\mathbb{E}_{\text{tail}}(X_2^2).$$

4.1.3. *The free de Finetti theorem.* We are now in position to state the main result of [**45**].

THEOREM 4.13 (Köstler-Speicher). *A family of non-commutative random variables $(X_i)_{i\in\mathbf{N}}$ is quantum exchangeable if and only if the variables are freely independent and identically distributed with respect to their tail algebra.*

PROOF. The proof is long and we cannot reproduce it in detail here. We will nevertheless try to highlight its main points, as well as the crucial features of the quantum permutation groups involved. The first step is to prove that quantum exchangeability is also satisfied in the conditional sense, meaning that for any $1 \leqslant i_1, \cdots, i_n \leqslant N$, $b_1, \cdots, b_{n-1} \in \mathbb{E}_{\text{tail}}$,

$$\begin{aligned}&\mathbb{E}_{\text{tail}}\left(X_{i_1}b_1\cdots b_{n-1}X_{i_n}\right)\otimes 1\\ &= \sum_{j_1,\cdots,j_n}\mathbb{E}_{\text{tail}}\left(X_{j_1}b_1\cdots b_{n-1}X_{j_n}\right)\otimes p_{j_1i_1}\cdots p_{j_ni_n}.\end{aligned}$$

The proof is not very difficult since it is enough to prove that the equation above holds with $\mathbb{E}$ replaced by τ. Then, we can approximate each b_i by polynomials in the X_i's with indices greater than N and exchangeability readily yields the result.

The second step uses only classical exchangeability. It consists in showing the following factorization property: for any $1 \leqslant i_1, \cdots, i_n \leqslant N$ such that there is $1 \leqslant k \leqslant n$ with i_k different from all the other indices, and for any $p_1, \cdots, p_n \in \mathcal{N}_{\text{tail}}\langle X\rangle$,

$$\begin{aligned}&\mathbb{E}_{\text{tail}}\left(p_1(X_{i_1})\cdots p_n(X_{i_n})\right)\\ &= \mathbb{E}_{\text{tail}}\left(p_1(X_{i_1})\cdots p_{k-1}(X_{i_{k-1}})\mathbb{E}_{\text{tail}}(p_k(X_{i_k}))p_{k+1}(X_{i_{k+1}})\cdots p_n(X_{i_n})\right).\end{aligned}$$

The proof relies on the observation that one can define an action of S_N on the von Neumann subalgebra generated by the variables $(X_i)_{i\in\mathbf{N}}$ (and we may even assume without loss of generality that the subalgebra in question is $\mathcal{N}$) by permuting the variables and that the fixed point subalgebra for that action coincide with $\mathcal{N}_{\text{tail}}$. One can then use von Neumann's ergodic theorem (see for instance [**44**]) to prove the result, together with the first step which proves in particular that the variables are conditionnally exchangeable.

We now enter the difficult part of the proof. Given $i_1 \neq \cdots \neq i_n$, we have to show that if for any $p_1, \cdots, p_n \in \mathcal{N}_{\text{tail}}$ such that $\mathbb{E}_{\text{tail}}(p_i(X_i)) = 0$ for all $1 \leqslant i \leqslant N$,

$$\mathbb{E}_{\text{tail}}\left(p_1(X_{i_1})\cdots p_n(X_{i_n})\right) = 0.$$

The trick is to consider the partition $\ker(i)$ on $\{1, \cdots, n\}$ where two elements k and l are in the same subset if and only if $i_k = i_l$. Observe that if $\ker i$ has n subsets, then by the factorization property the result holds. This suggests working by decreasing induction on the number of subsets of $\ker(i)$. Starting from the conditional quantum exchangeability formula, we can use the properties of the generators to simplify it. For instance, if $i_k \neq i_{k+1}$, then we have $p_{j_ki_k}p_{j_{k+1}i_{k+1}} = 0$ if $j_k = j_{k+1}$, so that we may sum only over tuples without any consecutive equal indices. But then, the expectation can be shown to be 0 by induction if $\ker(j)$ has more subsets than $\ker(i)$. As a consequence,

$$\begin{aligned}
&\mathbb{E}_{\text{tail}}\left(p_1(X_{i_1})\cdots p_n(X_{i_n})\right) \\
&= \sum_{j_1,\cdots,j_n=1}^{N} \mathbb{E}_{\text{tail}}\left(p_1(X_{j_1})\cdots p_n(X_{j_n})\right)\otimes p_{j_1i_1}\cdots p_{j_ni_n} \\
&= \left(\sum_{\ker(j)=\ker(i)} p_{j_1i_1}\cdots p_{j_ni_n}\right)\mathbb{E}_{\text{tail}}\left(p_1(X_{i_1})\cdots p_n(X_{i_n})\right)
\end{aligned}$$

where we used in the second line the fact that by (classical) conditional exchangeability, the expectation only depends on which indices are equal in the product, hence on $\ker(j)$. To conclude, it suffices to show that the element of $C(S_N^+)$ in parenthesis is not equal to 1, which can be done through explicit computation. Note that the image of that element in $C(S_N)$ is in fact equal to 1, hence classical exchangeability is not enough to recover the result. □

There exist many variations and extensions of the de Finetti theorem in the classical setting (see for instance [**25**]). In the quantum world as well, the previous result has been generalized in many ways. First, a more general proof, which works for a large family of compact quantum groups, was developped in [**7**], recovering in particular Theorem 4.5. Second, it is classically known that exchangeability is not exactly the condition needed for the theorem to work. A famous theorem of C. Ryll-Nardzewski [**61**] shows that one just needs *spreadability*. A quantum version of spreadability, together with a corresponding free de Finetti theorem was worked out in [**22**]. More results in a slightly different context, as well as a summary of non-commutative de Finetti theorems, can be found in [**9**].

Even though we have said, based on [**67**], that classical and free independences were the only two possible ways of defining natural notions of independence, it turns out that if one is ready to give away some assumption, then some extra notions of independence appear, called *boolean*, *monotone* and *anti-monotone*. De Finetti theorems exist in this context too, and are due to W. Liu in [**48**], later extended to Ryll-Nardzewski-type results in [**49**].

Let us conclude with a few words on the analogy between the classical and free de Finetti theorem. In the classical case, the structure of exchangeable random variables is completely elucidated. Indeed, they are simply mixtures of independent random variables. This is because the tail algebra is a well-understood object. In the free case, the tail algebra can be almost anything: it was proven by K. Dykema and C. Köstler in [**27**] that any countably generated von Neumann algebra is the tail algebra of an infinite sequence of quantum exchangeable non-commutative random variables. This means that the free de Finetti theorem does not tell us as much on quantum exchangeable random variables as its classical counterpart.

4.2. Random quantum transpositions. We will conclude this survey by another recent result concerning the probabilistic aspects of S_N^+. This is about the convergence of some specific random walk coming from the symetric group. There is a very rich theory concerning these in the classical case, and it was shown in [**31**] and [**33**] that some of the tools can be adapted to compact quantum groups.

4.2.1. *Card shuffling.* The classical permutation group provides a very convenient setting to formalize problems of shuffling, for instance for cards. Indeed, assume that we have a deck of N cards and that we want to shuffle it. The shuffling can be modeled by a probability distribution μ on S_N with the following interpretation:

- We number the cards of the deck in their initial position by integers from 1 to N from top to bottom;
- When shuffling once, we pick a permutation $\sigma \in S_N$ at random according to μ;
- The order of the cards after shuffling is $\sigma(1), \cdots, \sigma(N)$.

If we assume moreover that each repetition of the shuffle is independent from all the other ones, we end up after k shufflings with a permutation of the deck which is described by the product of k permutations chosen randomly according to μ. Therefore, the probability that this belongs to a given subset $A \subset S_N$ is

$$\mu^{*k}(A) = \mu^{\otimes k}\left(\{(\sigma_1, \cdots, \sigma_k) \mid \sigma_1 \cdots \sigma_k \in A\}\right).$$

The conclusion of the previous discussion is that the study of the shuffling of a deck of cards boils down to the study of a sequence $(\mu^{*k})_{k\in\mathbf{N}}$ of measures on S_N, which are all determined by the initial measure μ. Two questions then arise:

(1) Will the deck be eventually well-shuffled, or in other words, does $(\mu^{*k})_{k\in\mathbf{N}}$ converge to the uniform measure μ_{Uniform} on S_N ?
(2) If yes, how long do we have to shufle, or in other words, can we bound the distance between μ^{*k} and μ_{Uniform} explicitly in terms of k ?

In the setting of finite groups, the first question has a well-known answer.

THEOREM 4.14. *The sequence will converge to $\mu_{Uniform}$ if and only if its support is not contained in a subgroup of S_N, nor in a coset with respect to a normal subgroup.*

PROOF. The result is in fact true for any finite group, and even *mutatis mutandis* for any (locally) compact group (there is also a version for finite Markov chains, of which finite groups are a particular case). One way to prove it is through the Fourier transform. Given an irreducible representation ρ of S_N of dimension d, we set

$$\widehat{\mu}(\rho) = \sum_{\sigma\in S_N} \rho(g)\mu(g) \in M_d(\mathbf{C})$$

and see the Fourier transform $\widehat{\mu}$ of μ as a matrix-valued function on the set of equivalence classes of irreducible representations. It is then straightforward to check on the definition that

$$\widehat{\mu^{*k}} = \widehat{\mu}^k,$$

so that the problem boils down to the convergence of powers of matrices with norm at most 1. More precisely, assume that there exists an irreducible representation π such that $(\widehat{\mu}(\pi)^k)_{k\in\mathbf{N}}$ does not converge. Then, the matrix $\widehat{\mu}(\pi)$ has an eigenvalue λ of modulus one with eigenvector ξ, and simple computations show that this implies that the set $H_\lambda = \{g \in G \mid \pi(g)\xi = \lambda\xi\}$ is a coset with respect to the normal subgroup $H_1 = \{g \in G \mid \pi(g)\xi = \xi\}$ which contains the support of μ (see for instance [**38**] for a detailed proof). $\square$

Let us mention right away that the corresponding statement for compact quantum groups (of which the reader will be able to make sense later on) is known to be false. For instance, there are measure on compact quantum groups whose support is not contained in any proper subgroup but in a so-called *quasi-subgroups* (see [**30**] for details) still preventing the sequence from converging to the uniform measure (by which we mean the Haar state, already mentioned in Theorem 2.28 and Remark 2.29). In the case of finite quantum groups, a complete characterization based on generalizing the two conditions of Theorem 4.14 was given by J.-P. McCarthy in [**53**], but for the general compact case, it is open at the day of this writing.

As for the second question, it first requires specifying a distance between probability measures. One which is quite commonly used is the *total variation distance*, given by

$$d_{TV}(\mu, \nu) = \max_{A \subset S_N} |\mu(A) - \nu(A)|.$$

To show the kind of results one may expect using that distance, we will describe an example. We consider the uniform measure μ_{tr} on the set of transpositions and we would like to produce a random walk by composing randomly selected transpositions. This can obviously not converge to the uniform measure, since the permutations obtained will be even at even times and odd at odd times. One way to remedy this is to make the walk *lazy* by allowing it to stay where it is with small strictly positive probability. Concretely, this means that we consider the measure

$$\mu = \frac{1}{N}\delta_{\mathrm{id}} + \frac{N-1}{N}\mu_{tr}.$$

In terms of cards, this can be realized through the following process: spread the cards, randomly select one card (uniformly) and then a second card (uniformly and independently from the first one), then swap the two cards (this does nothing if the two cards are the same).

Because transpositions generate S_N and the only normal subgroup of S_N is the alternating group, the cosets of which have constant signature hence do not contain the support of μ_{tr}, we see that the sequence $(\mu^{*k})_{k\in\mathbf{N}}$ converges to the uniform measure μ_{Uniform} by Theorem 4.14. Concerning the precise estimate for the distance, P. Diaconis and M. Shahshahani obtained in [**26**] a remarkable result:

THEOREM 4.15 (Diaconis-Shahshahani). *Set $k_N = N\ln(N)/2$. For any $\epsilon > 0$, we have*

$$d_{TV}(\mu^{*(1-\epsilon)k_N}, \mu_{Uniform}) \underset{N\to\infty}{\longrightarrow} 1 \quad \& \quad d_{TV}(\mu^{*(1+\epsilon)k_N}, \mu_{Uniform}) \underset{N\to\infty}{\longrightarrow} 0.$$

This means that, asymptotically, the deck will be well shuffled after $N\ln(N)/2$ steps. However, these limits tell us more: the time needed around $N\ln(N)/2$ to go from a badly mixed deck (d_{TV} close to one) to a well-mixed one (d_{TV} close to zero) is negligible compared to $N\ln(N)/2$. Such a sharp behaviour is called a *cut-off phenomenon* and is often pictured as follows:

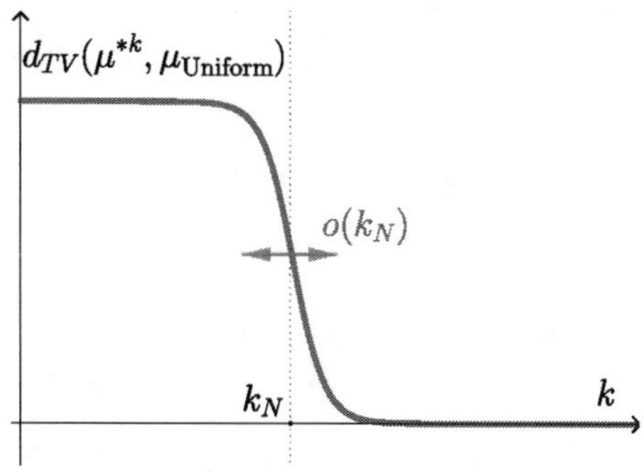

In view of this, it is natural to wonder whether the dependence on N of the "cut-off window" can be described more explicitly. This is indeed possible and the result is due to L. Teyssier in [**68**]:

THEOREM 4.16 (Teyssier). *For any $c \in \mathbf{R}$, we have*

$$d_{TV}(\mu^{*(k_N + cN/2)}, \mu_{Uniform}) \underset{N\to+\infty}{\longrightarrow} d_{TV}(\text{Poiss}(1+e^{-c}), \text{Poiss}(1)).$$

One may observe that the number of fixed points of a uniformly chosen permutation follows a Poisson distribution with parameter 1. This can be used to give an interpretation of Theorem 4.16. The measure μ^{*k} is supported on products of transpositions, and transpositions have lots of fixed points. Therefore, looking at the average number of fixed points of the permutations that we can reach through the random walk gives an indication of the distance to the uniform measure. Theorem 4.16 tells us that asymptotically, this exactly gives us the distance.

4.2.2. *A quantum random walk.* We will now explain how one may consider (and describe) a similar situation for the quantum permutation group. We first have to translate the previous constructions in terms of quantum groups. For this, recall that a Borel probability measure μ on a compact group G can be viewed through integration as a state on $C(G)$. We will therefore build a specific state φ on $C(S_N^+)$ playing the role of the random transposition probability measure on S_N. One idea to define such a state is the following remark: transpositions form a conjugacy class, hence one may define their uniform measure by taking the Dirac mass at any element and then averaging with respect to the adjoint action of the group. In other terms:

$$\mu_{tr} = \frac{1}{N!}\sum_{\sigma\in S_N}\delta_\tau(\sigma \bullet \sigma^{-1}), \tag{8}$$

where τ is any transposition. To use this idea in the quantum world, we will need to restate it. First, let us consider the map $\mathbb{E} : C(S_N) \to C(S_N)$ sending a function f to the function

$$\mathbb{E}(f) : x \mapsto \frac{1}{N!}\sum_{\sigma\in S_N} f(\sigma x\sigma^{-1}).$$

The range of that map is the subalgebra $C(S_N)_{\text{central}}$ of *central functions*, that is to say functions which are invariant under conjugation. Moreover, $\mathbb{E}$ is a conditional expectation in the sense already used in Section 4.1. Recall that this means that

- For any $f, h \in C(S_N)$ and $g \in C(S_N)_{\text{central}}$,
$$\mathbb{E}(fgh) = f\mathbb{E}(g)h;$$
- $\mathbb{E}(1) = 1$;
- $\|\mathbb{E}\| = 1$.

The point is that we can now restate Equation (8) as

$$\mu_{tr} = \delta_\tau \circ \mathbb{E}.$$

We should therefore try to build an analogue of $\mathbb{E}$ for S_N^+, but this requires another step of reformulation. Indeed, the algebra of central function admits in the case of groups an alternative description: it is the subalgebra generated by the characters of irreducible representations.

DEFINITION 4.17. Let $(\mathbb{G}, U)$ be a compact quantum group and let $\rho : \mathbf{C}^d \to \mathbf{C}^d \otimes C(\mathbb{G})$ be a finite-dimensional representation. The *character* of ρ is

$$\chi_\rho = \sum_{i=1}^{n} \widehat{\rho}_{ii} \in C(\mathbb{G}).$$

The sub-C*-algebra of $C(\mathbb{G})$ generated by the characters of all representations is called the *central subalgebra* and is denoted by $C(\mathbb{G})_{\text{central}}$.

To define a conditional expectation onto the central subalgebra, it would be convenient to have a basis of $C(S_N^+)$. It turns out that there are some which are very practical for computations.

THEOREM 4.18 (Woronowicz). *Let $(\mathbb{G}, U)$ be an orthogonal compact matrix quantum groups and let $\mathrm{Irr}(\mathbb{G})$ be a set of pairwise inequivalent irreducible representations such that any irreducible representation is equivalent to an element of $\mathrm{Irr}(\mathbb{G})$. Then, the family $(\widehat{\rho}_{ij})_{\rho\in\mathrm{Irr}(\mathbb{G}),1\leqslant i,j\leqslant\dim(\rho)}$ is linearly independent and spans a dense subspace of $C(\mathbb{G})$.*

For an irreducible representation ρ, we can now set

$$\mathbb{E}_{\text{central}}(\widehat{\rho}_{ij}) = \delta_{ij}\frac{\chi_\rho}{\dim(\rho)}$$

and this defines a linear map which is the identity on $C(\mathbb{G})_{\text{central}}$. Moreover, it can be proven by direct computation (using formulæ for products of coefficients of irreducible representations that we will not give) that

$$\mathbb{E}_{\text{central}}(\chi_{\rho'}\rho_{ij}\chi_{\rho''}) = \chi_{\rho'}\mathbb{E}_{\text{central}}(\rho_{ij})\chi_{\rho'},$$

so that $\mathbb{E}_{\text{central}}$ is indeed a conditional expectation.

Let us now sum everything up and describe our random transposition measure on S_N^+. We start with an arbitrary transposition τ and consider the Dirac mass δ_τ. This is not defined on $C(S_N^+)$ but on $\mathcal{F}(S_N)$. Nevertheless, it is easy to extend it by composing with the canonical surjection $\pi_{\text{ab}} : C(S_N^+) \to \mathcal{F}(S_N)$. Now, we simply compose it with the conditional expectation onto the central sublagebra, yielding the state

$$\varphi_{tr} = \delta_\tau \circ \pi \circ \mathbb{E}.$$

Even though the whole procedure seems to depend on τ, it does not. In fact, it is even possible to explicitly compute the state above. Let us define inductively a sequence of polynomials $(Q_n)_{n\in\mathbf{N}}$ by $Q_0(X) = 1$, $Q_1(X) = X - 1$ and

$$Q_1(X)Q_n(X) = Q_{n-1}(X) + Q_n(X) + Q_{n+1}(X).$$

REMARK 4.19. These can also be defined through dilated Chebyshev polynomials of the second kind. Indeed, setting $P_0(X) = 1$, $P_1(X) = X$ and $XP_n(X) = P_{n-1}(X) + P_{n+1}(X)$, we have $Q_n(X) = P_{2n}(\sqrt{X})$.

PROPOSITION 4.20. *Let χ_n denote the character of the n-th irreducible representation of S_N^+. Then,*

$$\varphi_{tr}(\chi_n) = Q_n(N-2).$$

PROOF. Let us start by some computations for small values of n. First, $\chi_0 = 1$ so that the result holds in that case. Next, the representation ρ_N splits by Theorem 3.23 as $\rho_0 \oplus \rho_1$, but this also gives, when restricted to S_N the natural representation on $\mathbf{C}^N$. As a consequence, $\pi(\chi_{\rho_0\oplus\rho_1}) = \pi(\chi_0 + \chi_1)$ is the character of that representation, which is nothing but the function Fix giving the number of fixed points of a permutation. Putting things together therefore yields

$$\varphi_{tr}(\chi_1) = \mathrm{Fix}(\tau) - \varphi_{tr}(\chi_0) = N - 2 - 1 = Q_1(N-2).$$

To conclude, observe that $\chi_{\rho^{\otimes n}} = \chi_\rho^n$ and that combining this with Theorem 3.23 yields

$$\chi_n = Q_n(\chi_0 \oplus \chi_1).$$

Because $\mathbb{E}$ is the identity on $C(S_N^+)_{\mathrm{central}}$, we then have

$$\begin{aligned}\pi \circ \mathbb{E}(\chi_n) &= \pi(\chi_n)\\ &= \pi(Q_n(\chi_0 \oplus \chi_1))\\ &= Q_n(\pi(\chi_0 \oplus \chi_1))\\ &= Q_n(\mathrm{Fix}).\end{aligned}$$

Applying δ_τ, which is an algebra homomorphism, then finishes the proof. □

Now that we have a state, we can start investigating the corresponding random walk. This requires defining a convolution operation on states, which is easily done:

DEFINITION 4.21. Let $\mathbb{G}$ be a compact quantum group and let φ, ψ be states on $C(\mathbb{G})$. Their *convolution product* is the state

$$\varphi * \psi = (\varphi \otimes \psi) \circ \Delta.$$

REMARK 4.22. It is an easy exercise to check that convolution is associative precisely because of the co-associativity of Δ. This allows to define unambiguously the sequence of convolution powers of a given state.

Let us now state a result analogous to Theorem 4.15.

THEOREM 4.23. *The random walk given by φ_{tr} has a cut-off at $N\ln(N)/2$.*

Of course, this does not mean much as long as we have not defined the limit of the random walk and the distance used. For the limit, we need a generalization of the uniform measure, and the natural candidate for an arbitrary compact group is the Haar measure. As explained in Remark 2.29, the generalization of that object is the Haar state of S_N^+.

We are left with the problem of defining the distance, but this is the easy part. Indeed, we are dealing with linear forms on $C(\mathbb{G})$, hence the simplest choice is the norm on the space $C(\mathbb{G})^*$ of continuous linear forms. Let us see what this gives us for S_N. If μ is a probability measure, then it is in particular a function on S_N. Moreover, its norm as a linear functional equals its 1-norm as a function. As it turns out,

$$\|\mu - \mu_{\text{Uniform}}\|_1 = 2d_{TV}(\mu, \mu_{\text{Uniform}}).$$

To see why, let us consider the set $A \subset S_N$ of all permutations σ such that $\mu(\sigma) \geqslant \mu_{\text{Uniform}}(\sigma)$. It is clear that the maximum in the definition of d_{TV} is attained at A as well as at its complement A^c. Therefore,

$$\begin{aligned} 2d_{TV}(\mu, \mu_{\text{Uniform}}) &= |\mu(A) - \mu_{\text{Uniform}}(A)| + |\mu(A^c) - \mu_{\text{Uniform}}(A^c)| \\ &= \sum_{\sigma\in A} \mu(\sigma) - \mu_{\text{Uniform}}(\sigma) - \sum_{\sigma\in A^c} \mu(\sigma) - \mu_{\text{Uniform}}(\sigma) \\ &= \sum_{\sigma\in S_N} |\mu(\sigma) - \mu_{\text{Uniform}}(\sigma)| \\ &= \|\mu - \mu_{\text{Uniform}}\|_1. \end{aligned}$$

Now that the statement of Theorem 4.23 makes sense, let us announce that we will not explain the proof since this is any ways a consequence of Theorem 4.24 below. But let us note a surprising fact: we have only considered a pure transposition walk – no laziness has been added – and yet we have convergence to the Haar measure! As already mentioned, this is impossible in S_N for elementary reasons. This is connected to a fact that was already observed from several points of view, namely that there is no natural analogue of the alternating group A_N inside S_N^+, or equivalently no nice quantum version of the signature of a permutation. A fancier way of putting it is: any permutation (classical or not) can be approximated by products of *quantum* transpositions.

As in the classical case, one would like to get a more precise description of the total variation distance around the cut-off point and since the quantum cut-off time is the same as the classical one, it is natural to conjecture that the correct second order is cN. An in fact, we have

THEOREM 4.24 (Freslon-Teyssier-Wang). *Set $k_N = N\ln(N)/2$. For any $c \in \mathbf{R}$,*

$$\left\|\varphi_{tr}^{*\lceil k_N + cN/2\rceil} - \varphi_{\text{Haar}}\right\|_{C(S_N^+)^*}$$

converges to

$$2d_{TV}\left(D_{\sqrt{1+e^{-c}}}\left(\text{Meix}^+\left(\frac{1-e^{-c}}{\sqrt{1+e^{-c}}}, \frac{-e^{-c}}{1+e^{-c}}\right)\right) * \delta_{e^{-c}}, \text{Meix}^+(1,0)\right)$$

where Meix$^+$ *are the free Meixner laws.*

The notations in the statement require explanations. The free Meixner law with parameter a, b has an asbolutely continuous part with density

$$\frac{\sqrt{4(1+b)-(t-a)^2}}{2\pi(bt^2+at+1)}\mathbf{1}_{[a-2\sqrt{1+b},a+2\sqrt{1+b}]}$$

with respect to the Lebesgue measure. It can also, depending on the values of a and b, have an atom. With the notations of Theorem 4.24, there is no atom for $c \geqslant 0$, and one atom at $(e^c + e^{-c})^2$ with mass $1 - e^{2c}$ for $c < 0$. These laws are normalized in the sense that they have mean 0 and variance 1 and the symbols $D_{\sqrt{1+e^{-c}}}$ and $*\delta_{e^{-c}}$ simply denote a dilation and a translation in order to make the mean equal to e^{-c} and the variance equal to $\sqrt{1+e^{-c}}$. One last comment: $\mathrm{Meix}^+(1,0)$ is the *free Poisson law*, an analogue of the classical Poisson distribution which plays an important role in free probability theory.

4.2.3. *A glimpse of the proof.* The proof of Theorem 4.24 is long and involves many computations. We will nevertheless try to describe its main ingredients. The first step is to consider the following power series, at least in a formal sense:

$$a_k = \sum_{n=0}^{+\infty} \frac{Q_n(N-2)^k}{Q_n(N-2)^{k-1}} \chi_{\rho_n}.$$

If that series converges in $C(S_N^+)$, then

$$\varphi_{\mathrm{Haar}}(a_k x) = \varphi_{\mathrm{tr}}^{*k}(x)$$

for all $x \in C(S_N^+)$. In other words, the state $\varphi_{\mathrm{tr}}^{*k}$ is then *absolutely continuous* with respect to the Haar state, with density a_k. There are some analytic subtleties at play when considering questions of this kind about non-commutative integration and non-commutative L^p-spaces, the reader will therefore have to trust us that everything works fine as soon as a_k is L^2-bounded in the sense that

$$\sum_{n=0}^{+\infty} \frac{Q_n(N-2)^{2k}}{Q_n(N-2)^{2k-2}} < +\infty.$$

Indeed, in that case, $\varphi_{\mathrm{tr}}^{*k}$ will have an extension to the von Neumann algebra $L^\infty(S_N^+)$ (an object which we have not defined). Then, the distance between $\varphi_{\mathrm{tr}}^{*k}$ and φ_{Haar} equals their distance as linear forms on $L^\infty(S_N^+)$, which itself is twice the L^1-distance between their densities. The conclusion of all this is that the problem boils down to the dominated convergence theorem for series in a Banach space and this can be handled if one is brave enough to carry out the calculations.

The strategy outlined above works as soon as $c > 0$. This is the result of an easy but tedious computation showing that in that case, L^2-boundedness will be achieved for k large enough. But on the opposite side, it can be checked that for $c < 0$, $\varphi_{\mathrm{tr}}^{k}$ will never be absolutely continuous. Then, the idea is first to reduce the problem to a commutative one. This is done first by observing that the C*-algebra $C(S_N^+)_{\mathrm{central}}$ is commutative, hence by the Gelfand-Naimark theorem it is isomorphic to the algebra of continuous functions on a compact space. One can even directly prove that the map sending χ_{ρ_n} to $Q_{n|[0,N]}$ provides an isomorphism with $C([0,N])$. As a consequence, there exists a unique probability measure μ_k on $[0,N]$ such that for all $f \in C([0,N])$,

$$\varphi_{\mathrm{tr}}^{*k}(f) = \int_0^N f(x)\mathrm{d}\mu_k(x).$$

Because $[0,N]$ is compact, the measure μ_k is completely determined by its moments and we know these explicitly. One may therefore try to find an explicit description of the measure from this, yielding

THEOREM 4.25. *There exist $0 < \alpha_k < 1$ and $\widetilde{N}(k) \in [0,N]$ such that for any $c < 0$ and $k_c = \lceil (N\ln(N) + cN)/2 \rceil$,*

$$m_{k_c}^{(N)} = \sum_{n=0}^{+\infty} \left[\varphi_{\mathrm{tr}}^{*k_c}(n)Q_n(N) - Q_n(\widetilde{N}(k_c))\right] Q_n \mathrm{d}\,\mathrm{Poiss}^+(1,1) + \alpha(k_c)\delta_{\widetilde{N}(k_c)}.$$

In other words, the failure of absolute continuity is due to a single atom which can be isolated. Once this is done, we have two separate problems. One is the convergence of the absolute part, which is done as in the case $c > 0$, and the other one is the convergence of the atom, which is straightforward.

The picky reader will object that the case $c = 0$ has not been treated. Fortunately, a simple continuity argument yields the result as a consequence of the previous two.

Acknowledgements

The author is thankful to the many people with whom he has discovered, learnt and studied quantum permutation groups. Special thanks are moreover due to M. Weber for discussions around this survey and many comments and suggestions on a preliminary version.

References

[1] A. Atserias, L. Mančinska, D. E. Roberson, R. Šámal, S. Severini, and A. Varvitsiotis, *Quantum and non-signalling graph isomorphisms*, J. Combin. Theory Ser. B **136** (2019), 289–328, DOI 10.1016/j.jctb.2018.11.002. MR3926289

[2] L. Babai and L. Kučera, *Canonical labelling of graphs in linear average time*, 20th Annual Symposium on Foundations of Computer Science (SFCS 1979), IEEE, 1979, pp. 39–46.

[3] T. Banica, *Symmetries of a generic coaction*, Math. Ann. **314** (1999), no. 4, 763–780, DOI 10.1007/s002080050315. MR1709109

[4] T. Banica, *Quantum automorphism groups of small metric spaces*, Pacific J. Math. **219** (2005), no. 1, 27–51, DOI 10.2140/pjm.2005.219.27. MR2174219

[5] T. Banica, J. Bichon, and B. Collins, *Quantum permutation groups: a survey*, Noncommutative harmonic analysis with applications to probability, Banach Center Publ., vol. 78, Polish Acad. Sci. Inst. Math., Warsaw, 2007, pp. 13–34, DOI 10.4064/bc78-0-1. MR2402345

[6] T. Banica, J. Bichon, and S. Natale, *Finite quantum groups and quantum permutation groups*, Adv. Math. **229** (2012), no. 6, 3320–3338, DOI 10.1016/j.aim.2012.02.012. MR2900441

[7] T. Banica, S. Curran, and R. Speicher, *De Finetti theorems for easy quantum groups*, Ann. Probab. **40** (2012), no. 1, 401–435, DOI 10.1214/10-AOP619. MR2917777

[8] T. Banica and R. Speicher, *Liberation of orthogonal Lie groups*, Adv. Math. **222** (2009), no. 4, 1461–1501, DOI 10.1016/j.aim.2009.06.009. MR2554941

[9] I. Baraquin, G. Cébron, U. Franz, L. Maassen, and M. Weber, *De Finetti theorems for the unitary dual group*, SIGMA Symmetry Integrability Geom. Methods Appl. **18** (2022), Paper No. 067, 29 pp., DOI 10.3842/SIGMA.2022.067. MR4480649

[10] E. Bédos, G. J. Murphy, and L. Tuset, *Co-amenability of compact quantum groups*, J. Geom. Phys. **40** (2001), no. 2, 130–153, DOI 10.1016/S0393-0440(01)00024-9. MR1862084

[11] J. Bichon, *Quantum automorphism groups of finite graphs*, Proc. Amer. Math. Soc. **131** (2003), no. 3, 665–673, DOI 10.1090/S0002-9939-02-06798-9. MR1937403

[12] J. Bichon, *Gerstenhaber-Schack and Hochschild cohomologies of Hopf algebras*, Doc. Math. **21** (2016), 955–986. MR3548138

[13] J. Bichon, A. De Rijdt, and S. Vaes, *Ergodic coactions with large multiplicity and monoidal equivalence of quantum groups*, Comm. Math. Phys. **262** (2006), no. 3, 703–728, DOI 10.1007/s00220-005-1442-2. MR2202309

[14] J. Bichon, U. Franz, and M. Gerhold, *Homological properties of quantum permutation algebras*, New York J. Math. **23** (2017), 1671–1695. MR3741856

[15] B. Blackadar, *Operator algebras: Theory of C^*-algebras and von Neumann algebras*, Encyclopaedia of Mathematical Sciences, vol. 122, Springer-Verlag, Berlin, 2006. Operator Algebras and Non-commutative Geometry, III, DOI 10.1007/3-540-28517-2. MR2188261

[16] F. P. Boca, *Ergodic actions of compact matrix pseudogroups on C^*-algebras: Recent advances in operator algebras (Orléans, 1992)*, Astérisque **232** (1995), 93–109. MR1372527

[17] A. Chirvasitu, *Quantum rigidity of negatively curved manifolds*, Comm. Math. Phys. **344** (2016), no. 1, 193–221, DOI 10.1007/s00220-015-2553-z. MR3493142

[18] A. Chirvasitu, *Generic quantum metric rigidity*, Int. Math. Res. Not. **2021** (2021), no. 18, 14379–14397.

[19] A. Chirvasitu, *Quantum isometries and loose embeddings*, J. Geom. Phys. **161** (2021), 104089.

[20] A. Chirvasitu and D. Goswami, *Existence and rigidity of quantum isometry groups for compact metric spaces*, Comm. Math. Phys. **380** (2020), no. 2, 723–754, DOI 10.1007/s00220-020-03849-3. MR4170290

[21] A. Chirvasitu, *Topological generation results for free unitary and orthogonal groups*, Internat. J. Math. **31** (2020), no. 1, 2050003, 13 pp., DOI 10.1142/S0129167X20500032. MR4060567

[22] S. Curran, *A characterization of freeness by invariance under quantum spreading*, J. Reine Angew. Math. **659** (2011), 43–65, DOI 10.1515/CRELLE.2011.066. MR2837010

[23] B. de Finetti, *La prévision : ses lois logiques, ses sources subjectives* (French), Ann. Inst. H. Poincaré **7** (1937), no. 1, 1–68. MR1508036

[24] P. Diaconis, *Group representations in probability and statistics*, Institute of Mathematical Statistics Lecture Notes—Monograph Series, vol. 11, Institute of Mathematical Statistics, Hayward, CA, 1988. MR964069

[25] P. Diaconis and D. Freedman, *A dozen de Finetti-style results in search of a theory* (English, with French summary), Ann. Inst. H. Poincaré Probab. Statist. **23** (1987), no. 2, suppl., 397–423. MR898502

[26] P. Diaconis and M. Shahshahani, *Generating a random permutation with random transpositions*, Z. Wahrsch. Verw. Gebiete **57** (1981), no. 2, 159–179, DOI 10.1007/BF00535487. MR626813

[27] K. J. Dykema and C. Köstler, *Tail algebras of quantum exchangeable random variables*, Proc. Amer. Math. Soc. **142** (2014), no. 11, 3853–3863, DOI 10.1090/S0002-9939-2014-12116-2. MR3251725

[28] P. Erdős and A. Rényi, *Asymmetric graphs*, Acta Math. Acad. Sci. Hungar. **14** (1963), 295–315, DOI 10.1007/BF01895716. MR156334

[29] U. Franz, A. Kula, and A. Skalski, *Lévy processes on quantum permutation groups*, Noncommutative analysis, operator theory and applications, Oper. Theory Adv. Appl., vol. 252, Birkhäuser/Springer, [Cham], 2016, pp. 193–259, DOI 10.1007/978-3-319-29116-1_11. MR3526959

[30] U. Franz and A. Skalski, *On idempotent states on quantum groups*, J. Algebra **322** (2009), no. 5, 1774–1802, DOI 10.1016/j.jalgebra.2009.05.037. MR2543634

[31] A. Freslon, *Cut-off phenomenon for random walks on free orthogonal quantum groups*, Probab. Theory Related Fields **174** (2019), no. 3-4, 731–760, DOI 10.1007/s00440-018-0863-8. MR3980303

[32] A. Freslon, F. Taipe, and S. Wang, *Tannaka-Krein reconstruction and ergodic actions of easy quantum groups*, Comm. Math. Phys. **399** (2023), no. 1, 105–172, DOI 10.1007/s00220-022-04555-y. MR4567370

[33] A. Freslon, L. Teyssier, and S. Wang, *Cutoff profiles for quantum Lévy processes and quantum random transpositions*, Probab. Theory Related Fields **183** (2022), no. 3-4, 1285–1327, DOI 10.1007/s00440-022-01121-4. MR4453326

[34] A. Freslon and M. Weber, *On the representation theory of partition (easy) quantum groups*, J. Reine Angew. Math. **720** (2016), 155–197, DOI 10.1515/crelle-2014-0049. MR3565972

[35] D. Goswami, *Existence and examples of quantum isometry groups for a class of compact metric spaces*, Adv. Math. **280** (2015), 340–359, DOI 10.1016/j.aim.2015.03.024. MR3350223

[36] D. Goswami, *Non-existence of genuine (compact) quantum symmetries of compact, connected smooth manifolds*, Adv. Math. **369** (2020), no. 1, 107181.

[37] D. Goswami and A. Skalski, *On two possible constructions of the quantum semigroup of all quantum permutations of an infinite countable set*, Operator algebras and quantum groups, Banach Center Publ., vol. 98, Polish Acad. Sci. Inst. Math., Warsaw, 2012, pp. 199–214, DOI 10.4064/bc98-0-7. MR3059661

[38] U. Grenander, *Probabilities on algebraic structures*, Courier Corporation, 2008.

[39] D. Gromada, *Quantum symmetries of Cayley graphs of abelian groups*, Glasg. Math. J. **65** (2023), no. 3, 655–686, DOI 10.1017/s0017089523000198. MR4653714

[40] E. Hewitt and L. J. Savage, *Symmetric measures on Cartesian products*, Trans. Amer. Math. Soc. **80** (1955), 470–501, DOI 10.2307/1992999. MR76206

[41] H. Huang, *Faithful compact quantum group actions on connected compact metrizable spaces*, J. Geom. Phys. **70** (2013), 232–236, DOI 10.1016/j.geomphys.2013.03.027. MR3054297

[42] H. Huang, *Invariant subsets under compact quantum group actions*, J. Noncommut. Geom. **10** (2016), no. 2, 447–469, DOI 10.4171/JNCG/239. MR3519043

[43] L. Junk, S. Schmidt, and M. Weber, *Almost all trees have quantum symmetry*, Arch. Math. (Basel) **115** (2020), no. 4, 367–378, DOI 10.1007/s00013-020-01476-x. MR4145763

[44] C. Köstler, *A noncommutative extended de Finetti theorem*, J. Funct. Anal. **258** (2010), no. 4, 1073–1120, DOI 10.1016/j.jfa.2009.10.021. MR2565834

[45] C. Köstler and R. Speicher, *A noncommutative de Finetti theorem: invariance under quantum permutations is equivalent to freeness with amalgamation*, Comm. Math. Phys. **291** (2009), no. 2, 473–490, DOI 10.1007/s00220-009-0802-8. MR2530168

[46] J. Kustermans and S. Vaes, *Locally compact quantum groups* (English, with English and French summaries), Ann. Sci. École Norm. Sup. (4) **33** (2000), no. 6, 837–934, DOI 10.1016/S0012-9593(00)01055-7. MR1832993

[47] D. Kyed, *L^2-Betti numbers of coamenable quantum groups*, Münster J. Math. **1** (2008), 143–179. MR2502497

[48] W. Liu, *A noncommutative de Finetti theorem for boolean independence*, J. Funct. Anal. **269** (2015), no. 7, 1950–1994, DOI 10.1016/j.jfa.2015.07.007. MR3378866

[49] W. Liu, *Extended de Finetti theorems for boolean independence and monotone independence*, Trans. Amer. Math. Soc. **370** (2018), no. 3, 1959–2003, DOI 10.1090/tran/7034. MR3739198

[50] L. Lovász, *Operations with structures*, Acta Math. Acad. Sci. Hungar. **18** (1967), 321–328, DOI 10.1007/BF02280291. MR214529

[51] M. Lupini, L. Mančinska, and D. E. Roberson, *Nonlocal games and quantum permutation groups*, J. Funct. Anal. **279** (2020), no. 5, 108592, 44 pp., DOI 10.1016/j.jfa.2020.108592. MR4097284

[52] L. Mančinska and D. E. Roberson, *Quantum isomorphism is equivalent to equality of homomorphism counts from planar graphs*, 2020 IEEE 61st Annual Symposium on Foundations of Computer Science, IEEE Computer Soc., Los Alamitos, CA, [2020], pp. 661–672, DOI 10.1109/FOCS46700.2020.00067. MR4232075

[53] J. P. McCarthy, *The ergodic theorem for random walks on finite quantum groups*, Comm. Algebra **49** (2021), no. 9, 3850–3871, DOI 10.1080/00927872.2021.1908551. MR4290115

[54] P.-A. Meyer, *Quantum probability for probabilists*, Lecture Notes in Mathematics, vol. 1538, Springer-Verlag, Berlin, 1993, DOI 10.1007/978-3-662-21558-6. MR1222649

[55] S. Neshveyev and L. Tuset, *Compact quantum groups and their representation categories*, Cours Spécialisés [Specialized Courses], vol. 20, Société Mathématique de France, Paris, 2013. MR3204665

[56] A. Nica and R. Speicher, *Lectures on the combinatorics of free probability*, London Mathematical Society Lecture Note Series, vol. 335, Cambridge University Press, Cambridge, 2006, DOI 10.1017/CBO9780511735127. MR2266879

[57] K. R. Parthasarathy, *An introduction to quantum stochastic calculus*, Monographs in Mathematics, vol. 85, Birkhäuser Verlag, Basel, 1992, DOI 10.1007/978-3-0348-8641-3. MR1164866

[58] C. Pinzari and J. E. Roberts, *A duality theorem for ergodic actions of compact quantum groups on C^*-algebras*, Comm. Math. Phys. **277** (2008), no. 2, 385–421, DOI 10.1007/s00220-007-0371-7. MR2358289

[59] D. E. Roberson and S. Schmidt, *Solution group representations as quantum symmetries of graphs*, J. Lond. Math. Soc. (2) **106** (2022), no. 4, 3379–3410. MR4524201

[60] L. Rollier and S. Vaes, *Quantum automorphism groups of connected locally finite graphs and quantizations of discrete groups*, arXiv preprint (2022).

[61] C. Ryll-Nardzewski, *On stationary sequences of random variables and the de Finetti's equivalence*, Colloq. Math. **4** (1957), 149–156, DOI 10.4064/cm-4-2-149-156. MR88823

[62] S. Schmidt, *The Petersen graph has no quantum symmetry*, Bull. Lond. Math. Soc. **50** (2018), no. 3, 395–400, DOI 10.1112/blms.12154. MR3829728

[63] S. Schmidt, *On the quantum symmetry of distance-transitive graphs*, Adv. Math. **368** (2020), 107150.

[64] S. Schmidt, *Quantum automorphisms of folded cube graphs* (English, with English and French summaries), Ann. Inst. Fourier (Grenoble) **70** (2020), no. 3, 949–970. MR4117054

[65] S. Schmidt, *Quantum isomorphic strongly regular graphs from the E_8 root system*, arXiv preprint (2022).

[66] M. Schürmann, *White noise on bialgebras*, Lecture Notes in Mathematics, vol. 1544, Springer-Verlag, Berlin, 1993, DOI 10.1007/BFb0089237. MR1238942

[67] R. Speicher, *On universal products*, Free probability theory (Waterloo, ON, 1995), Fields Inst. Commun., vol. 12, Amer. Math. Soc., Providence, RI, 1997, pp. 257–266. MR1426844

[68] L. Teyssier, *Limit profile for random transpositions*, Ann. Probab. **48** (2020), no. 5, 2323–2343, DOI 10.1214/20-AOP1424. MR4152644

[69] T. Timmermann, *An invitation to quantum groups and duality: From Hopf algebras to multiplicative unitaries and beyond*, EMS Textbooks in Mathematics, European Mathematical Society (EMS), Zürich, 2008, DOI 10.4171/043. MR2397671

[70] D. V. Voiculescu, K. J. Dykema, and A. Nica, *Free random variables: A noncommutative probability approach to free products with applications to random matrices, operator algebras and harmonic analysis on free groups*, CRM Monograph Series, vol. 1, American Mathematical Society, Providence, RI, 1992, DOI 10.1090/crmm/001. MR1217253

[71] C. Voigt, *Infinite quantum permutations*, Adv. Math. **415** (2023), Paper No. 108887, 37 pp., DOI 10.1016/j.aim.2023.108887. MR4543448

[72] S. Wang, *Quantum symmetry groups of finite spaces*, Comm. Math. Phys. **195** (1998), no. 1, 195–211, DOI 10.1007/s002200050385. MR1637425

[73] M. Weber, *Quantum permutation matrices*, Complex Anal. Oper. Theory **17** (2023), no. 3, Paper No. 37, 26 pp., DOI 10.1007/s11785-023-01335-x. MR4564553

[74] S. L. Woronowicz, *Compact matrix pseudogroups*, Comm. Math. Phys. **111** (1987), no. 4, 613–665. MR901157

[75] S. L. Woronowicz, *Tannaka-Kreĭn duality for compact matrix pseudogroups. Twisted* SU(N) *groups*, Invent. Math. **93** (1988), no. 1, 35–76, DOI 10.1007/BF01393687. MR943923

[76] S.L. Woronowicz, *Compact quantum groups*, Symétries quantiques (Les Houches, 1995) (1998), 845–884.

Email address: amaury.freslon@universite-paris-saclay.fr

Laboratoire de Mathématique d'Orsay, CNRS, Université Paris-Saclay, 91405 Orsay, France.

Contemporary Mathematics
Volume **798**, 2024
https://doi.org/10.1090/conm/798/15985

A note on the quotient of a locally JC-algebra by a closed Jordan ideal

Oleg Friedman and Alexander A. Katz

ABSTRACT. In the sequel some locally JC-algebra versions of results of Apostol and Effros-Størmer about quotient by a closed Jordan ideal are established.

1. Introduction

The real Jordan analogues of C^*-algebras, so called JB**-algebras,** were first introduced in 1978 by Alfsen, Shultz and Størmer in [**1**]. One of the main results of the aforementioned paper stated that modulo factorization over a unique Jordan ideal each JB-algebra is isomerically isomorphic to a JC-algebra, i.e. an operator norm closed Jordan subalgebra of the Jordan algebra of all bounded self-adjoint operators with symmetric multiplication acting on a complex Hilbert space $\mathcal{H}$. The development of the subject is reflected in the monographs [**9**] and [**4**], which we will use for further references on the general theory of JB-algebras. In particular, in 1967, in their paper [**11**] Effros and Størmer, using second dual techniques, have shown that the quotient of a JC-algebra by a closed Jordan ideal is isometrically Jordan isomorphic to a JC-algebra. This result is of interest because it is known (see for example [**5**]) that, in general, Jordan homomorphic images of special Jordan algebras can fail to be special.

The Hausdorff projective limits of projective families of Banach algebras as natural locally-convex generalizations of Banach algebras have been studied sporadically by many authors since 1952, when they were first introduced by Arens [**3**] and Michael [**18**]. The Hausdorff projective limits of projective families of C^*-algebras were first mentioned by Arens [**3**]. They have since been studied under various names by many authors. Development of the subject is reflected in the monograph of Fragoulopoulou [**12**]. We will follow Inoue [**10**] in the usage of the name **locally** C^***-algebras** for these algebras. In particular, in 1971 Apostol in his paper [**2**] has shown that a bounded part $b(\mathcal{A})$ of a unital locally C^*-algebra $\mathcal{A}$ is a C^*-algebra, and the natural projections in the Arens-Michael decomposition of a locally C^*-algebra into a projective limit of a projective family of C^*-algebra are all surjections.

2020 *Mathematics Subject Classification.* Primary 46L70, 46H70; Secondary 17C65.
Key words and phrases. Locally JC-algebras, locally Hilbert spaces, Jordan ideals.
This paper is in final form and no version of it will be submitted for publication elsewhere.

The Jordan analogues of locally C^*-algebras- so called **locally JB-algebras** (*locally JC-algebras*) as projective limits of projective families of JB-algebras (JC-algebras) were first introduced in 2006 by Katz and Friedman (see [**15**]).

The aim of the present paper is the obtain analogues the aforementioned theorems of Apostol and Effros-Størmer in the case of locally JC-algebras.

2. Preliminaries

2.1. Locally Hilbert Spaces. By definition, $\{\mathcal{H}_\lambda\}_{\lambda\in\Lambda}$ is a *strictly inductive system of Hilbert spaces* if

i). $(\Lambda; \preceq)$ is a directed poset;

ii). $\{\mathcal{H}_\lambda\}_{\lambda\in\Lambda}$ is a net of Hilbert spaces $(\mathcal{H}_\lambda; \langle\cdot,\cdot\rangle_{\mathcal{H}_\lambda})$, $\lambda\in\Lambda$;

iii). for each $\lambda,\mu\in\Lambda$ with $\lambda\preceq\mu$ we have $\mathcal{H}_\lambda\subseteq\mathcal{H}_\mu$;

iv). for each $\lambda,\mu\in\Lambda$ with $\lambda\preceq\mu$ the inclusion map $J_{\mu,\lambda}\colon\mathcal{H}_\lambda\to\mathcal{H}_\mu$ is isometric, that is, $\langle x,y\rangle_{\mathcal{H}_\lambda}=\langle x,y\rangle_{\mathcal{H}_\mu}$, for all $x,y\in\mathcal{H}_\lambda$.

As it is shown in [**6**], for any strictly inductive system of Hilbert spaces $\{\mathcal{H}_\lambda\}_{\lambda\in\Lambda}$, its inductive limit

$$\mathcal{H}=\varinjlim_{\lambda\in\Lambda}\mathcal{H}_\lambda$$

is a Hausdorff locally convex space.

A *locally Hilbert space* (see [**10**][**6**]) is, by definition, the inductive limit

$$\mathcal{H}=\varinjlim_{\lambda\in\Lambda}\mathcal{H}_\lambda=\bigcup_{\lambda\in\Lambda}\mathcal{H}_\lambda,$$

of a strictly inductive system $\{\mathcal{H}_\lambda\}_{\lambda\in\Lambda}$ of Hilbert spaces. We note that a locally Hilbert space is a special type of locally convex space and, in general, not a Hilbert space. It is clear that a locally Hilbert space is uniquely determined by the strictly inductive system of Hilbert spaces.

2.2. Locally Bounded Operators. Let

$$\mathcal{H}=\varinjlim_{\lambda\in A}\mathcal{H}_\lambda$$

and

$$\mathcal{K}=\varinjlim_{\lambda\in A}\mathcal{K}_\lambda$$

be two locally Hilbert spaces generated by strictly inductive systems of Hilbert spaces $(\{\mathcal{H}_\lambda\}_{\lambda\in\Lambda};\{J^{\mathcal{H}}_{\mu,\lambda}\}_{\lambda\preceq\mu})$ and, respectively, $(\{\mathcal{K}_\lambda\}_{\lambda\in\Lambda};\{J^{\mathcal{K}}_{\mu,\lambda}\}_{\lambda\preceq\mu})$, indexed on the same directed poset Λ. A linear map

$$T\colon\mathcal{H}\to\mathcal{K}$$

is called a *locally bounded operator* if T is a continuous coherent linear map and adjointable, more precisely,

i). There exists a net of operators $\{T_\lambda\}_{\lambda\in\Lambda}$, with $T_\lambda\in\mathcal{B}(\mathcal{H}_\lambda,\mathcal{K}_\lambda)$ such that

$$TJ^{\mathcal{H}}_\lambda=J^{\mathcal{K}}_\lambda T_\lambda$$

for all $\lambda\in\Lambda$.

ii). The net of operators $\{T^*_\lambda\}_{\lambda\in\Lambda}$ is coherent as well, that is,

$$T^*_\mu J^{\mathcal{K}}_{\mu,\lambda}=J^{\mathcal{H}}_{\mu,\lambda}T^*_\lambda,$$

for all $\lambda,\mu\in\Lambda$ such that $\lambda\le\mu$.

We denote by $\mathcal{B}_{loc}(\mathcal{H},\mathcal{K})$ the collection of all locally bounded operators

$$T\colon \mathcal{H}\to\mathcal{K}$$

It is easy to see that $\mathcal{B}_{loc}(\mathcal{H},\mathcal{K})$ is a vector space.

2.3. ***JC*-algebras.** A *Jordan Banach algebra* is a real Jordan algebra $(\mathcal{A},\circ)$ equipped with a complete norm satisfying

$$\|a\circ b\|\le\|a\|\,\|b\|\,,$$

$a,b\in\mathcal{A}$ A *JB-algebra* is a Jordan Banach algebra in which the norm satisfies the following two additional properties for $a,b\in\mathcal{A}$,

(*i*) $\|a^2\|=\|a\|^2$;

(*ii*) $\|a^2\|\le\|a^2+b^2\|$.

Examples of JB-algebras are *JC-algebras*, i.e. the norm closed , Jordan algebras of self-adjoint operators on a complex Hilbert space, and the exceptional M_3^8 consisting of all Hermitian 3×3 matrices over Cayley numbers.

Notice, that an associative JB-algebra can be realized as a self-adjoint part of a commutative C^*-algebra.

The self-adjoint part $B(H)_{sa}$ of all bounded operators $B(H)$ on a complex Hilbert space H is a Jordan algebra under the (non-associative) product

$$a\circ b=\frac{1}{2}(ab+ba).$$

JC-algebra is an ordered space with order inherited from that of $B(H)_{sa}$ via the cone $B(H)_+$ of positive operators on H.

A *Jordan ideal* J in a Jordan algebra ($\mathcal{A}$,$\circ$) is a linear subspace of $\mathcal{A}$ such that if $a\in\mathcal{A}$, and $b\in J$, $a\circ b\in J$.

3. Theorems of Apostol for locally *JC*-algebras

Let now $\mathcal{A}_\lambda$, $\lambda\in\Lambda$, be a special JB-algebra, otherwise speaking JC-algebra, thus (see, for example [**9**]), there exists a complex Hilbert space $\mathcal{G}_\lambda$, such that $\mathcal{A}_\lambda$ is isometrically Jordan isomorphic to a norm closed Jordan subalgebra of the JC-algebra $\mathcal{B}(\mathcal{G}_\lambda)_{sa}$ of all self-adjoint bounded linear operators acting on $\mathcal{G}_\lambda$. Let $\mathcal{A}$ be locally JC-algebra which is a projective limit

$$\mathcal{A}=\varprojlim_{\lambda\in\Lambda}\mathcal{A}_\lambda$$

of the projective family of JC-algebras $\mathcal{A}_\lambda$, $\lambda\in\Lambda$. It is easy to see that $\mathcal{A}$ is unital iff each $\mathcal{A}_\lambda$ is unital (see [**15**]). The following version of Gelfand-Naimark theorem is valid:

THEOREM 1 (Gelfand-Naimark). *Any locally JC-algebra $\mathcal{A}$ can be coherently identified with some concrete operator locally JC-algebra, more precisely,*

$$\mathcal{A}=\varprojlim_{\lambda\in\Lambda}\mathcal{A}_\lambda,$$

where $\mathcal{A}_\lambda$, $\lambda\in\Lambda$, is a projective family of JC-algebras over some directed poset Λ, then there exists a complex locally Hilbert space

$$\mathcal{H}=\varinjlim_{\lambda\in\Lambda}\mathcal{H}_\lambda,$$

and a faithful coherent Jordan representation

$$\sigma : \mathcal{A} \to \mathcal{B}_{loc}(\mathcal{H})_{sa}.$$

PROOF. As it was pointed above, for each $\mu \in \Lambda$, there exists a Hilbert space $\mathcal{G}_\mu$ and a faithful Jordan morphism

$$\rho_\mu : \mathcal{A}_\mu \to \mathcal{B}(\mathcal{H}_\mu)_{sa}.$$

Now, for each $\lambda \in \Lambda$, let us consider a Hilbert space

$$\mathcal{H}_\lambda = \bigoplus_{\mu \preceq \lambda} \mathcal{G}_\mu.$$

By identifying $\mathcal{H}_\lambda$ with the subspace $\mathcal{H}_\lambda \oplus 0$ of $\mathcal{H}_\eta$ for any $\lambda \preceq \eta$, we can see that $\mathcal{H}_\lambda$, $\lambda \in \Lambda$, is a strictly inductive family of Hilbert spaces. Let

$$\mathcal{H} = \varinjlim_{\lambda \in \Lambda} \mathcal{H}_\lambda,$$

be a locally Hilbert space which is the inductive limit of the inductive family of $\mathcal{H}_\lambda, \lambda \in \Lambda$. Thus, for each $\lambda \in \Lambda$ we can define

$$\sigma_\lambda : \mathcal{A}_\lambda \to \mathcal{B}(\mathcal{H}_\lambda),$$

as

$$\sigma_\lambda(a_\lambda) = \bigoplus_{\mu \preceq \lambda} \rho_\mu(a_\mu),$$

where

$$a_\mu = \pi_\mu(a),$$

$\mu \in \Lambda$, $a \in \mathcal{A}$. One can see that each morphism $\sigma_\lambda, \lambda \in \Lambda$ is faithful. This the family of morphisms $\sigma_\lambda, \lambda \in \Lambda$, correctly defines a coherent faithful Jordan representation

$$\sigma : A \to \mathcal{B}_{loc}(\mathcal{H})_{sa},$$

of $\mathcal{A}$ on $\mathcal{H}$, such that for each $\lambda \in \Lambda$,

$$\sigma_\lambda \circ \pi_\lambda = \pi_\lambda \circ \sigma.$$

□

Let $\mathcal{A}$ be a locally JC-algebra, and $S(\mathcal{A})$ be the family of JC-seminorms that define the topology τ_S on $\mathcal{A}$. Let

$$b(\mathcal{A}) = \{a \in \mathcal{A} : \sup_{p \in S(\mathcal{A})} p(a) < \infty\}.$$

We equip $b(\mathcal{A})$ with the norm

$$\|a\|_b = \sup_{p \in S(\mathcal{A})} p(a), \ a \in b(\mathcal{A}).$$

It is easy to see that $b(\mathcal{A})$ is a Jordan subalgebra of $\mathcal{A}$. $b(\mathcal{A})$ is called the *bounded part* of $\mathcal{A}$.

For what follows we need the following lemmata:

LEMMA 1. *Let $(\mathcal{A}, \tau_S)$ be a unital locally JC-algebra with a unit $\mathbf{1}_A$. Let $z \in (\mathcal{A}, \tau_S)^+$, and*

$$x = z \circ (\mathbf{1}_\mathcal{A} + z)^{-1}, \ y = z \circ (\mathbf{1}_\mathcal{A} + z)^{-2}.$$

Then

$$p(x) = p(z)(1 + p(z))^{-1}, \text{ and } p(y) \leq \frac{1}{4},$$

for each $p \in S(\mathcal{A})$.

PROOF. We consider the associative subalgebra of $\mathcal{A}$ generated by $\mathbf{1}_{\mathcal{A}}$ and z. This subalgebra contains the elements x and y. Using the results from [**13**] we turn our attention to a locally multiplicatively convex algebra of real valued functions on compactly generated space. After that we proceed as in [**2**] to reach the conclusions. □

LEMMA 2. *Let* $(\mathcal{A}, \tau_S)$ *be a unital locally* JC*-algebra with a unit* $\mathbf{1}_A$*, and* $b(\mathcal{A})$ *be its bounded part. Let* $a \in \mathcal{A}$ *and* $t > 0$*. Then* $a \circ (\mathbf{1}_{\mathcal{A}} + ta^2)^{-1}$ *belongs to* $b(\mathcal{A})$.

PROOF. It is easy to see that $a^2 \geq \mathbf{0}$, so that $\mathbf{1}_{\mathcal{A}} + ta^2 > 0$ for any $t > 0$, therefore $(\mathbf{1}_{\mathcal{A}} + ta^2)^{-1}$ exists in $\mathcal{A}$. Also, considering an associative subalgebra generated by $\mathbf{1}_{\mathcal{A}}$ and a, we get

$$\begin{gathered} p(a \circ (\mathbf{1}_{\mathcal{A}} + ta^2)^{-1})^2 = p((a \circ ((\mathbf{1}_{\mathcal{A}} + ta^2)^{-1})^2) \\ = p((a \circ (\mathbf{1}_{\mathcal{A}} + ta^2)^{-1} \circ (a \circ (\mathbf{1}_{\mathcal{A}} + ta^2)^{-1}) \\ = p((\mathbf{1}_{\mathcal{A}} + ta^2)^{-1} \circ a^2 \circ (\mathbf{1}_{\mathcal{A}} + ta^2)^{-1}) = p(a^2 \circ (\mathbf{1}_{\mathcal{A}} + ta^2)^{-2}) \\ = \frac{1}{t} p(ta^2 \circ (\mathbf{1}_{\mathcal{A}} + ta^2)^{-2}) \leq \frac{1}{4t}, \end{gathered}$$

for any $p \in S(\mathcal{A})$. Thus, $a \circ (\mathbf{1}_{\mathcal{A}} + ta^2)^{-1}$ belongs to $b(\mathcal{A})$. □

Now we are ready to consider Apostol theorems for locally JC-algebras:

THEOREM 2 (Apostol). *Let* $(\mathcal{A}, \tau_S)$ *be a unital locally* JC*-algebra. Then* $b(\mathcal{A})$ *is a unital* JC*-algebra, dense in* $\mathcal{A}$.

PROOF. Let $a \in \mathcal{A}$ and $t > 0$. From Lemma 2 above it follows that $a \circ (\mathbf{1}_{\mathcal{A}} + ta^2)^{-1}$ belongs to $b(\mathcal{A})$. One can see that

$$a - a \circ (\mathbf{1}_{\mathcal{A}} + ta^2)^{-1} = ta \circ a^2 \circ (\mathbf{1}_{\mathcal{A}} + ta^2)^{-1}.$$

Therefore, for any $p \in S(\mathcal{A})$,

$$\begin{aligned} p(a - a \circ (\mathbf{1}_{\mathcal{A}} + ta^2)^{-1}) &= p(ta \circ a^2 \circ (\mathbf{1}_{\mathcal{A}} + ta^2)^{-1}) \\ &\leq t \cdot p(a^2) p(a \circ (\mathbf{1}_{\mathcal{A}} + ta^2)^{-1}) \\ &\leq t \cdot p(a)^2 \cdot \frac{1}{2\sqrt{t}} = \frac{1}{2}\sqrt{t} \cdot p(a)^2. \end{aligned}$$

Thus, when $t \to 0$,

$$p(a - a \circ (\mathbf{1}_{\mathcal{A}} + ta^2)^{-1}) \to 0,$$

for any $p \in S(\mathcal{A})$, which proves the density of $b(\mathcal{A})$ in $(\mathcal{A}, \tau_S)$.

It is easy to see that $b(\mathcal{A})$ is a pre-JC-algebra. We show that completeness of $b(\mathcal{A})$ in $\|.\|_b$. Let a_n be any Cauchy sequence in $b(\mathcal{A})$. Thus there exists

$$M > 0,$$

such that

$$\|a_n\|_b < M,$$

for all $n \in \mathbb{N}$. The topology generated by $\|.\|_b$ on $b(\mathcal{A})$ is stronger than the topology τ_S, thus, the sequence a_n is a Cauchy sequence in $(\mathcal{A}, \tau_S)$. Thus, there exists such $a \in \mathcal{A}$, so that

$$a_n \xrightarrow{\tau_S} a.$$

It implies that

$$p(a) = \lim_{n\to\infty} p(a_n) \le \lim_{n\to\infty} \|a_n\|_b < M,$$

for all $p \in S(\mathcal{A})$, thus, we conclude that $a \in b(\mathcal{A})$. It remains to show that

$$a_n \xrightarrow{\|\cdot\|_b} a.$$

Let $\varepsilon > 0$ be arbitrary. Due to the sequence a_n being Cauchy, there exists $N(\varepsilon) \in \mathbb{N}$, such that when $l, m > N(\varepsilon)$,

$$\|a_l - a_m\|_b < \varepsilon.$$

Therefore, for any $n, m > N(\varepsilon)$,

$$p(a - a_n) = \lim_{m\to\infty} p(a_m - a_n) \le \lim_{m\to\infty} \|a_m - a_n\|_b < \varepsilon,$$

for any $p \in S(\mathcal{A})$, which completes the proof. □

Let $(\mathcal{A}, \tau_S)$ be a locally JC-algebra. For $p \in S(\mathcal{A})$, we define

$$\mathcal{N}_p = \{a \in \mathcal{A} : p(a) = 0\}.$$

THEOREM 3 (Apostol). *Let $(\mathcal{A}, \tau_S)$ be a locally JC-algebra. Then for every $p \in S(\mathcal{A})$, the pre-JC-algebra $(\mathcal{A}, p)/\mathcal{N}_p$ is complete, hence JC-algebra. Moreover, we have that*

$$(\mathcal{A}, p)/\mathcal{N}_p = \mathcal{A}_p,$$

for any $p \in S(\mathcal{A})$, where $\mathcal{A}_p$ is the completion of $(\mathcal{A}, p)/\mathcal{N}_p$ under the quotient norm $\|.\|_p$, such that

$$\|a_p\|_p = p(a),$$

for each $a \in \mathcal{A}$ and $a_p \in (\mathcal{A}, p)/\mathcal{N}_p$.

PROOF. Using standard arguments, without the loss of generality we may assume that $\mathcal{A}$ is unital. Let now

$$J_p = b(\mathcal{A}) \cap \mathcal{N}_p = \{a \in b(\mathcal{A}) : p(a) = 0\},$$

for any $p \in S(\mathcal{A})$. J_p is a closed Jordan ideal in $b(\mathcal{A})$. Therefore, from [**11**] it follows that $b(\mathcal{A})/J_p$ is a JC-algebra. Let us denote the quotient JC-norm on $b(\mathcal{A})/J_p$ by $\|.\|_p^b$. Due to the fact that a surjective Jordan isomorphism between two JC-algebra is an isometry (see, for example [**9**]),

$$\|a + J_p\|_p^b = p(a),$$

for any $a \in \mathcal{A}$, and $p \in S(\mathcal{A})$. We consider a mapping

$$\varphi : b(\mathcal{A})/J_p \to (\mathcal{A}, p)/\mathcal{N}_p,$$

as

$$\varphi(a + J_p) = a_p,$$

where

$$a_p = a + \mathcal{N}_p,$$

$p \in S(\mathcal{A})$. It is obvious that φ is an isometry. We show that φ is a surjection. Let $a_p \in (\mathcal{A}, p)/\mathcal{N}_p$ be arbitrary. From continuity of natural projection

$$\pi_p : (\mathcal{A}, \tau_S) \to (\mathcal{A}, p)/\mathcal{N}_p,$$

and Lemma 2 it follows that

$$a_p = \pi_p(a) = \lim_{t\to 0}\pi_p(a \circ (\mathbf{1}_{\mathcal{A}} + ta^2)^{-1})$$

$$= \lim_{t\to 0}(a \circ (\mathbf{1}_{\mathcal{A}} + ta^2)^{-1} + \mathcal{N}_p) = \lim_{t\to 0}\varphi(a \circ (\mathbf{1}_{\mathcal{A}} + ta^2)^{-1} + J_p).$$

Thus, we can conclude that $\varphi(b(\mathcal{A})/J_p)$ is dense in $(\mathcal{A}, p)/\mathcal{N}_p$ for every $p \in S(\mathcal{A})$. The algebra $\varphi(b(\mathcal{A})/J_p)$ is complete for every $p \in S(\mathcal{A})$, and thus the algebra $(\mathcal{A}, p)/\mathcal{N}_p$ is complete. □

4. Quotients of a locally JC-algebra by closed Jordan ideals

Now we establish an extension of Effros-Størmer to locally JC-algebras.

THEOREM 4 (Effros-Størmer). *Let $(\mathcal{A}, \tau_S)$ be a locally JC-algebra, and J be a τ_S-closed Jordan ideal in $\mathcal{A}$. Then $(\mathcal{A}, \tau_S)/J$ is a locally JC-algebra, and in particular, each continuous homomorphic image of a locally JC-algebra in a locally JB-algebra is a locally JC-algebra.*

PROOF. Let $(\mathcal{A}, \tau_S)$ be a locally JC-algebra, and let

$$(\mathcal{A}, \tau_S) \cong \varprojlim_{\lambda\in\Lambda} \mathcal{A}_\lambda,$$

be its Arens-Michael decomposition as a projective limit of a projective family of JC-algebras $\mathcal{A}_\lambda$, $\lambda \in \Lambda$. Let

$$\pi_\lambda : \mathcal{A} \to \mathcal{A}_\lambda$$

be the natural projection from $\mathcal{A}$ onto $\mathcal{A}_\lambda, \lambda \in \Lambda$. For each $\lambda \in \Lambda$, we denote by J_λ the projection of J onto J_λ, i.e.

$$\pi_\lambda(J) = J_\lambda.$$

From the fact that J is a Jordan ideal of $\mathcal{A}$ it follows that J_λ is a Jordan ideal of $\mathcal{A}_\lambda$ for each $\lambda \in \Lambda$. From continuity of π_λ it follows that each J_λ is closed in the JC-norm of $\mathcal{A}_\lambda, \lambda \in \Lambda$. Thus, from **[11]** it follows that for each $\lambda \in \Lambda$, J_λ is a JC-algebra. One can see that the family $J_\lambda, \lambda \in \Lambda$, is projective, and

$$(J, \tau_S) \cong \varprojlim_{\lambda\in\Lambda} J_\lambda,$$

$\lambda \in \Lambda$. From that it follows that

$$(\mathcal{A}, \tau_S)/J \cong \varprojlim_{\lambda\in\Lambda} \mathcal{A}_\lambda/J_\lambda.$$

And, since the kernel of a continuous homomorphism to a locally JB-algebra is a Jordan ideal, the homomorphic image of a locally JC-algebra is again a locally JC-algebra. □

5. Ideals in locally JC-algebras and locally C^*-algebras they generate

Let $(\mathcal{A}, \tau_S)$ be a unital locally JC-algebra acting on a locally Hilbert space $\mathcal{H}$, i.e. $(\mathcal{A}, \tau_S)$ is a locally JC-subalgebra of a locally JC-algebra $\mathcal{B}_{loc}(\mathcal{H})_{sa}$. Let S be a subspace of $\mathcal{A}$. Let us denote by $C^*_{loc}(\mathcal{S})$ the intersection of all locally C^*-subalgebras of $\mathcal{B}_{loc}(\mathcal{H})$, that contain S.

The following theorem is valid:

THEOREM 5 (Effros-Størmer). *Let $(\mathcal{A}, \tau_S)$ be a unital locally JC-subalgebra in $\mathcal{B}_{loc}(\mathcal{H})_{sa}$, and J be a τ_S-closed Jordan ideal in $\mathcal{A}$. Then $\mathcal{A}/J$ is topologically Jordan isomorphic with a locally JC-algebra, $J = C^*_{loc}(J) \cap \mathcal{A}$, and $C^*_{loc}(J)$ is an ideal in $C^*_{loc}(\mathcal{A})$.*

PROOF. Let $(\mathcal{A}, \tau_S)$ be a locally JC-subalgebra in $\mathcal{B}_{loc}(\mathcal{H})_{sa}$, where

$$\mathcal{H} = \varinjlim_{\lambda \in \Lambda} \mathcal{H}_\lambda,$$

$$\mathcal{B}_{loc}(\mathcal{H}) \cong \varprojlim_{\lambda \in \Lambda} \mathcal{B}(\mathcal{H}_\lambda),$$

$$\pi_\lambda : \mathcal{B}_{loc}(\mathcal{H}) \to \mathcal{B}(\mathcal{H}_\lambda)$$

be the natural projection from $\mathcal{B}_{loc}(\mathcal{H})$ onto $\mathcal{B}(\mathcal{H}_\lambda), \lambda \in \Lambda$.and let

$$(\mathcal{A}, \tau_S) \cong \varprojlim_{\lambda \in \Lambda} \mathcal{A}_\lambda,$$

be its Arens-Michael decomposition as a projective limit of a projective family of JC-algebras $\mathcal{A}_\lambda$, $\lambda \in \Lambda$, where each $\mathcal{A}_\lambda$.acts on $\mathcal{H}_\lambda$ Let

$$\pi_\lambda : \mathcal{A} \to \mathcal{A}_\lambda$$

be the natural projection from $\mathcal{A}$ onto $\mathcal{A}_\lambda, \lambda \in \Lambda$.

It has been already established in Theorem 4 above that

$$(\mathcal{A}, \tau_S)/J \cong \varprojlim_{\lambda \in \Lambda} \mathcal{A}_\lambda / J_\lambda,$$

where $\mathcal{A}_\lambda / J_\lambda$ is a projective family of JC-algebras, $\lambda \in \Lambda$.

Let us denote the intersection of all C^*-subalgebras in $\mathcal{B}(\mathcal{H}_\lambda)$ that contain a set $S_\lambda \subset \mathcal{B}(\mathcal{H}_\lambda)_{sa}$ by $C^*(S_\lambda)$. One can see that for each $\lambda \in \Lambda$,

$$\pi_\lambda(C^*_{loc}(\mathcal{A})) = C^*(\mathcal{A}_\lambda), \quad \pi_\lambda(C^*_{loc}(J)) = C^*(J_\lambda),$$

and

$$C^*_{loc}(\mathcal{A}) \cong \varprojlim_{\lambda \in \Lambda} C^*(\mathcal{A}_\lambda), \quad C^*_{loc}(J) \cong \varprojlim_{\lambda \in \Lambda} C^*(J_\lambda).$$

From Theorem 2 in [**11**] it follows, that for each $\lambda \in \Lambda$,

$$J_\lambda = C^*(J_\lambda) \cap \mathcal{A}_\lambda,$$

and $C^*(J_\lambda)$ is an ideal in $C^*(\mathcal{A}_\lambda)$. From that it follows that

$$J = C^*_{loc}(J) \cap \mathcal{A},$$

and $C^*_{loc}(J)$ is an ideal in $C^*_{loc}(\mathcal{A})$. □

6. Characterization of Jordan ideals

At last, we obtain the following characterization of closed Jordan ideals in locally JC-algebras:

THEOREM 6 (Effros-Størmer). *Let $(\mathcal{A}, \tau_S)$ be a unital locally JC-algebra, and J be a τ_S-closed subspace of $\mathcal{A}$ that contains $\mathbf{1}_{\mathcal{A}}$. The following properties are equivalent:*

i). J is a Jordan ideal in $\mathcal{A}$;

ii). If $a \in \mathcal{A}$, and $b \in J$, then $aba \in J$.

PROOF. Let J be a Jordan ideal in $\mathcal{A}$, $a \in \mathcal{A}$, and $b \in J$. From the equation

$$aba = \frac{1}{2}\left[(a(ba+ab) + (ba+ab)a - (a^2b + ba^2)\right]$$

it follows that $aba \in J$.

Conversely, let for each $a \in \mathcal{A}$, and each $b \in J$, the element $aba \in J$. Then

$$a \circ b = \frac{1}{2}(ab+ba) = \frac{1}{2}\left((a+\mathbf{1}_{\mathcal{A}})b(a+\mathbf{1}_{\mathcal{A}}) - aba - b\right).$$

Due to the assumptions, $a \circ b \in J$. □

References

[1] E. M. Alfsen, F. W. Shultz, and E. Størmer, *A Gel′fand-Neumark theorem for Jordan algebras*, Advances in Math. **28** (1978), no. 1, 11–56, DOI 10.1016/0001-8708(78)90044-0. MR482210

[2] C. Apostol, *b^*-algebras and their representation*, J. London Math. Soc. (2) **3** (1971), 30–38, DOI 10.1112/jlms/s2-3.1.30. MR284816

[3] R. Arens, *A generalization of normed rings*, Pacific J. Math. **2** (1952), 455–471. MR51445

[4] E. M. Alfsen and F. W. Shultz, *Geometry of state spaces of operator algebras*, Mathematics: Theory & Applications, Birkhäuser Boston, Inc., Boston, MA, 2003, DOI 10.1007/978-1-4612-0019-2. MR1947002

[5] P. M. Cohn, *On homomorphic images of special Jordan algebras*, Canad. J. Math. **6** (1954), 253–264, DOI 10.4153/cjm-1954-026-9. MR60496

[6] A. Gheondea, *On locally Hilbert spaces*, Opuscula Math. **36** (2016), no. 6, 735–747, DOI 10.7494/OpMath.2016.36.6.735. MR3571419

[7] A. Gheondea, *Operator models for Hilbert locally C^*-modules*, Oper. Matrices **11** (2017), no. 3, 639–667, DOI 10.7153/oam-11-43. MR3655673

[8] A. Gheondea, *The spectral theorem for locally normal operators*, Opuscula Math. **38** (2018), no. 5, 597–621, DOI 10.7494/opmath.2018.38.5.597. MR3818589

[9] H. Hanche-Olsen and E. Størmer, *Jordan operator algebras*, Monographs and Studies in Mathematics, vol. 21, Pitman (Advanced Publishing Program), Boston, MA, 1984. MR755003

[10] A. Inoue, *Locally C^*-algebra*, Mem. Fac. Sci. Kyushu Univ. Ser. A **25** (1971), 197–235, DOI 10.2206/kyushumfs.25.197. MR305089

[11] E. G. Effros and E. Størmer, *Jordan algebras of self-adjoint operators*, Trans. Amer. Math. Soc. **127** (1967), 313–316, DOI 10.2307/1994650. MR206733

[12] M. Fragoulopoulou, *Topological algebras with involution*, North-Holland Mathematics Studies, vol. 200, Elsevier Science B.V., Amsterdam, 2005. MR2172581

[13] O. Friedman and A. A. Katz, *On Gelfand-Naimark type theorems for unital abelian complex and real locally C^*-, and locally JB-algebras*, Topological algebras and their applications, De Gruyter Proc. Math., De Gruyter, Berlin, 2018, pp. 73–108. MR3793865

[14] D. J. Karia and Y. M. Parmar, *Operators on locally Hilbert space*, J. Anal. **23** (2015), 59–73. MR3491647

[15] A. A. Katz and O. Friedman, *On projective limits of real C^*- and Jordan operator algebras*, Vladikavkaz. Mat. Zh. **8** (2006), no. 2, 33–38, DOI 10.21314/jor.2006.128. MR2434387

[16] A. A. Katz and O. Friedman, *On a Gelfand-Naimark type theorem for locally JB-algebras*, Indian J. Math. **55** (2013), no. Fifth Dr. George Bachman Memorial Conference, suppl., 103–116. MR3310060

[17] A. A. Katz and O. Friedman, *On universal representations and universal enveloping locally* C*-*algebras for locally JB-algebras*, Topics in functional analysis and algebra, Contemp. Math., vol. 672, Amer. Math. Soc., Providence, RI, 2016, pp. 185–204, DOI 10.1090/conm/672/13472. MR3546701

[18] E. A. Michael, *Locally multiplicatively-convex topological algebras.* (English) Mem. Amer. Math. Soc. Vol. 11 (1952), 79 pp.

[19] K. McCrimmon, *A taste of Jordan algebras.* (English) Universitext. Springer-Verlag, New York (2004), 562 pp.

[20] G. J. Murphy, *C^*-algebras and operator theory.* (English) Academic Press, Inc., Boston, MA, (1990), 286 pp.

[21] D. M. Topping, *Jordan algebras of self-adjoint operators.* (English) Mem. Amer. Math. Soc. Vol. 53 (1965), 48 pp.

Department of Mathematics, Lander College for Men, Touro University, 75-31 150th Street, Kew Gardens Hills, NY 11367, USA

Email address: oleg.friedman@touro.edu

Department of Mathematics and Computer Science, St. John's College of LAS, St. John's University, 8000 Utopia Parkway, SJH-334-G, Queens, NY 11439, USA

Email address: katza@stjohns.edu

Contemporary Mathematics
Volume **798**, 2024
https://doi.org/10.1090/conm/798/15986

Villadsen Idempotents

Cristian Ivanescu and Dan Kucerovsky

Abstract. C*-algebras are rings, sometimes nonunital, obeying certain axioms that ensure a very well-behaved representation theory upon Hilbert space. Moreover, there are some well-known features of the representation theory leading to subtle questions about norms on tensor products of C*-algebras, and thus to the subclass of nuclear C*-algebras. The question whether all separable nuclear C*-algebras satisfy the Universal Coefficient Theorem (UCT) remains one of the most important open problems in the structure and classification theory of such algebras. One of the ways to test the UCT conjecture depends on finding C*-algebras that behave as idempotents under the tensor product, and satisfy certain additional properties. Briefly put, if there exists a simple, separable, and nuclear C*-algebra that is an idempotent under the tensor product, satisfies a certain technical property, and is not one of the already known such elements $\{O_\infty, O_2, \mathrm{UHF}_\infty, J, Z, \mathbb{C}, \mathcal{K}\}$ then the UCT fails. Although we do not disprove the UCT in this publication, we do find new idempotents in the class of Villadsen algebras.

1. Idempotents and their relation to the classification of C*-algebras

It is known that in many cases, C*-algebras can be classified by invariants coming from K-theory. Thus it is of interest to compute K-theory groups of C*-algebras, and the main computational tool for K-theoretical questions is possibly the so-called UCT theorem due to Rosenberg and Schochet [**15**]. It has been conjectured that the UCT may hold for all nuclear separable C*-algebras, but on the other hand, it has been pointed out that this conjecture could be disproven by finding a new example of a tensor idempotent having the strongly self-absorbing property. A natural place to look for such an idempotent is in the famous class of algebras studied by Villadsen, because after all, this class has been the main source of counterexamples to many questions about C*-algebras. However, it has also been conjectured that the Villadsen algebras might be classifiable by using sufficiently many invariants. We remain agnostic with respect to these conjectures, but it is generally recognized that the classification program makes key use of tensor idempotents with the strongly self-absorbing property. Thus, whether one hopes to disprove the UCT conjecture, or to classify the Villadsen algebras, it seems very relevant to first investigate the idempotents that may exist in the Villadsen class.

2020 *Mathematics Subject Classification.* Primary 46L05; Secondary 46L35.
Key words and phrases. UCT conjecture, C*-algebras, tensor products, Grothendieck rings.

DEFINITION 1. A unital C*-algebra $\mathcal{A}$ is *strongly self-absorbing* if there is an isomorphism $\phi : \mathcal{A} \to \mathcal{A} \otimes \mathcal{A}$ so that ϕ is approximately unitarily equivalent to the *-homomorphism $a \to a \otimes 1_{\mathcal{A}}$.

All strongly self-absorbing C*-algebras must be simple, nuclear, and $\mathcal{Z}$-stable. The following is a list of the known strongly self-absorbing C*-algebras: the Jiang-Su algebra $\mathcal{Z}$, UHF algebras of infinite type, the Cuntz algebra $\mathcal{O}_\infty$, tensor products of UHF algebras of infinite type, $\mathcal{O}_\infty$, and $\mathcal{O}_2$. If all nuclear C*-algebras satisfy the UCT, this is a complete list of strongly self-absorbing C*-algebras. It is a fact that the strongly self-absorbing property implies $\mathcal{Z}$-stability, and that most Villadsen algebras are not $\mathcal{Z}$-stable. We therefore suggest some possible replacements for the strongly self-absorbing property, and we find idempotents having these properties. The Villadsen algebras are conjectured classifiable by sufficiently many invariants, however there is no known corresponding idempotent. We construct a Villadsen idempotent in the following.

2. Villadsen's algebras of the first type

We start by recalling the class of inductive limits introduced by Villadsen [**20**, pgs. 110-111], generalized slightly as in [**18**].

DEFINITION 2. Let X be a compact Hausdorff space and $n, m, k \in \mathbb{N}$. Consider a unital diagonal $*$-homomorphism

$$\phi : M_n \otimes C(X) \to M_k \otimes C(X^{\times m})$$

where the entries on the diagonal are either of the form $f \mapsto f(y)$ for a point $y \in X$ *or* are of the form $f \mapsto f \circ \pi_i$ where π_i is one of the canonical projection maps $\pi_i \colon X^{\times m} \to X$. This map will be called a *Villadsen connecting map*, or simply a connecting map if no ambiguity can arise.

DEFINITION 3. Let X be a compact Hausdorff space, and let $(n_i)_{i=1}^{\infty}$ and $(m_i)_{i=1}^{\infty}$ be sequences of natural numbers with $n_1 = 1$. Fix a compact Hausdorff space X, and put $X_i = X^{\times n_i}$. A unital C^*-algebra A is a Villadsen algebra if it can be written as an inductive limit algebra

$$A \cong \lim_{i \to \infty} \left(M_{m_i} \otimes C(X^{\times n_i}), \phi_i \right)$$

where each ϕ_i is a Villadsen connecting map.

Recall that in Definition 2, we are allowed to choose evaluation points $y_i \in X$. Thus, with each Villadsen algebra, there is associated a sequence y_i of evaluation points, $y_i \in X$. Villadsen algebras as defined above need not be simple as C*-algebras, but as pointed out by Villadsen [ibid.], if the sequence of evaluation points y_i happens to have dense range in X, then the inductive limit algebra can be shown to be simple. In this case, the algebra obtained is actually independent of the choice of evaluation points, provided that they are dense (See [**4**].)

The Hilbert cube $\mathfrak{U}$ is the universal space in the class of metric spaces with a countable base. Define a topological manifold to be a separable metric space which has an open cover by open subsets of $[0,1]^n$ for some $n = 1, 2, 3, \cdots, \infty$. When $n = \infty$, the term Hilbert cube manifold is also used. Call a space X *n-homogeneous* if for every natural number n and any two sets $F, G \subset X$ with cardinality $|F| = |G| = n$, there is a homeomorphism $h : X \to X$ such that $h(\{F]\}) = \{G\}$. Also, if h can be chosen to extend any previously given bijection between F and G, then X is called *strongly n-homogeneous.* A space U is said to be *ω-homogeneous* if given (a_i) and (b_i) that are countable dense in U, there is a homeomorphism h of U onto U preserving the range of these sequences, so that $h(\{a_i\}) = \{b_i\}$. Cantor showed that the real line is ω-homogeneous, then Fort showed that the Hilbert cube is ω-homogeneous, and then Ungar generalized to the (possibly infinite-dimensional) locally Euclidean case. The following summarizes some known results from their papers [**2**, **6**, **19**].

THEOREM 4 (Bennett-Fort-Unger). *The Hilbert cube, and more generally, all connected n-dimensional topological manifolds with a countable base are ω-homogeneous. This is equivalent to strong M-homogenity for* all *finite M.*

It seems interesting to see how the ω-homogenity of the Hilbert cube manifests in the Villadsen construction; it shows the independence of the choice of (dense) set of evaluation points, and more importantly, it allows us to construct a Villadsen algebra that is an idempotent in the sense discussed previously.

2.1. The basic lemma. Let us first consider how base space homeomorphisms interact with the Villadsen connecting maps in Definition 2.

LEMMA 5. *Suppose that $\phi : M_n \otimes C(X) \to M_k \otimes C(X^{\times m})$ is a Villadsen connecting map with evaluation points (y_i). Let $h\colon X \to X$ be a given self-homeomorphism of X. Let us denote the pullback action of h on $M_n \otimes C(X)$ by $\hat{h}$. Then*

$$\left(\bigoplus_m \hat{h}^{-1}\right) \circ \phi \circ \hat{h}$$

is a Villadsen connecting map with evaluation points $(h(y_i))$.

PROOF. Recall that this diagonal connecting map has two types of entries on the diagonal: they may be either $f \mapsto f(y_i)$ for a given evaluation point $y_i \in X$ or $f \mapsto f \circ \pi_i$ where π_i is one of the canonical projection maps $\pi_i\colon X^{\times m} \to X$. Let us consider the two cases separately. In the first case, we first pull back f by the given homeomorphism h, and evaluate at y_i, giving $f(h(y_i))$. Since this has already been evaluated, when we subsequently apply the pullback action of $\hat{h}^{-1}$, nothing further happens. In other words, $\hat{h}^{-1}$ acts trivially on constant functions in $M_n \otimes C(X)$.

In the second case, we first pull back f by the given homeomorphism h, and then subsequently apply the pullback action of h^{-1}, and that gives us f back again. This completes the proof that the map obtained is a Villadsen connecting map , and the first paragraph already showed that the evaluation points are of the form $(h(y_i))$, as was to be shown. □

We can use the basic lemma to show the algebra obtained in the limit is independent of the choice of basepoint (this is already known [**4**].)

COROLLARY 6. *Let X denote a separable compact topological manifold. Villadsen algebras V_1 and V_2 with the same initial space X, differing only in their choice of evaluation points, are isomorphic provided that the two sets of evaluation points are both dense.*

PROOF. According to Definition 3, we are thus given two countable inductive limits with the same objects:

$$A_1 \xrightarrow{\phi_1} A_2 \xrightarrow{\phi_2} A_3 \longrightarrow \qquad \text{with} \lim_{\to} A_i =: V_1$$

$$A_1 \xrightarrow{\psi_1} A_2 \xrightarrow{\psi_2} A_3 \longrightarrow \qquad \text{with} \lim_{\to} A_i =: V_2.$$

The difference between these limits is that there may be different evaluation points x_i and x_i' used in the two cases. Given that the cardinalities of the sets $\{x_i\}$ and $\{x_i'\}$ are equal, and observing that the cardinality is at most $\aleph_0$, we use the strong n-homogenity of Theorem 4 to find a sequence of homeomorphisms, $h_i : X \to X$ that intertwine the evaluation points of ϕ_i and ψ_i. By Lemma 5, these homeomorphisms h_i will conjugate ϕ_i and ψ_i, up to an amplification. In other words, they induce automorphisms $\hat{h}_i : M_n(C(X)) \to M_n(C(X))$ whose amplifications intertwine the above two inductive limits:

$$\begin{array}{ccccccc} A_1 & \xrightarrow{\phi_1} & A_2 & \xrightarrow{\phi_2} & A_3 & \longrightarrow & \cdots \\ \big\downarrow{\scriptstyle \hat{h}_1} & & \big\downarrow{\scriptstyle \hat{h}_1 \circ \hat{h}_2} & & \big\downarrow{\scriptstyle \hat{h}_1 \circ \hat{h}_2 \circ \hat{h}_3} & & \\ A_1 & \xrightarrow{\psi_1} & A_2 & \xrightarrow{\psi_2} & A_3 & \longrightarrow & \cdots \end{array}$$

on larger and larger finite sets (of points), becoming dense in the limit. □

In the case of simple Villadsen algebras with the Hilbert cube as base space, all of the base spaces are homeomorphic to the Hilbert cube, and self–homeomorphisms of the Hilbert cube are equivalent to different choices of (dense) evaluation points, but as we have seen, the algebra obtained does not change. Generally, homeomorphisms of topological manifolds need not be connected to the identity homeomorphism, but the Hilbert cube is contractible. So in the special case of the Hilbert cube, the homeomorphisms constructed are connected to the identity, and therefore the induced automorphisms are also connected to the identity. Therefore we can conclude that

COROLLARY 7. *Self-homeomorphisms of the Hilbert cube induce automorphisms (connected to the identity) of simple Hilbert cube Villadsen algebras.*

3. On Villadsen idempotents.

The essential idea is very simple: consider a sequential inductive limit of Villadsen's form, where each object is the tensor square of the preceding object, and each connecting map is, up to a perturbation, the corresponding doubling of the previous connecting map. Succinctly put, the perturbations are chosen large enough so that we obtain a simple (Villadsen) C*-algebra, but small enough to obtain the desired idempotence property. There is of course nothing about this that requires Villadsen's form of inductive limit, but we may as well use a specific case for illustrating the concept.

It is a general fact in analysis that repeated squaring will, when it converges, produce an idempotent. We may as well choose the initial object to be approximately an idempotent, although it is not necessary to do so. Let us thus choose initial object $A_1 := M_2(C(\mathfrak{U}))$, and let $\phi_{\{y\}}$ denote a Villadsen connecting map with one evaluation point, $y \in \mathfrak{U}$. Let $h_n : \mathfrak{U} \to \mathfrak{U}$ be a sequence of homeomorphisms such that:

- The orbit of the point $y \in \mathfrak{U}$ under the h_n is dense in $\mathfrak{U}$
- The homeomorphisms $h_n : \mathfrak{U} \to \mathfrak{U}$ satisfy $\operatorname{dist}(h_n, h_{n+1}) < \frac{1}{n}$.

Now, the first property insures that, in view of Lemma 5, the inductive limit

$$V_2 \cong \varinjlim_{i\to\infty} \left(A_1^{\otimes^{2^{i-1}}}, \bigoplus_{2^i} \hat{h}_i^{-1} \circ \phi_{\{y\}} \circ \hat{h}_i \right) \tag{1}$$

has a dense set of evaluation points $\{h_i(y)\}_{i=1}^{\infty}$ and is therefore simple. The second property insures the process converges, and as pointed out before, we then automatically obtain the desired idempotence property. To see this in detail, we shall compare the obtained inductive limit decomposition of V_2 with that of $V_2 \otimes V_2$. We exhibit both limits in the rows of the diagram below, where to reduce the size of the diagram, we temporarily denote the objects of the previous inductive limit (in Eq. (1)) by A_i and denote the connecting maps by ϕ_i. Then we have:
(2)

$$\begin{array}{ccccccccccc}
A_1 & \xrightarrow{\phi_1} & A_2 & \xrightarrow{\phi_2} & A_3 & \xrightarrow{\phi_3} & A_4 & \xrightarrow{\phi_4} & A_5 \cdots & \xrightarrow[\lim]{} & V_2 \\
 & & \downarrow & & \downarrow & & \downarrow & & \downarrow & & \\
 & & A_1 \otimes A_1 & \xrightarrow{\phi_1 \oplus \phi_1} & A_2 \otimes A_2 & \xrightarrow{\phi_2 \oplus \phi_2} & A_3 \otimes A_3 & \xrightarrow{\phi_3 \oplus \phi_3} & A_4 \otimes A_4 \cdots & \xrightarrow[\lim]{} & V_2 \otimes V_2
\end{array}$$

The vertical arrows are the natural isomorphisms of $A_i \otimes A_i$ with A_{i+1}, and the diagram is asymptotically commutative because of how the homeomorphisms were chosen. This shows that $V_2 \cong V_2 \otimes V_2$, as claimed. Setting the initial object A_1 equal to $M_p(C(\mathfrak{U}))$ we obtain from the above construction a family of idempotents, V_p, thus proving:

THEOREM 8. *There is an idempotent in the class of Villadsen algebras of the first kind, and it is simple, separable and nuclear.*

Certainly, the class of C*-algebras which are V_p*-stable,* i.e. the class of C*-algebras W for which $W \otimes V_p \cong W$, would seem to be interesting. The product of all of the V_p need not even be Villadsen, but only locally AH. (See [**3**].) We denote this product by $V_\infty := \bigotimes_p V_p$, where the tensor product ranges over all primes p.

The usual concept of a homotopy of compact spaces in topology, under the Gelfand functor, gives rise to the following definition or lemma:

DEFINITION 9. Let $\psi_0, \psi_1 : A \to B$ be unital *-homomorphisms of unital C*-algebras. We say that they are *homotopic* if there is a point-norm continuous path of unital *-homomorphisms $\Psi_t : A \to B$ such that $\Psi_0 = \psi_0$ and $\Psi_1 = \psi_1$. This is equivalent to being given a unital $*$-homomorphism $\Psi : A \to C([0,1]) \otimes B$ such that $(\pi_0 \otimes \mathrm{Id}) \circ \Psi = \psi_0$ and $(\pi_1 \otimes \mathrm{Id}) \circ \Psi = \psi_1$, where π_t denotes the natural evaluation map on $C([0,1])$. If the path of unital *-homomorphisms $\Psi_t : A \to B$ is operator norm continuous, then we say there is a *norm homotopy.*

Homotopies are well behaved under tensor products:

LEMMA 10. *If f_1 is (norm) homotopic to f_0, and g_1 is (norm) homotopic to g_0, then $f_1 \otimes g_1$ is (norm) homotopic to $f_0 \otimes g_0$.*

PROOF. If $\Psi_1 \colon A_1 \to C([0,1]) \otimes B_1$ and $\Psi_0 \colon A_0 \to C([0,1]) \otimes B_0$ are homotopies, their tensor product is a *-homomorphism from $A_1 \otimes A_0$ to $C(\square, B_1 \otimes B_0)$, where $\square$ denotes the unit square. Restricting the functions in $C(\square, B_1 \otimes B_0)$ to the diagonal of the unit square, we obtain at one endpoint $f_1 \otimes g_1$ and at the other endpoint $f_0 \otimes g_0$. The given topologies are inherited by the tensor product map, so if we begin with norm homotopies, we end up with a norm homotopy. $\square$

DEFINITION 11. Let us say that a unital C*-algebra A has *(norm) homotopic half flip* with respect to a given tensor product if the tensor factor inclusion maps $\phi\colon a \mapsto a \otimes 1_A$ and $\psi\colon a \mapsto 1_A \otimes a$ are (norm) homotopic as maps $\phi, \psi\colon A \to A \otimes A$. Let us say that a unital C*-algebra A has *(norm) homotopic flip* with respect to a given tensor product if the identity map $\mathrm{Id}\colon A \otimes A \to A \otimes A$ and the flip map $\sigma\colon A \otimes A \to A \otimes A$ are norm homotopic.

We have the following sufficient condition for the norm homotopic half flip property:

PROPOSITION 12. *A unital C*-algebra A has the norm homotopic half flip property if the automorphism group of $A \otimes A$ is norm contractible.*

PROOF. Being contractible implies path-connectedness, so the flip automorphism $\sigma\colon A{\otimes}A \to A{\otimes}A$ is connected to the trivial automorphism $\mathrm{Id}\colon A{\otimes}A \to A{\otimes}A$ by a norm continuous path of automorphisms, denoted $\alpha_t|_{t\in[0,1]}$. But then the given tensor factor inclusion maps ϕ and ψ are also connected by such a path, namely $\alpha_t \circ \phi$, where $t \in [0,1]$ is the parameter. $\square$

THEOREM 13. *The Villadsen idempotents V_p have the norm homotopic flip property.*

PROOF. Recall that the objects A_i of a Villadsen inductive limit of our form were chosen to be $M_n(C(\mathfrak{U}))$. Therefore, $V_p \otimes V_p$ is itself a Villadsen algebra, with a decomposition of the form

$$A_1 \otimes A_1 \xrightarrow{\phi_1 \oplus \phi_1} A_2 \otimes A_2 \xrightarrow{\phi_2 \oplus \phi_2} A_3 \otimes A_3 \xrightarrow{\phi_3 \oplus \phi_3} A_4 \otimes A_4 \cdots .$$

In particular, a self-homotopy of $\mathfrak{U} \times \mathfrak{U}$ will not change the algebra obtained, it will just permute the choice of basepoints, which we are free to do, by Corollary 7. The flip σ_{M_k} on a tensor product of matrix algebras is inner, and the flip $\sigma_{\mathfrak{U}}$ on $C(\mathfrak{U}) \otimes C(\mathfrak{U}) \cong C(\mathfrak{U})$ is homotopically inner by Proposition 12. Now, letting the matrix flip act on the first factor of $(M_n \otimes M_n) \otimes (C(\mathfrak{U}) \otimes C(\mathfrak{U}))$, and the flip $\sigma_{\mathfrak{U}}$ act on the second factor, we have maps $\sigma_i\colon A_i \otimes A_i \to C([0,1]) \otimes (A_i \otimes A_i)$ implementing the homotopically inner flip on $A_i \otimes A_i$ as below

$$\begin{array}{ccccccc} A_1 \otimes A_1 & \xrightarrow{\phi_1 \oplus \phi_1} & A_2 \otimes A_2 & \xrightarrow{\phi_2 \oplus \phi_2} & A_3 \otimes A_3 & \xrightarrow{\phi_3 \oplus \phi_3} & A_4 \otimes A_4 \cdots \\ \downarrow{\scriptstyle \pi_t \circ \sigma_1} & & \downarrow{\scriptstyle \pi_t \circ \sigma_2} & & \downarrow{\scriptstyle \pi_t \circ \sigma_3} & & \downarrow{\scriptstyle \pi_t \circ \sigma_4} \\ A_1 \otimes A_1 & \xrightarrow{\phi_1 \oplus \phi_1} & A_2 \otimes A_2 & \xrightarrow{\phi_2 \oplus \phi_2} & A_3 \otimes A_3 & \xrightarrow{\phi_3 \oplus \phi_3} & A_4 \otimes A_4 \cdots \end{array}$$

where π_t denotes the natural evaluation map on $C([0,1])$. To assure that the above diagram is sufficiently commutative, we need to choose the matrix flip homotopies $\sigma_{M_k} : M_k \otimes M_k \to C([0,1]) \otimes (M_k \otimes M_k)$ to be consistent with each other. This is straightforward because the matrix flip is inner, and the connecting maps respect the tensor product. Then the vertical maps above are the tensor product of a homotopy of matrix algebras and a norm homotopy of the commutative C*-algebra $C(\mathfrak{U} \times \mathfrak{U})$, as in Lemma 10. This proves the norm homotopic flip property. □

THEOREM 14 ([**12**, Theorem 3.2]). *Let A be a C^*-algebra that is separable and simple. An automorphism of A that is norm homotopic to the identity is approximately inner.*

This lets us improve norm homotopic flip to approximately inner flip:

PROPOSITION 15. *Let A be a separable unital simple C^*-algebra with norm homotopic flip. Then the algebra A also has approximately inner flip.*

PROOF. We are given that the flip automorphism $\sigma\colon A \otimes A \to A \otimes A$ is connected to the identity automorphism $\mathrm{Id}\colon A \otimes A \to A \otimes A$ by a norm continuous path of automorphisms. We are to show that the flip automorphism is in fact an approximately inner automorphism of $A \otimes A$, and this follows from Theorem 14. □

COROLLARY 16. *The tensor idempotents V_p have approximately inner flip.*

We recall the known result that if an algebra A has approximately inner flip, then its infinite tensor product becomes $\mathcal{Z}$-stable and also has the strongly self-absorbing property [**18**, Prop 1.9]:

COROLLARY 17. *The infinite tensor product $\bigotimes V_p$ is strongly self-absorbing.*

This provides an interesting collection of idempotents. As previously mentioned, we can replace the Hilbert cube $\mathfrak{U}$ by an infinite-dimensional sphere. We omit the details, and give just the first step:

LEMMA 18. *The infinite-dimensional sphere S^∞ has the ω-homogeneity property, and the associated abelian C^*-algebra $C(S^\infty)$ is an idempotent.*

PROOF. Modelling S^∞ by the unitary group of separable Hilbert space, we have $S^\infty \times S^\infty \cong S^\infty$, and the ω-homogeneity property is a corollary of Theorem 4. □

4. Villadsen algebras of the second kind

Villadsen's second construction adds line bundle coefficients to the connecting map, and we now generalize our basic lemmas accordingly.

DEFINITION 19. Let X be a compact Hausdorff space and $n, m, k \in \mathbb{N}$. Let p_i be a given finite set of orthogonal projections $p_i \in C(X) \otimes \mathcal{K}$. Consider a unital diagonal $*$-homomorphism

$$\phi : \mathcal{K} \otimes C(X) \to \mathcal{K} \otimes \mathcal{K} \otimes C(X^{\times m})$$

where the entries on the diagonal are either of the form $f \mapsto p_i \otimes f(y)$ for a point $y \in X$ *or* are of the form $f \mapsto p_i \otimes f \circ \pi_i$ where π_i is one of the canonical projection maps $\pi_i \colon X^{\times m} \to X$. This map will be called a *Villadsen connecting map with coefficients*, or simply a connecting map if no ambiguity can arise.

Let us consider how base space homeomorphisms interact with the Villadsen connecting maps when there are coefficients, as in Definition 19.

LEMMA 20. *Suppose that $\phi\colon \mathcal{K} \otimes C(X) \to \mathcal{K} \otimes \mathcal{K} \otimes C(X^{\times m})$ is a Villadsen connecting map with evaluation points (y_i) and coefficient projections p_i. Let $h\colon X \to X$ be a given self-homeomorphism of X. Let us denote the pullback action of h on $M_n \otimes C(X)$ by $\hat{h}$. Then*

$$\left(\bigoplus_m \hat{h}^{-1}\right) \circ \phi \circ \hat{h}$$

is a Villadsen connecting map with evaluation points $(h(y_i))$ and coefficient projections $\hat{h} \circ p_i$. If the given homeomorphism h happens to be connected to the identity, then $\hat{h} \circ p_i$ is implemented by applying an inner automorphism to p_i.

PROOF. The first part is proven just as in Lemma 5. For the statement on inner automorphisms, recall that if two projections are linked by a homotopy (*i.e.* are linked by a norm-continuous path of projections) then the homotopy is implemented by unitaries, see for example Corollary 5.2.9 in [**21**]. The Corollary is for single projections, not families, but we can as well take a finite orthogonal sum and apply the Corollary to $\sum p_i$ and $\hat{h}\left(\sum p_i\right)$. $\square$

Villadsen's second construction uses projective spaces. Following this method, beginning with infinite dimensional projective space as the initial object, we can obtain an additional Villadsen idempotent W_p in the second class of Villadsen's algebras:

THEOREM 21. *There is an idempotent in the class of Villadsen algebras of the second kind, and it is simple, separable and nuclear.*

PROOF. Let us choose initial object $A_1 := \mathcal{K} \otimes C(\mathbb{CP}^\infty)$, and let $\phi_{\{y\}}$ denote a Villadsen connecting map with one evaluation point, $y \in \mathbb{CP}^\infty$. We may for example choose the coefficient vector bundle to be the tautological line bundle over $\mathbb{CP}^\infty$. Since $\mathbb{CP}^\infty$ has only one nontrivial homotopy group, that one being equal to $\mathbb{Z}$, the self-homeomorphisms are topologically classified by their action on $\mathbb{Z}$, in other words, they may act either as 1 or as -1. (Compare [**13**].) Let $h_n \colon \mathbb{CP}^\infty \to \mathbb{CP}^\infty$ be a sequence of homeomorphisms in the topological component of the identity, and approaching the identity, such that:

- The orbit of the point $y \in \mathbb{CP}^\infty$ under the h_n is dense in $\mathbb{CP}^\infty$
- The homeomorphisms $h_n \colon \mathbb{CP}^\infty \to \mathbb{CP}^\infty$ satisfy $\operatorname{dist}(h_n, h_{n+1}) < \frac{1}{n}$.

Now, the first property insures that, in view of Lemma 5, the inductive limit

$$W_2 \cong \varinjlim_{i\to\infty} \left(A_1^{\otimes^{2^{i-1}}}, \bigoplus_{2^i} \hat{h}_i^{-1} \circ \phi_{\{y\}} \circ \hat{h}_i \right) \tag{3}$$

has a dense set of evaluation points $\{h_i(y)\}_{i=1}^\infty$ and is therefore simple. The second property insures the process converges, and as pointed out before, we then automatically obtain the desired idempotence property. This can be seen in detail by following through the argument from the top of page 213. □

REMARK 1. We remark that Villadsen's argument from [**20**] goes through to show that the Villadsen idempotent W_2 of the second kind has stable rank > 1 and therefore this idempotent is quite different from the previous family of idempotents V_p of the first kind. The homotopic flip property is thus not to be expected, and indeed, the previous argument for homotopic flip was based on contractibility and does not apply in the case of $\mathbb{CP}^\infty$.

Appendix A. Idempotents from the algebraic point of view

It is very interesting that the UCT, and the classification program for C*-algebras depends in such a crucial way upon the existence of idempotents. Let us present an algebraic perspective on the classification program, inspired by the Grothedieck ring construction from modern algebraic geometry.

Given any two nuclear C*-algebras A and B we may form a direct sum, denoted $A \oplus B$, or a tensor product, denoted $A \otimes B$. Under these operations, a given set of C*-algebras will generate a ring. Since $A \otimes B$ is isomorphic to $B \otimes A$, this is a commutative ring up to isomorphism of C*-algebras. A C*-algebra A is an *idempotent* if $A \otimes A \cong A$. Each such idempotent generates a ring ideal, and such ideals are of great help in understanding the structure of an abstract ring. Indeed the classification program for C*-algebras has for the most part proceeded by focusing on one such ideal and then finding invariants that classify the algebras within each such ideal. Thus it appears that the structure of this abstract ring has loosely guided the classification program for C*-algebras [**24**]. The two best known C*-algebraic idempotents are probably $\mathcal{K}$, the compact operators, and $\mathcal{Z}$, the Jiang-Su algebra. From a historical point of view, $\mathcal{K}$, the compact operators, is a very important idempotent, one of the very first operator algebras to be studied, and indeed the ideal generated, the stable nuclear C*-algebras, are of vast importance. However, certain idempotents seem more suited for the classification program than others, and at present it is believed that the so called strongly self absorbing idempotents are suitable.

It has been shown [**22**] that all so-called strongly self-absorbing simple idempotents have the property of absorbing $\mathcal{Z}$, meaning that $W \otimes \mathcal{Z} \cong W$ for such unital idempotents W. Kirchberg has shown that O_2 absorbs all the known unital simple separable and nuclear idempotents [**10**], meaning that for such idempotents $W \otimes O_2 \cong O_2$. In some cases, though, tensoring two idempotents simply gives a new idempotent, as is the case with nearly every tensor product with $\mathcal{K}$, for example. The complex numbers, $\mathbb{C}$, regarded as a C*-algebra, certainly are an idempotent, and also provide a multiplicative unit for a tensor product ring. Eventually, in the classification program, it was observed that in a given classifiable class of C*-algebras, such as, say, the AF algebras, the projectionless simple algebras, or the purely infinite simple algebras, it seems interesting to single out the algebras having the same invariants as the complex numbers, $\mathbb{C}$. In most cases, such algebras turned out to be self-absorbing, i.e. idempotent in the sense that, for example, $\mathcal{Z} \otimes \mathcal{Z}$ is isomorphic to $\mathcal{Z}$. In fact, these algebras have a stronger property of being strongly self-absorbing.

The list of special idempotents that have this kind of connection with the classification program is: $\mathcal{Z}$, the Jiang-Su algebra; O_2 and O_∞, the Cuntz algebras;

UHF algebras with all primes of infinite multiplicity; and of course $\mathbb{C}$ the algebra of complex numbers. The nonunital Jacelon–Razak algebra, J, is not strongly self-absorbing, but has received considerable attention. (See **[14]** and the references there). An idempotent need not be simple as a C*-algebra. An especially interesting example of an idempotent that is not simple as a C*-algebra is $C([0,1])^\infty$, in other words, $C(\mathfrak{U})$ where $\mathfrak{U}$ is the Hilbert cube (*i.e.* Tychonoff cube) of weight $\aleph_0$. We remark that in the category of C*-algebras the answer to Ulam's problem is negative: in other words, it is not true that $A \otimes A \cong B \otimes B$ implies that A is isomorphic to B. There is a somewhat involved counterexample in the unital and commutative case already **[7]**.

The following Hasse diagram summarizes the known relationships between simple idempotents:

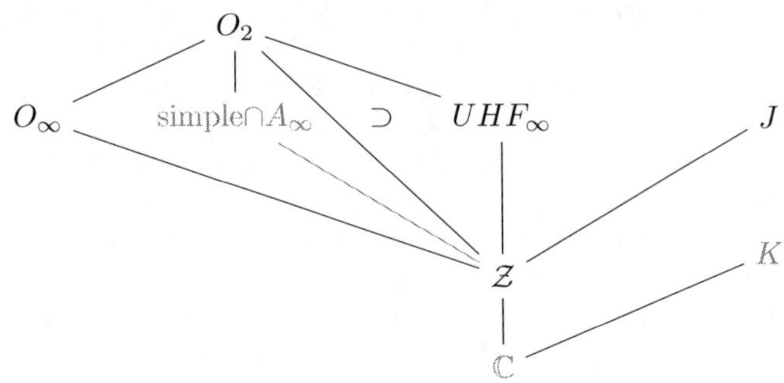

The ideals that have been classified are shown in black, the others in grey. The Villadsen algebras are conjectured classifiable by sufficiently many invariants, however there is no already known idempotent associated with them.

References

[1] A. V. Arhangel'skii and J. van Mill, *Topological homogeneity*, Recent progress in general topology. III, Atlantis Press, Paris, 2014, pp. 1–68, DOI 10.2991/978-94-6239-024-9_1. MR3204728

[2] R. Bennett, *Countable dense homogeneous spaces*, Fund. Math. **74** (1972), no. 3, 189–194, DOI 10.4064/fm-74-3-189-194. MR301711

[3] M. Dădărlat and S. Eilers, *Approximate homogeneity is not a local property*, J. Reine Angew. Math. **507** (1999), 1–13, DOI 10.1515/crll.1999.015. MR1670254

[4] G. A. Elliott, C. G. Li, Z. Niu, *A remark on Villadsen algebras*, arXiv:2209.10649

[5] I. Farah, *Calkin algebra, Kazhdan's property (T), strongly self-absorbing C*-algebras*, arXiv:2208.12301

[6] M. K. Fort Jr., *Homogeneity of infinite products of manifolds with boundary*, Pacific J. Math. **12** (1962), 879–884. MR145499

[7] R. H. Fox, *On a problem of S. Ulam concerning Cartesian products*, Fund. Math. **34** (1947), 278–287, DOI 10.4064/fm-34-1-278-287. MR27502

[8] V. L. Hansen, *Some theorems on direct limits of expanding sequences of manifolds*, Math. Scand. **29** (1971), 5–36, DOI 10.7146/math.scand.a-11031. MR319206

[9] O.-H. Keller, *Die Homoiomorphie der kompakten konvexen Mengen im Hilbertschen Raum* (German), Math. Ann. **105** (1931), no. 1, 748–758, DOI 10.1007/BF01455844. MR1512740

[10] E. Kirchberg, The classification of purely infinite C*-algebras using Kasparov's theory, unpublished manuscript

[11] V. L. Klee Jr., *Some topological properties of convex sets*, Trans. Amer. Math. Soc. **78** (1955), 30–45, DOI 10.2307/1992947. MR69388

[12] H. Lin, *Stable approximate unitary equivalence of homomorphisms*, J. Operator Theory **47** (2002), no. 2, 343–378. MR1911851

[13] C. A. McGibbon, *Self-maps of projective spaces*, Trans. Amer. Math. Soc. **271** (1982), no. 1, 325–346, DOI 10.2307/1998769. MR648096

[14] N. Nawata, *A characterization of the Razak-Jacelon algebra*, preprint, arXiv:2008.10235v3

[15] J. Rosenberg and C. Schochet, *The Künneth theorem and the universal coefficient theorem for Kasparov's generalized K-functor*, Duke Math. J. **55** (1987), no. 2, 431–474, DOI 10.1215/S0012-7094-87-05524-4. MR894590

[16] I. R. Shafarevich, *On some infinite-dimensional groups. II* (Russian), Izv. Akad. Nauk SSSR Ser. Mat. **45** (1981), no. 1, 214–226, 240. MR607583

[17] H. Thiel and E. Vilalta, *Covering dimension of Cuntz semigroups*, Adv. Math. **394** (2022), Paper No. 108016, 44 pp., DOI 10.1016/j.aim.2021.108016. MR4355725

[18] A. S. Toms and W. Winter, *The Elliott conjecture for Villadsen algebras of the first type*, J. Funct. Anal. **256** (2009), no. 5, 1311–1340, DOI 10.1016/j.jfa.2008.12.015. MR2490221

[19] G. S. Ungar, *Countable dense homogeneity and n-homogeneity*, Fund. Math. **99** (1978), no. 3, 155–160, DOI 10.4064/fm-99-3-155-160. MR482626

[20] J. Villadsen, *Simple C^*-algebras with perforation*, J. Funct. Anal. **154** (1998), no. 1, 110–116, DOI 10.1006/jfan.1997.3168. MR1616504

[21] N. E. Wegge-Olsen, *K-theory and C^*-algebras: A friendly approach*, Oxford Science Publications, The Clarendon Press, Oxford University Press, New York, 1993. MR1222415

[22] W. Winter, *Strongly self-absorbing C^*-algebras are $\mathcal{Z}$-stable*, J. Noncommut. Geom. **5** (2011), no. 2, 253–264, DOI 10.4171/JNCG/74. MR2784504

[23] W. Winter, *Structure of nuclear C^*-algebras: from quasidiagonality to classification and back again*, Proceedings of the International Congress of Mathematicians—Rio de Janeiro 2018. Vol. III. Invited lectures, World Sci. Publ., Hackensack, NJ, 2018, pp. 1801–1823. MR3966830

[24] W. Winter, *Localizing the Elliott conjecture at strongly self-absorbing C^*-algebras*, J. Reine Angew. Math. **692** (2014), 193–231, DOI 10.1515/crelle-2012-0082. MR3274552

Department of Mathematics and Statistics, Grant MacEwan University, 10700 104 Ave NW Edmonton, AB T5J 4S2

Email address: `ivanescuc@macewan.ca`

Department of Mathematics and Statistics, University of New Brunswick at Fredericton, P. O. Box 4400, Fredericton NB E3B 5A3

Email address: `dkucerov@unb.ca`

Contemporary Mathematics
Volume **798**, 2024
https://doi.org/10.1090/conm/798/15987

A Remark on the Kadison-Singer Transform

Dan Kucerovsky

Abstract. We interpret the recently proven Kadison-Singer conjecture as a transform, giving it a geometric interpretation, and in particular we show that it gives in a natural way a product operation on infinite-dimensional projective space.

1. Introduction

We interpret the recently proven Kadison-Singer conjecture as a transform, giving it a geometric flavour. The embryonic form of this conjecture is generally agreed to be due to the physicist P.A.M. Dirac, who wanted to be able to simultaneously diagonalize all elements of a maximal family of self-adjoint operators (masa), acting on a Hilbert space $\mathcal{H}$. Dirac's formulation of the conjecture (see Appendix) was simple, clear, and beautiful. Technical issues with non-existence of eigenvectors appeared to scuttle Dirac's original conjecture, but eventually Kadison and Singer found a reasonable form of the conjecture, and showed that the conjecture fails for certain maximal families called "continuous masas." For, in particular, the most immediately evident type of masa, the "discrete masa," the problem remained open until recently.

Question 1. Do pure states of maximal abelian self-adjoint algebras of operators on a Hilbert space extend *uniquely* to the algebra of all bounded operators?

Kadison and Singer played key roles in both the formulation of the general conjecture and the disproof of the continuous case. They showed that the answer was negative in the case that the given masa was diffuse (we say a masa is *diffuse* when it is algebraically isomorphic to $L^\infty([0,1],\mu)$ where μ is Lebesgue measure). On the other hand, it was shown in [**11**] that the answer is positive for so-called discrete masas of $B(\mathcal{H})$. A masa is called *discrete* when it is algebraically isomorphic to $L^\infty(\mathbb{N},c)$, where c is counting measure.

Projective spaces are classic in algebraic geometry, and to a lesser extent, in physics. The most common are the finite-dimensional ones like $\mathbb{CP}^n$, but for example [**14**] defines the infinite-dimensional $\mathbb{CP}^\infty$ as a direct limit of these as n approaches infinity, using the fact that $\mathbb{CP}^n$ sits inside $\mathbb{CP}^{n+1}$ as a subspace. Alternatively, take a complex Hilbert space $\mathcal{H}$ with countably infinite dimension and form the space of all 1-dimensional linear subspaces in $\mathcal{H}$. In physics, this is the space

2020 *Mathematics Subject Classification.* Primary 46L05; Secondary 46L35.
Key words and phrases. Kadison-Singer conjecture, C*-algebras, projective space.

of all pure states of the quantum system whose Hilbert space is $\mathcal{H}$. In these special cases, one can define a minimal projection to be a projection on a 1-dimensional subspace of our Hilbert space. The definition of the term pure state in physics is not quite the same as in mathematics, and we will be using the mathematical definition here. See [**16**, pg. 144-5] for a discussion of the equivalence of the definitions.

We will show that the positive answer to the Kadison–Singer conjecture implies the existence of a convolution product of a certain form on an infinite-dimensional $\mathbb{CP}^\infty$ space. Motivated by the above physical considerations, we define *infinite-dimensional* $\mathbb{CP}^\infty$ *space* to be the set of extremal rays in the positive cone of linear functionals on the compact operators $\mathcal{K}(\mathcal{H})$, where we denote the usual separable Hilbert space by $\mathcal{H}$. Since this is not quite the usual definition, it might be reasonable to call it the pure state model of $\mathbb{CP}^\infty$. We will show that these rays uniquely determine nonzero pure states on a discrete masa M of $B(\mathcal{H})$.

THEOREM 1. *Let f denote a nonzero positive linear functional on the compact operators $\mathcal{K}(\mathcal{H})$. It is an extremal ray in the positive cone of linear functionals if and only if it is pure when extended to $B(\mathcal{H})$ and then restricted to a masa there.*

PROOF. Given a nonzero positive linear functional f on the compact operators $\mathcal{K}(\mathcal{H})$, we denote by $\pi_f\colon \mathcal{K} \to B(\mathcal{H})$ the associated representation. If f is extremal, this representation will be irreducible [**13**, Th. 3.13.2.i], and the representation naturally extends to an irreducible representation of $\mathcal{M}(\mathcal{K})$ on $B(\mathcal{H})$. But it is then an irreducible representation of one algebra of bounded operators $B(\mathcal{H}')$ on another, and is therefore implemented by a unitary operator [**3**]:

$$\begin{aligned} \pi_f\colon \mathcal{M}(\mathcal{K}) &\to B(\mathcal{H}) \\ x &\mapsto UxU^*. \end{aligned}$$

If it were the case that π_f restricted to a masa $M \subset B(\mathcal{H})$ was no longer irreducible, then this would mean, by [**13**, Th. 3.13.2.ii], that the commutant of the range of $\pi_f(M)$ consists of more than just the scalar multiples of 1 in $B(\mathcal{H})$. Therefore, there would be an element $y \in B(\mathcal{H})$ that commutes with the elements of $\pi_f(M)$. But then, conjugating by the unitary $U \in B(\mathcal{H})$, we would have that the elements of M commute with U^*yU, and in particular, we would then have an element $U^*yU \in B(\mathcal{H})$ that commutes with every element of the masa M. This would contradict the maximality of the masa. Therefore, if f is extremal on the compact operators $\mathcal{K}(\mathcal{H})$, as assumed, then the process of extending the representation to $B(\mathcal{H})$ and restricting to the masa M does give an irreducible representation. It then follows that $f|_M$ is extremal, and is therefore a multiple of a pure state (but being unital, it is itself a pure state). The converse follows from the conjecture itself. □

2. The transform

THEOREM 2. *There is a one to one correspondence between:*

i. *Extremal rays of the positive cone of linear functions on $\mathcal{K}(\mathcal{H})$,*
ii. *Normal states of $B(\mathcal{H})$, and*
iii. *Normal characters of the discrete masa $L^\infty(\mathbb{N}, c)$.*

PROOF. That elements as in *i* map to elements as in *ii* comes from, for example, [**13**, thm. 3.6.4]. Then that *ii* maps to *iii* comes from Theorem 1 and the fact that weak continuity of a state on $B(\mathcal{H})$ implies its weak continuity when restricted to the discrete masa $L^\infty(\mathbb{N}, c)$. The character that we restricted can be reconstructed by (uniquely) extending it back to $B(\mathcal{H})$ using the positive solution to the Kadison-Singer conjecture, and this shows that elements of *iii* map to elements of *i*. □

The above is thus the basic form of our transform result, where positive linear functions on $\mathcal{K}$ are transformed to characters on $L^\infty(\mathbb{N}, c)$ and vice versas. Although this reminds one of Pontryagin's duality theory, the underlying machinery is very different. We now give an application of this fundamental transformation. Since the Kadison-Singer transform takes pure states on the compact operators to characters on a masa, this provides us with a convolution-like operation on the pure states. Just as in Pontryagin's theory, we take products of characters pointwise. The following definition or lemma defines the product operation:

DEFINITION 3. *Let f_1 and f_2 denote points of $\mathbb{CP}^\infty$, in the pure state model. Let $\hat{f}_i$ denote the normal character corresponding to f_i when restricted to the discrete masa M. Then there is a convolution product $f_1 \Box f_2$ defined by*

$$\widehat{(f_1 \Box f_2)} := \widehat{(f_1)}\widehat{(f_2)}.$$

To bring this result to bear, let us introduce a tensor product decomposition on our Hilbert space, $\mathcal{H} = \mathcal{H}_1 \otimes \mathcal{H}_2$. Then we have compact operators $\mathcal{K}(\mathcal{H}_1 \otimes \mathcal{H}_2) \cong \mathcal{K}(\mathcal{H}_1) \otimes \mathcal{K}(\mathcal{H}_2)$ and $B(\mathcal{H}_1 \otimes \mathcal{H}_2) \cong B(\mathcal{H}_1) \otimes_{min} B(\mathcal{H}_2)$. Recalling that tensor products of Hilbert spaces are unique, we may consider diagonal operators on Hilbert spaces, and in terms of the diagonal algebras, we have $L^\infty(\mathbb{N} \times \mathbb{N}, c) \cong L^\infty(\mathbb{N}, c) \otimes L^\infty(\mathbb{N}, c)$. Since this is a tensor product of abelian C*-algebras, nuclearity implies that there is only one unique C*-tensor product. The properties of tensor products then give the following unique decompositions:

THEOREM 4. *Let us consider $\mathcal{K}(\mathcal{H}_1) \otimes \mathcal{K}(\mathcal{H}_2)$. There are one to one correspondences between:*

i. *Pairs of pure normal quasi-states $\mathbb{P}\mathcal{K}(\mathcal{H}_1) \times \mathbb{P}\mathcal{K}(\mathcal{H}_2)$ and the pure normal quasi-states $\mathbb{P}(\mathcal{K}(\mathcal{H}_1) \otimes \mathcal{K}(\mathcal{H}_2))$*
ii. *Pairs of pure normal states $\mathbb{P}B(\mathcal{H}_1) \times \mathbb{P}B(\mathcal{H}_2)$ and the pure normal quasi-states $\mathbb{P}(B(\mathcal{H}_1 \otimes \mathcal{H}_2))$*
iii. *Pairs of normal characters on $L^\infty(\mathbb{N} \times \mathbb{N}, c)$, and the normal characters of $L^\infty(\mathbb{N}, c)$,*

PROOF. The correspondence between quasi-states, states, and characters follows from Theorem 2. In addition, we remark that normality is equivalent to weak continuity, and that the normality of a tensor product of states follows from checking continuity on elementary tensors. □

In terms of projective space, we can restate property *i* in the above list in the following striking way that we feel gives insight into the Kadison–Singer transform as we have interpreted it here:

COROLLARY 5. *There exists a product $\Box \colon \mathbb{CP}^\infty \times \mathbb{CP}^\infty \to \mathbb{CP}^\infty$. Moreover, the corresponding decomposition $x = x_1 \Box x_2$ is unique.*

PROOF. We have already shown above the existence of this product. To show uniqueness of the decomposition, we recall that the Kadison-Singer conjecture provides a unique character corresponding to each state on $\mathcal{K}(\mathcal{H}_1) \otimes \mathcal{K}(\mathcal{H}_2)$. But such a character is after all a point in the Gelfand spectrum of the algebra of diagonal operators (on the tensor product), namely $L^\infty(\mathbb{N}, c) \otimes L^\infty(\mathbb{N}, c)$. We remark that this is a tensor product of abelian, hence nuclear, C*-algebras; and thus this tensor product is in effect unique. Since this algebra has Gelfand spectrum $\beta\mathbb{N} \times \beta\mathbb{N}$, it follows that the character can be written $a \otimes b$ where a is in the Gelfand spectrum of the first factor $\beta\mathbb{N}$ and b is in the Gelfand spectrum of the second factor $\beta\mathbb{N}$. □

Of course, there are already known product operations on the direct limit model of $\mathbb{CP}^\infty$, one is due to the H-space structure and the other is due to the existence of a Segre–type embedding. The relationships between these products is not immediately clear.

3. Towards Hilbert bundles on $\mathbb{CP}^\infty$

Recall that a Hilbert module over a C*-algebra A is defined in much the same way as a separable Hilbert space, but the scalar field of the Hilbert space is replaced by the C*-algebra A. See [**9**] for more information. In particular, Hilbert A-modules have an A-valued inner product. Choosing a pure state ρ of A, we can then apply the pure state to the Hilbert module inner product, to obtain a classical scalar-valued inner product. The family of inner products so obtained can be used to generate a Hilbert bundle using a construction similar to the well-known GNS construction. Moreover, when the pure states separate the points of the C*-algebra, we may reconstruct the Hilbert module from the family of inner products.

A map from one Hilbert space to another will be termed holomorphic if it is continuous and Gâteaux differentiable (meaning that the restriction to each one-dimensional affine complex subspace should be holomorphic as a function of one complex variable).

THEOREM 6. *There is a one-to-one correspondence between isomorphism classes of countably generated Hilbert modules over $\mathcal{K}$ and a certain class of locally trivial holomorphic Hilbert bundles over $\mathbb{CP}^\infty$ with respect to the pure state model.*

PROOF. The noncommutative Serre–Swan theorem [**5**] makes the category of Hilbert C*-modules over a given C*-algebra equivalent to a category of (locally trivial) Hilbert bundles over the space of pure states of the C*-algebra with the zero functional adjoined. Thus, in particular, in the case of the compact operators, $\mathcal{K}$, we have that countably generated Hilbert modules over $\mathcal{K}$ are equivalent to holomorphic Hilbert bundles over $\mathbb{CP}^\infty \cup \{0\}$. Restricting the base space to $\mathbb{CP}^\infty$, we obtain a holomorphic Hilbert bundle over $\mathbb{CP}^\infty$. □

The Hilbert bundles obtained in the above theorem consist of locally trivial continuous fields of Hilbert spaces, possibly finite-dimensional, over the base space $\mathbb{CP}^\infty$. The sections are Gâteaux differentiable, and uniformly continuous. Let us now take a closer look at the isomorphism classes of countably generated Hilbert modules over $\mathcal{K}$.

THEOREM 7. *There exist only countably many isomorphism classes of Hilbert bundles as in* 6

PROOF. Since the proof of Theorem 6 was based on an equivalence of categories, it is clear that isomorphism classes of Hilbert bundles correspond to isomorphism classes of Hilbert modules over $\mathcal{K}$, at least in the countably generated case. But the compact operators, $\mathcal{K}$, are strongly Morita equivalent to $\mathbb{C}$, and this equivalence takes Hilbert modules over $\mathcal{K}$ to Hilbert modules over $\mathbb{C}$. But these are nothing other than classical Hilbert spaces, and are classified by their dimension. (From another viewpoint, we can say that since the compact operators have stable rank 1, we are in effect computing their Cuntz semigroup, and this is just the natural numbers $\mathbb{N}$.) □

It is clear that Hilbert bundles over $P(\mathcal{K})$ cannot be isomorphic if they have different fibre dimension at some point of the base space. Moreover, since the base space is connected, the fibre dimension is the same at all points of the base space. From the Hilbert module point of view, the Hilbert modules of the form $\mathcal{K}^n$, together with the standard Hilbert module $\mathcal{H}_{\mathcal{K}}$, are sufficient to exhaust the isomorphism classes in the above theorem. Under the previously mentioned GNS construction, the Hilbert modules $\mathcal{K}^n$ are mapped to Hilbert bundles having fibre $\mathbb{C}^n$ and the standard Hilbert module $\mathcal{H}_{\mathcal{K}}$ is mapped to a Hilbert bundle with fibre ℓ^2. Therefore, the fibre dimension of the Hilbert bundle is an explicit invariant implementing the classification. Thus we have the corollary:

COROLLARY 8. *Hilbert bundles over $P(\mathcal{K})$ are classified, up to isomorphism, by their fibre dimension, taking values in $\mathbb{N} \cup \{\infty\}$.*

Appendix A. Dirac on his theory and self-adjoint evaluations

A state of a von Neumann algebra $\mathcal{A}$ is a linear functional f on $\mathcal{A}$ for which $f(I) = 1$ and $f(T) \geq 0$ whenever T is a positive operator. The set of states of $\mathcal{A}$ is a convex subset of the dual space of $\mathcal{A}$, which is compact in the $w*$-topology. By the Kreĭn–Mil'man theorem, this convex set is the closed convex hull of its extreme points. The extremal elements in the space of states are by definition the *pure states* (of $\mathcal{A}$).

Then today's form of the Kadison–Singer conjecture is:: *Does every pure state on the algebra of bounded diagonal operators on $B(\mathcal{H})$ have a unique extension to a (pure) state on all of $B(\mathcal{H})$, the von Neumann algebra of all bounded linear operators on a separable infinite-dimensional $B(\mathcal{H})$?*

However, it is likely that this mathematically sophisticated question came from simpler and more physical considerations introduced by, most probably, P.A.M. Dirac, who wanted to be able to simultaneously diagonalize a family of commuting self-adjoint operators. His general idea was that self-adjoint operators represent physically meaningful quantities, but that to actually evaluate or observe the system, more needs to be done. He wished to be able to find a simultaneous basis, which he called a *representation,* for a compatible family of *observables* (*i.e.*, a commutative family of self-adjoint operators). The following gives the flavour; and in particular, paragraph iv can be understood in terms of a mathematical operation, namely, the simultaneous diagonalization of a family of self-adjoint operators:

> [Dirac's conjecture [**4**, pp. 74–75]]: To introduce a representation in practice
>
> i. We look for observables which we would like to have diagonal either because we are interested in their probabilities or for reasons of mathematical simplicity;
> ii. We must see that they all commute—a necessary condition since diagonal matrices always commute;

iii. We then see that they form a complete commuting set, and if not we add some more commuting observables to make them into a complete commuting set;
iv. [...] We may count the representation as being completely determined by the observables that are diagonal in it.

Set in a historical context, the above suggestions are not unreasonable, since during the same period, David Hilbert was studying the eigenvalues of integral operators by viewing the operators as infinite matrices [**7**]. However, the formulation by Dirac, in the setting of infinite matrices, did not see much more development, and it seems plausible that using the language of pure states represents a necessary concession from the physics of the problem towards the mathematics of the problem. In either case, Kadison and Singer published the first paper on the problem of the unique extension of pure states[**8**], and came to a negative conclusion. This was based on earlier work of von Neumann[**15**], who found what in today's language would be called a conditional expectation from $B(\mathcal{H})$ to a masa in $B(\mathcal{H})$. In the case of a diffuse masa, this conditional expectation was not unique, and that turned out to solve the problem in this case (negatively). After that, only the case of discrete masas was left open. The problem seemed likely to have counterexamples, and indeed, a counterexample was found under some set–theoretical hypotheses[**1**]. Moreover, a closely related conjecture (Anderson's conjecture[**2**]) that is only very slightly stronger than the Kadison–Singer conjecture was settled negatively. There was sufficient confusion about the exact status of the Kadison–Singer conjecture so that Math Reviews commissioned a special review in order to clarify the situation [**6**]. We would like to jokingly suggest that after that, most people thought it could not be done, but then Marcus, Spielman, and Srivastava [**11**] went ahead and did it.

References

[1] C. Akemann and N. Weaver, *$\mathcal{B}(H)$ has a pure state that is not multiplicative on any masa*, Proc. Natl. Acad. Sci. USA **105** (2008), no. 14, 5313–5314, DOI 10.1073/pnas.0801176105. MR2403097

[2] J. Anderson, *Extreme points in sets of positive linear maps on $\mathcal{B}(\mathcal{H})$*, J. Functional Analysis **31** (1979), no. 2, 195–217, DOI 10.1016/0022-1236(79)90061-2. MR525951

[3] O. Bratteli, P. E. T. Jorgensen, and G. L. Price, *Endomorphisms of $\mathcal{B}(\mathcal{H})$*, Quantization, nonlinear partial differential equations, and operator algebra (Cambridge, MA, 1994), Proc. Sympos. Pure Math., vol. 59, Amer. Math. Soc., Providence, RI, 1996, pp. 93–138, DOI 10.1090/pspum/059/1392986. MR1392986

[4] P. A. M. Dirac, *The Principles of Quantum Mechanics*, Oxford, at the Clarendon Press, 1947. 3d ed. MR23198

[5] G. A. Elliott and K. Kawamura, *A Hilbert bundle characterization of Hilbert C^*-modules*, Trans. Amer. Math. Soc. **360** (2008), no. 9, 4841–4862, DOI 10.1090/S0002-9947-08-04600-X. MR2403706

[6] R. S. Doran, review of "Akemann, Charles; Weaver, Nik $B(\mathcal{H})$ has a pure state that is not multiplicative on any masa." in MR2403097

[7] D. Hilbert, *Grundzüge einer allgemeinen Theorie der linearen Integralgleichungen* (German), Chelsea Publishing Co., New York, 1953. MR56184

[8] R. V. Kadison and I. M. Singer, *Extensions of pure states*, Amer. J. Math. **81** (1959), 383–400, DOI 10.2307/2372748. MR123922

[9] G. G. Kasparov, *Hilbert C^*-modules: theorems of Stinespring and Voiculescu*, J. Operator Theory **4** (1980), no. 1, 133–150. MR587371

[10] O.-H. Keller, *Die Homoiomorphie der kompakten konvexen Mengen im Hilbertschen Raum* (German), Math. Ann. **105** (1931), no. 1, 748–758, DOI 10.1007/BF01455844. MR1512740

[11] A. W. Marcus, D. A. Spielman, and N. Srivastava, *Interlacing families II: Mixed characteristic polynomials and the Kadison-Singer problem*, Ann. of Math. (2) **182** (2015), no. 1, 327–350, DOI 10.4007/annals.2015.182.1.8. MR3374963

[12] E. Mezzetti, *Introduction To Algebraic Geometry*, Note del corso di Instituzioni di Geometria Superiore 2, Trieste, Italy, 2005.

[13] G. K. Pedersen, *C^*-algebras and their automorphism groups*, London Mathematical Society Monographs, vol. 14, Academic Press, Inc. [Harcourt Brace Jovanovich, Publishers], London-New York, 1979. MR548006

[14] I. R. Šafarevič, *On some infinite-dimensional groups, II*; Math USSR Izv. **18** (1982) 185–194

[15] J. v. Neumann, *On rings of operators. III*, Ann. of Math. (2) **41** (1940), 94–161, DOI 10.2307/1968823. MR898

[16] V. S. Varadarajan, *Geometry of quantum theory. Vol. II*, The University Series in Higher Mathematics, Van Nostrand Reinhold Co., New York-Toronto-London, 1970. MR471675

Department of Mathematics and Statistics, University of New Brunswick at Fredericton, P. O. Box 4400, Fredericton NB E3B 5A3

Email address: dkucerov@unb.ca

Contemporary Mathematics
Volume **798**, 2024
https://doi.org/10.1090/conm/798/15988

Shafarevich-Tate groups of abelian varieties

Igor V. Nikolaev

ABSTRACT. The Shafarevich-Tate group $Ш(\mathscr{A})$ measures the failure of the Hasse principle for an abelian variety $\mathscr{A}$. Using a correspondence between the abelian varieties and the higher dimensional non-commutative tori, we prove that $Ш(\mathscr{A}) \cong Cl\ (\Lambda) \oplus Cl\ (\Lambda)$ or $Ш(\mathscr{A}) \cong (\mathbf{Z}/2^k\mathbf{Z}) \oplus Cl_{\ \mathbf{odd}}\ (\Lambda) \oplus Cl_{\ \mathbf{odd}}\ (\Lambda)$, where $Cl\ (\Lambda)$ is the ideal class group of a ring Λ associated to the K-theory of the non-commutative tori and 2^k divides the order of $Cl\ (\Lambda)$. The case of elliptic curves with complex multiplication is considered in detail.

1. Introduction

The study of Diophantine equations is the oldest part of mathematics. If such an equation has an integer solution, then using the reduction modulo any prime p, one gets a solution of the equation lying in the finite field $\mathbf{F}_p$ and a solution in the field of real numbers $\mathbf{R}$. The equation is said to satisfy the Hasse principle, if the converse is true. For instance, the quadratic equations satisfy the Hasse principle, while the equation $x^4 - 17 = 2y^2$ has a solution over $\mathbf{R}$ and each $\mathbf{F}_p$, but no rational solutions. Measuring the failure of the Hasse principle is a difficult open problem of number theory.

Let $\mathscr{A}_K$ be an abelian variety over the number field K which we assume to be simple, i.e. the $\mathscr{A}_K$ has no proper sub-abelian varieties over K. Denote by K_v the completion of K at the (finite or infinite) place v. Consider the Weil-Châtelet group $WC(\mathscr{A}_K)$ of the abelian variety $\mathscr{A}_K$ and the group homomorphism:

$$WC(\mathscr{A}_K) \to \prod_v WC(\mathscr{A}_{K_v}). \tag{1.1}$$

The Shafarevich-Tate group $Ш(\mathscr{A}_K)$ of $\mathscr{A}_K$ is defined as the kernel of homomorphism (1.1). The variety $\mathscr{A}_K$ satisfies the Hasse principle, if and only if, the group $Ш(\mathscr{A}_K)$ is trivial. Little is known about the $Ш(\mathscr{A}_K)$ in general. The existing methods include an evaluation of the analytic order of $Ш(\mathscr{A}_K)$ based on the second part of the Birch and Swinnerton-Dyer Conjecture and an exact calculation of the p-part of $Ш(\mathscr{A}_K)$, see e.g. [Rubin 1989] [**R**].

The aim of our note is calculation of the group $Ш(\mathscr{A}_K)$ based on a functor F between the n-dimensional abelian varieties $\mathscr{A}_K$ and the $2n$-dimensional non-commutative tori $\mathcal{A}_\Theta$, i.e. the C^*-algebras generated by the unitary operators

2020 *Mathematics Subject Classification.* Primary 11G10; Secondary 46L85.
Key words and phrases. Abelian varieties, non-commutative tori.

$U_1, \dots, U_{2n}$ satisfying the commutation relations $\{U_jU_i = e^{2\pi i\theta_{ij}}U_iU_j \mid 1 \le i,j \le 2n\}$ described by a skew-symmetric matrix

$$\Theta = \begin{pmatrix} 0 & \theta_{12} & \dots & \theta_{1,2n} \\ -\theta_{12} & 0 & \dots & \theta_{2,2n} \\ \vdots & \vdots & \ddots & \vdots \\ -\theta_{1,2n} & -\theta_{2,2n} & \dots & 0 \end{pmatrix} \in GL_{2n}(\mathbf{R}). \tag{1.2}$$

The functor maps isomorphic abelian varieties $\mathscr{A}$ and $\mathscr{A}'$ to the Morita equivalent algebras $\mathcal{A}_\Theta = F(\mathscr{A})$ and $\mathcal{A}'_\Theta = F(\mathscr{A}')$, see e.g. [**N**, Section 1.3]. Restricting F to the simple abelian varieties $\mathscr{A}_K$, one gets the non-commutative tori $\mathcal{A}_{\Theta(k)} = F(\mathscr{A}_K)$, where $\Theta(k)$ is the matrix (1.2) defined over a number field $k \subset \mathbf{R}$.

Roughly speaking, the idea is this. To calculate the $\text{Ш}(\mathscr{A}_K)$, we recast (1.1) in terms of the K-theory of algebra $\mathcal{A}_{\Theta(k)} = F(\mathscr{A}_K)$. Recall that an isomorphism class of $\mathcal{A}_{\Theta(k)}$ is defined by the triple $\left(K_0(\mathcal{A}_{\Theta(k)}), K_0^+(\mathcal{A}_{\Theta(k)}), \Sigma(\mathcal{A}_{\Theta(k)})\right)$ consisting of the K_0-group, the positive cone K_0^+ and the scale Σ of the algebra $\mathcal{A}_{\Theta(k)}$ [Blackadar 1986] [**B**, Section 6], see also Section 2.3. It is proved, that $\Sigma(\mathcal{A}_{\Theta(k)})$ is a torsion group, such that $WC(\mathscr{A}_K) \cong \Sigma(\mathcal{A}_{\Theta(k)})$ (lemma 3.1). The RHS of (1.1) corresponds to the crossed product C^*-algebra $\mathcal{A}_{\Theta(k)} \rtimes_{L_v} \mathbf{Z}$, where the notation is explained in Section 2.4.2. It is proved, that $\prod_v WC(\mathscr{A}_{K_v}) \cong \prod_v K_0\left(\mathcal{A}_{\Theta(k)} \rtimes_{L_v} \mathbf{Z}\right)$ (corollary 3.4). Thus (1.1) can be written in the form:

$$\Sigma(\mathcal{A}_{\Theta(k)}) \to \prod_v K_0\left(\mathcal{A}_{\Theta(k)} \rtimes_{L_v} \mathbf{Z}\right). \tag{1.3}$$

Both sides of (1.3) are functions of a single positive matrix $B \in GL(2n, \mathbf{Z})$, see definition 2.1. However, the LHS of (1.3) depends on the similarity class of B, while the RHS of (1.3) depends on the characteristic polynomial of B. This observation is critical, since it puts the elements of $\text{Ш}(\mathscr{A}_K)$ into a one-to-one correspondence with the similarity classes of matrices having the same characteristic polynomial. The latter is an old problem of linear algebra and it is known, that the number of such classes is finite. They correspond to the ideal classes $Cl\ (\Lambda)$ of an order Λ in the field $k = \mathbf{Q}(\lambda_B)$, where λ_B is the Perron-Frobenius eigenvalue of matrix B [Latimer & MacDuffee 1933] [**LM**]. Let $Cl\ (\Lambda) \cong \left(\mathbf{Z}/2^k\mathbf{Z}\right) \oplus Cl_{\textbf{odd}}(\Lambda)$ for $k \ge 0$. Using the Atiyah pairing between the K-theory and the K-homology, one gets the following result.

THEOREM 1.1. *The Shafarevich-Tate group of an abelian variety $\mathscr{A}_K$ is a finite group given by the formulas:*

$$\text{Ш}(\mathscr{A}_K) \cong \begin{cases} Cl\ (\Lambda) \oplus Cl\ (\Lambda), & \text{if } k \text{ is even}, \\ \left(\mathbf{Z}/2^k\mathbf{Z}\right) \oplus Cl_{\textbf{odd}}\ (\Lambda) \oplus Cl_{\textbf{odd}}\ (\Lambda), & \text{if } k \text{ is odd}. \end{cases} \tag{1.4}$$

The article is organized as follows. Preliminary facts can be found in Section 2. The proof of theorem 1.1 is given in Section 3. As an illustration, we consider the abelian varieties with complex multiplication in Section 4.

2. Preliminaries

We briefly review the abelian varieties, the Shafarevich-Tate groups and the n-dimensional non-commutative tori. For a detailed exposition, we refer the reader to [Mumford 1983] [**M2**, Chapter 2, §4], [Cassels 1966] [**C**], [Blackadar 1986] [**B**, Section 6] and [**N**, Section 3.4.1], respectively.

2.1. Abelian varieties. Let $\mathbf{C}^n$ be the n-dimensional complex space and Λ be a lattice in the underlying $2n$-dimensional real space $\mathbf{R}^{2n}$. The complex torus $\mathbf{C}^n/\Lambda$ is said to be an n-dimensional (principally polarized) abelian variety $\mathscr{A}$, if it admits an embedding into a projective space. In other words, the set $\mathscr{A}$ is both an additive abelian group and a complex projective variety. Recall that the Siegel upper half-space $\mathbb{H}_n$ consists of the symmetric $n \times n$ matrices with complex entries τ_i, such that: $\mathbb{H}_n := \left\{\tau = (\tau_i) \in \mathbf{C}^{\frac{n(n+1)}{2}} \mid \Im(\tau_i) > 0\right\}$. The points of $\mathbb{H}_n$ parametrize the n-dimensional abelian varieties., i.e. $\mathscr{A} \cong \mathscr{A}_\tau$ for some $\tau \in \mathbb{H}_n$. Denote by $Sp\ (2n, \mathbf{R})$ the $2n$-dimensional symplectic group, i.e. a subgroup of the linear group defined by the equation:

$$\begin{pmatrix} A & B \\ C & D \end{pmatrix}^T \begin{pmatrix} 0 & I \\ -I & 0 \end{pmatrix} \begin{pmatrix} A & B \\ C & D \end{pmatrix} = \begin{pmatrix} 0 & I \\ -I & 0 \end{pmatrix}, \tag{2.1}$$

where A, B, C, D are the $n \times n$ matrices with real entries and I is the identity matrix. It is well known, that the abelian varieties $\mathscr{A}_\tau$ and $\mathscr{A}_{\tau'}$ are isomorphic, whenever

$$\tau' = \frac{A\tau + B}{C\tau + D}, \quad \text{where} \quad \begin{pmatrix} A & B \\ C & D \end{pmatrix} \in Sp\ (2n, \mathbf{Z}). \tag{2.2}$$

2.2. Shafarevich-Tate group. Let $K \subset \mathbf{C}$ be a number field and let $\mathscr{A}_K$ be an abelian variety over K. The principal homogeneous space of $\mathscr{A}_K$ is a variety $\mathscr{C}$ over K with a map $\mu : \mathscr{C} \times \mathscr{A}_K \to \mathscr{C}$ satisfying (i) $\mu(x, 0) = x$, (ii) $\mu(x, a + b) = \mu(\mu(x, a), b)$ for all $x \in \mathscr{C}$ and $a, b \in \mathscr{A}_K$ and (iii) for all $x \in \mathscr{C}$ the map $a \mapsto \mu(x, a)$ is a K-isomorphism between $\mathscr{A}_K$ and $\mathscr{C}$. The homogeneous spaces $(\mathscr{C}, \mu)$ and $(\mathscr{C}', \mu')$ are equivalent, if there is a K-isomorphism $\mathscr{C} \to \mathscr{C}'$ compatible with the maps μ and μ'. If $\mathscr{C}$ has K-points, then the equivalence class $(\mathscr{C}, \mu)$ is said to be trivial. The non-trivial classes $(\mathscr{C}, \mu)$ correspond to the varieties $\mathscr{C}$ without K-points, but with the L-points, where L is a Galois extension of K. The Weil-Châtelet group $WC(\mathscr{A}_K)$ is an additive abelian group of the equivalence classes $(\mathscr{C}, \mu)$, where the trivial class corresponds to the zero element of the group, see [Cassels 1966] [**C**] for the details. The $WC(\mathscr{A}_K)$ is a torsion group, i.e. each element of the group has finite order. It is not hard to see, that if $K \subset K_v$ is an extension of K by completion at the place v, then there exists a natural group homomorphism:

$$WC(\mathscr{A}_K) \to WC(\mathscr{A}_{K_v}). \tag{2.3}$$

The kernel of (2.3) consists of those homogeneous spaces having the K_v-points, but no K-points. The Shafarevich-Tate group $Ш(\mathscr{A}_K)$ of $\mathscr{A}_K$ is defined as the kernel of homomorphism $WC(\mathscr{A}_K) \to \prod_v WC(\mathscr{A}_{K_v})$. Thus the abelian variety $\mathscr{A}_K$ satisfies the Hasse principle, if and only if, the group $Ш(\mathscr{A}_K)$ is trivial.

2.3. K-theory of C^*-algebras. The C^*-algebra $\mathcal{A}$ is an algebra over $\mathbf{C}$ with a norm $a \mapsto ||a||$ and an involution $a \mapsto a^*$ such that it is complete with respect to the norm and $||ab|| \leq ||a||\,||b||$ and $||a^*a|| = ||a||^2$ for all $a, b \in \mathcal{A}$. Any commutative C^*-algebra is isomorphic to the algebra $C_0(X)$ of continuous complex-valued functions on some locally compact Hausdorff space X. Any other algebra $\mathcal{A}$ can be thought of as a non-commutative topological space. By $M_\infty(\mathcal{A})$ one understands the algebraic direct limit of the C^*-algebras $M_n(\mathcal{A})$ under the embeddings $a \mapsto \mathbf{diag}(a, 0)$. The direct limit $M_\infty(\mathcal{A})$ can be thought of as the C^*-algebra of infinite-dimensional matrices whose entries are all zero except for a finite number of the non-zero entries taken from the C^*-algebra $\mathcal{A}$. Two projections $p, q \in M_\infty(\mathcal{A})$ are equivalent, if there exists an element $v \in M_\infty(\mathcal{A})$, such that $p = v^*v$ and $q = vv^*$. The equivalence class of projection p is denoted by $[p]$. We write $V(\mathcal{A})$ to denote all equivalence classes of projections in the C^*-algebra $M_\infty(\mathcal{A})$, i.e. $V(\mathcal{A}) := \{[p] \;:\; p = p^* = p^2 \in M_\infty(\mathcal{A})\}$. The set $V(\mathcal{A})$ has the natural structure of an abelian semi-group with the addition operation defined by the formula $[p] + [q] := \mathbf{diag}(p, q) = [p' \oplus q']$, where $p' \sim p$, $q' \sim q$ and $p' \perp q'$. The identity of the semi-group $V(\mathcal{A})$ is given by $[0]$, where 0 is the zero projection. By the K_0-group $K_0(\mathcal{A})$ of the unital C^*-algebra $\mathcal{A}$ one understands the Grothendieck group of the abelian semi-group $V(\mathcal{A})$, i.e. a completion of $V(\mathcal{A})$ by the formal elements $[p] - [q]$. The image of $V(\mathcal{A})$ in $K_0(\mathcal{A})$ is a positive cone $K_0^+(\mathcal{A})$ defining the order structure $\leq$ on the abelian group $K_0(\mathcal{A})$. The pair $(K_0(\mathcal{A}), K_0^+(\mathcal{A}))$ is known as a dimension group of the C^*-algebra $\mathcal{A}$. The scale $\Sigma(\mathcal{A})$ is the image in $K_0^+(\mathcal{A})$ of the equivalence classes of projections in the C^*-algebra $\mathcal{A}$. The $\Sigma(\mathcal{A})$ is a generating, hereditary and directed subset of $K_0^+(\mathcal{A})$, i.e. (i) for each $a \in K_0^+(\mathcal{A})$ there exist $a_1, \dots, a_r \in \Sigma(\mathcal{A})$ such that $a = a_1 + \dots + a_r$; (ii) if $0 \leq a \leq b \in \Sigma(\mathcal{A})$, then $a \in \Sigma(\mathcal{A})$ and (iii) given $a, b \in \Sigma(\mathcal{A})$ there exists $c \in \Sigma(\mathcal{A})$, such that $a, b \leq c$. Each scale can always be written as $\Sigma(\mathcal{A}) = \{a \in K_0^+(\mathcal{A}) \mid 0 \leq a \leq u\}$, where u is an order unit of $K_0^+(\mathcal{A})$. The pair $(K_0(\mathcal{A}), K_0^+(\mathcal{A}))$ and the triple $(K_0(\mathcal{A}), K_0^+(\mathcal{A}), \Sigma(\mathcal{A}))$ are invariants of the Morita equivalence and isomorphism class of the C^*-algebra $\mathcal{A}$, respectively.

2.4. Higher dimensional non-commutative tori.

2.4.1. *Definition and properties.* The even-dimensional non-commutative torus $\mathcal{A}_\Theta$ is the universal C^*-algebra generated by the unitary operators $U_1, \dots, U_{2n}$ and satisfying the relations $\{U_jU_i = e^{2\pi i\theta_{ij}}U_iU_j \mid 1 \leq i, j \leq 2n\}$, where $\Theta = (\theta_{ij})$ is a skew-symmetric real matrix (1.2). Let $SO(n, n;\ \mathbf{Z})$ be a subgroup of the group $GL(2n, \mathbf{Z})$ preserving the quadratic form $x_1x_{n+1} + x_2x_{n+2} + \dots + x_nx_{2n}$. The matrix $\begin{pmatrix} A & B \\ C & D \end{pmatrix} \in SO(n, n;\ \mathbf{Z})$, if and only if,

$$A^TD + C^TB = I, \quad A^TC + C^TA = 0 = B^TD + D^TB, \tag{2.4}$$

where $A, B, C, D \in GL(n, \mathbf{Z})$ and I is the identity matrix. The noncommutative tori $\mathcal{A}_\Theta$ and $\mathcal{A}_{\Theta'}$ are Morita equivalent, whenever

$$\Theta' = \frac{A\Theta + B}{C\Theta + D}, \quad \text{where} \quad \begin{pmatrix} A & B \\ C & D \end{pmatrix} \in SO(n, n;\ \mathbf{Z}). \tag{2.5}$$

The reader can verify, that conditions (2.4) follow from the matrix equation (2.1), but not vice versa. In other words, one gets an inclusion $Sp\ (2n, \mathbf{Z}) \subseteq SO(n, n;\ \mathbf{Z})$. This observation prompts the construction of a functor F from the n-dimensional abelian varieties $\mathscr{A}$ to the $2n$-dimensional non-commutative tori $\mathcal{A}_\Theta$, such that if $\mathscr{A}$ and $\mathscr{A}'$ are isomorphic abelian varieties, the $\mathcal{A}_\Theta = F(\mathscr{A})$ and $\mathcal{A}'_\Theta = F(\mathscr{A}')$ will be the Morita equivalent non-commutative tori, see (2.2) and (2.5). The restriction of F to the simple abelian varieties $\mathscr{A}_K$ corresponds to the non-commutative tori $\mathcal{A}_{\Theta(k)} = F(\mathscr{A}_K)$, where $\Theta(k)$ is the matrix over a number field k of deg $(k|\mathbf{Q}) = 2n$.

2.4.2. *Crossed product* $\mathcal{A}_{\Theta(k)} \rtimes_{L_v} \mathbf{Z}$. Let $\mathcal{A}_\Theta$ be a $2n$-dimensional non-commutative torus endowed with the canonical trace $\tau : \mathcal{A}_\Theta \to \mathbf{C}$. Since $K_0(\mathcal{A}_\Theta) \cong \mathbf{Z}^{2^{2n-1}}$ and $\tau_* : K_0(\mathcal{A}_\Theta) \to \mathbf{R}$ is a homomorphism induced by τ, one gets a $\mathbf{Z}$-module $\Lambda := \tau_*(K_0(\mathcal{A}_\Theta)) \subset \mathbf{R}$ of rank 2^{2n-1}. The generators of Λ belong to the ring $\mathbf{Z}[\theta_{ij}]$. In what follows, we assume that $\theta_{ij} \in k$. In this case one gets the algebraic constraints and the rank of Λ is equal to $2n$ [**N**, Remark 6.6.1]. In other words,

$$\Lambda \cong \mathbf{Z} + \mathbf{Z}\theta_1 + \cdots + \mathbf{Z}\theta_{2n-1}, \quad \text{where } \theta_i \in k. \tag{2.6}$$

DEFINITION 2.1. We shall write $B \in GL(2n, \mathbf{Z})$ to denote a positive matrix, such that $(1, \theta_1, \ldots, \theta_{2n-1})$ is the normalized Perron-Frobenius eigenvector of B.

Let $\pi(n)$ be an integer-valued function defined in [**N**, Section 6.5.3]. Consider the characteristic polynomial $Char\left(B^{\pi(p)}\right) = x^{2n} - a_1 x^{2n-1} - \cdots - a_{2n-1}x - 1$ and a matrix

$$L_v = \begin{pmatrix} a_1 & 1 & \ldots & 0 & 0 \\ a_2 & 0 & 1 & 0 & 0 \\ \vdots & \vdots & \ddots & \vdots & \vdots \\ a_{2n-1} & 0 & \ldots & 0 & 1 \\ p & 0 & \ldots & 0 & 0 \end{pmatrix}, \tag{2.7}$$

where p is the prime underlying v. The L_v defines an endomorphism of the algebra $\mathcal{A}_{\Theta(k)}$ by its action on the generators $U_1, \ldots, U_{2n}$. We consider the crossed product C^*-algebra $\mathcal{A}_{\Theta(k)} \rtimes_{L_v} \mathbf{Z}$ associated to such an action.

REMARK 2.2. There exits an isomorphism:

$$K_0\left(\mathcal{A}_{\Theta(k)} \rtimes_{L_v} \mathbf{Z}\right) \cong \mathscr{A}_{\mathbf{F}_p}, \tag{2.8}$$

where $\mathscr{A}_{\mathbf{F}_p}$ is a localization of the $\mathscr{A}_K$ at the prime ideal $\mathscr{P} \subset K$ over p [**N**, Section 6.6]. Formula (2.8) hints that the crossed product $\mathcal{A}_{\Theta(k)} \rtimes_{L_v} \mathbf{Z}$ is a recast of the variety $\mathscr{A}_{K_v}$.

3. Proof of theorem 1.1

We refer the reader to Section 1 for an outline of the proof. The detailed argument is given below. We split the proof in a series of lemmas.

LEMMA 3.1. *The scale* $\Sigma\left(\mathcal{A}_{\Theta(k)}\right)$ *has the structure of a torsion group, so that* $WC(\mathscr{A}_K) \cong \Sigma\left(\mathcal{A}_{\Theta(k)}\right)$, *where* $\mathcal{A}_{\Theta(k)} = F(\mathscr{A}_K)$.

PROOF. (i) For the sake of clarity, we restrict to the case $n = 1$, i.e. when the variety $\mathscr{A}$ is an elliptic curve $\mathscr{E}$. In this case matrix (1.2) can be written as:

$$\Theta = \begin{pmatrix} 0 & \theta \\ -\theta & 0 \end{pmatrix}. \tag{3.1}$$

We shall denote the corresponding non-commutative torus by $\mathcal{A}_\theta$. One gets deg $(k|\mathbf{Q}) = 2$, i.e. k is a real quadratic field. Therefore $F(\mathscr{E}_K) = \mathcal{A}_\theta$, where $\theta \in k$ is a quadratic irrationality.

To prove that the scale $\Sigma(\mathcal{A}_\theta)$ has the structure of a torsion group, we shall use the Minkowski question-mark function $?(x)$ [Minkowski 1904] [**M1**, p. 172]. Such a function is known to map quadratic irrational numbers of the unit interval to the rational numbers of the same interval preserving their natural order. This observation implies $\Sigma(\mathcal{A}_\theta) \subset \mathbf{Q}/\mathbf{Z}$, i.e. the scale $\Sigma(\mathcal{A}_\theta)$ is a subgroup of the torsion group $\mathbf{Q}/\mathbf{Z}$. Let us pass to a detailed argument.

Let τ be the canonical trace on the algebra $\mathcal{A}_\theta$. Since $K_0(\mathcal{A}_\theta) \cong \mathbf{Z}^2$, one gets $\tau_*(K_0(\mathcal{A}_\theta)) = \mathbf{Z} + \mathbf{Z}\theta$ and $\tau_*(K_0^+(\mathcal{A}_\theta)) = \{m + n\theta \geq 0 \mid m, n \in \mathbf{Z}\}$. It is known, that $\tau_*(u) = 1$, where $u \in K_0^+(\mathcal{A}_\theta)$ is an order unit. Therefore the traces on the scale $\Sigma(\mathcal{A}_\theta)$ are given by the formula:

$$\tau_*(\Sigma(\mathcal{A}_\theta)) = [0,1] \cap \mathbf{Z} + \mathbf{Z}\theta. \tag{3.2}$$

Recall that the Minkowski question-mark function is defined by the convergent series

$$?(x) := a_0 + 2\sum_{k=1}^{\infty} \frac{(-1)^{k+1}}{2^{a_1+\cdots+a_k}}, \tag{3.3}$$

where $x = [a_0, a_1, a_2, \dots]$ is the continued fraction of the irrational number x. The $?(x) : [0,1] \to [0,1]$ is a monotone continuous function with the following properties: (i) $?(0) = 0$ and $?(1) = 1$; (ii) $?(\mathbf{Q}) = \mathbf{Z}[\frac{1}{2}]$ are dyadic rationals and (iii) $?(\mathcal{Q}) = \mathbf{Q} - \mathbf{Z}[\frac{1}{2}]$, where $\mathcal{Q}$ are quadratic irrational numbers [Minkowski 1904] [**M1**, p. 172].

Consider subset (3.2) of the interval $[0,1]$. Since θ is a quadratic irrationality, we conclude that the set $\tau_*(\Sigma(\mathcal{A}_\theta)) \subset \mathcal{Q}$. By property (iii) of the Minkowski question-mark function, one gets an inclusion $?(\tau_*(\Sigma(\mathcal{A}_\theta))) \subset \mathbf{Q}/\mathbf{Z}$. Thus we constructed a one-to-one map:

$$\Sigma(\mathcal{A}_\theta) \to \mathcal{Y} \subset \mathbf{Q}/\mathbf{Z}, \tag{3.4}$$

where $\mathcal{Y} := ?(\tau_*(\Sigma(\mathcal{A}_\theta)))$. It follows from (3.4), that $\Sigma(\mathcal{A}_\theta)$ is a torsion group.

REMARK 3.2. Formula (3.4) is part of a duality between the K-theory of noncommutative tori and the Galois cohomology of abelian varieties. Such a study is beyond the scope of present paper.

(ii) Let us show, that $WC(\mathcal{E}_K) \cong \Sigma(\mathcal{A}_\theta)$, where $\cong$ is an isomorphism of the torsion groups. Indeed, let $\mathcal{C}$ be the principal homogeneous space of the elliptic curve $\mathcal{E}_K$. It is known, that $\mathcal{C} \cong \mathcal{E}_K'$ is the twist of $\mathcal{E}_K$, i.e. an isomorphic but not K-isomorphic elliptic curve $\mathcal{E}_K'$. It follows, that the $\mathcal{A}_\theta$ and $\mathcal{A}_\theta' = F(\mathcal{E}_K')$ are Morita equivalent, but not isomorphic C^*-algebras. This would imply $\left(K_0(\mathcal{A}_\theta), K_0^+(\mathcal{A}_\theta)\right) \cong \left(K_0(\mathcal{A}_\theta'), K_0^+(\mathcal{A}_\theta')\right)$, but $\left(K_0(\mathcal{A}_\theta), K_0^+(\mathcal{A}_\theta), \Sigma(\mathcal{A}_\theta)\right) \not\cong \left(K_0(\mathcal{A}_\theta'), K_0^+(\mathcal{A}_\theta'), \Sigma(\mathcal{A}_\theta')\right)$. In other words, the principal homogeneous spaces of elliptic curve $\mathcal{E}_K$ are classified by the scales of the algebra $\mathcal{A}_\theta$.

On the other hand, each element of $K_0^+(\mathcal{A}_\theta)$ can be taken for an order unit u of the dimension group $\left(K_0(\mathcal{A}_\theta), K_0^+(\mathcal{A}_\theta)\right)$. Since $\Sigma(\mathcal{A}_\theta) = \{a \in K_0^+(\mathcal{A}_\theta) \mid 0 \leq a \leq u\}$, we conclude that the elements of $K_0^+(\mathcal{A}_\theta)$ classify all scales of the algebra $\mathcal{A}_\theta$. We can always restrict to the generating subset $\Sigma(\mathcal{A}_\theta) \subset K_0^+(\mathcal{A}_\theta)$, since all other elements of the positive cone $K_0^+(\mathcal{A}_\theta)$ correspond to the finite unions $\mathcal{C}_1 \cup \cdots \cup \mathcal{C}_k$ of the generating homogeneous spaces $\mathcal{C}_i$. It remains to notice, that equivalence classes of the principal homogeneous spaces $(\mathscr{C},\mu)$ coincide with the isomorphism classes of the algebra $\mathcal{A}_\theta$. In other words, one gets an isomorphism of the torsion groups:

$$WC(\mathscr{E}_K)\cong\Sigma(\mathcal{A}_\theta). \tag{3.5}$$

(iii) The general case $n>1$ is proved by an adaption of the argument for the case $n=1$. Notice that one must use the Perron-Frobenius n-dimensional continued fractions and a higher-dimensional analog of the Minkowski question-mark function. The details are left to the reader.

Lemma 3.1 is proved. □

LEMMA 3.3. $WC(\mathscr{A}_{K_v})\cong K_0\left(\mathcal{A}_{\Theta(k)}\rtimes_{L_v}\mathbf{Z}\right)$, *where* $\mathcal{A}_{\Theta(k)}=F(\mathscr{A}_K)$.

PROOF. The abelian group $WC(\mathscr{A}_{K_v})$ is known to be finite, see e.g. [Borel & Serre 1964] [**BS**]. Specifically,

$$WC(\mathscr{A}_{K_v})\cong(\widehat{\mathscr{A}_{K_v}})^*, \tag{3.6}$$

where $\widehat{\mathscr{A}_{K_v}}$ are the rational points of K_v on the dual abelian variety of $\mathscr{A}_{K_v}$ and $(\widehat{\mathscr{A}_{K_v}})^*$ is the character group of $\widehat{\mathscr{A}_{K_v}}$ [Tate 1958] [**T**].

Recall that $K_0\left(\mathcal{A}_{\Theta(k)}\rtimes_{L_v}\mathbf{Z}\right)\cong\mathscr{A}(\mathbf{F}_p)$, where $\mathcal{A}_{\Theta(k)}=F(\mathscr{A}_K)$ and $\mathscr{A}(\mathbf{F}_p)$ is a localization of the abelian variety $\mathscr{A}_K$ at the prime ideal $\mathscr{P}$ over prime p, see remark 2.2. In view of the Hensel Lemma, the points of $\mathscr{A}(\mathbf{F}_p)$ are the rational points of $\mathscr{A}_{K_v}$, where v is the place over prime p. Since the torsion points of $\mathscr{A}_{K_v}$ and the torsion points of $\widehat{\mathscr{A}_{K_v}}$ coincide, we conclude that $\widehat{\mathscr{A}_{K_v}}\cong\mathscr{A}(\mathbf{F}_p)$. On the other hand, since finite abelian groups are Pontryagin self-dual, one gets $(\widehat{\mathscr{A}_{K_v}})^*\cong\mathscr{A}(\mathbf{F}_p)$. Therefore isomorphism (3.6) can be written as:

$$WC(\mathscr{A}_{K_v})\cong K_0\left(\mathcal{A}_{\Theta(k)}\rtimes_{L_v}\mathbf{Z}\right). \tag{3.7}$$

Lemma 3.3 is proved. □

COROLLARY 3.4. $\prod_v WC(\mathscr{A}_{K_v})\cong\prod_v K_0\left(\mathcal{A}_{\Theta(k)}\rtimes_{L_v}\mathbf{Z}\right)$.

PROOF. One takes the direct sum of finite groups $WC(\mathscr{A}_{K_v})$ over all non-archimedean places v of the field K. The corollary follows from isomorphism (3.7). □

LEMMA 3.5.

$$Ш(\mathscr{A}_K)\cong\begin{cases} Cl\ (\Lambda)\oplus Cl\ (\Lambda), & \text{if } k \text{ is even},\\ (\mathbf{Z}/2^k\mathbf{Z})\oplus Cl_{\ \mathbf{odd}}\ (\Lambda)\oplus Cl_{\ \mathbf{odd}}\ (\Lambda), & \text{if } k \text{ is odd}.\end{cases}$$

PROOF. For the sake of clarity, let us outline the main idea. In view of lemma 3.1 and corollary 3.4, one can replace (1.1) by a group homomorphism:

$$\Sigma\left(\mathcal{A}_{\Theta(k)}\right)\to\prod_v K_0\left(\mathcal{A}_{\Theta(k)}\rtimes_{L_v}\mathbf{Z}\right). \tag{3.8}$$

To calculate the kernel $Ш(\mathscr{A}_K)$ of (3.8), recall that the algebra $\mathcal{A}_{\Theta(k)}$ at the LHS of (3.8) depends on the similarity class of a single matrix $B\in GL(2n,\mathbf{Z})$, see definition 2.1. On the other hand, the RHS of (3.8) depends solely on the characteristic polynomial of B. Roughly speaking, the problem of kernel of (3.8) reduces to an old question of the linear algebra: how many non-similar matrices with the same characteristic polynomial are there in $GL(2n, \mathbf{Z})$? This problem was solved by [Latimer & MacDuffee 1933] [**LM**]. Their theorem says that the similarity classes of matrices $B \in GL(2n, \mathbf{Z})$ with the fixed polynomial $Char\ (B)$ are in one-to-one correspondence with the ideal classes $Cl\ (\Lambda)$ of an order Λ of a number field k generated by the eigenvalues of B. Coupled with the fact, that the scale $\Sigma\left(\mathcal{A}_{\Theta(k)}\right)$ contains the scales corresponding to all similarity classes, the Latimer-MacDuffee Theorem implies that the (half of) kernel of (3.8) is isomorphic to the group $Cl\ (\Lambda)$. The rest of the proof follows from the Atiyah pairing between the K-theory and the K-homology. We pass to a step by step argument.

(i) We use notation of Sections 2.3 and 2.4.2. Recall that the positive cone $K_0^+(\mathcal{A}_{\Theta(k)})$ and the scale $\Sigma\left(\mathcal{A}_{\Theta(k)}\right)$ can be recovered from (2.6) by solving the inequality $\Lambda \geq 0$ and $0 \leq \Lambda \leq 1$, respectively. Since $\theta_i \in k$ are components of the normalized Perron-Frobenius eigenvector of a matrix $B \in GL(2n, \mathbf{Z})$, we conclude that the similarity class of B defines the algebra $\mathcal{A}_{\Theta(k)}$ up to an isomorphism, and vice versa. The set of all pairwise non-similar matrices with the characteristic polynomial $Char\ (B)$ will be denoted by $B_1, \ldots, B_h$.

(ii) Let us show, that $\Lambda_1 \subset \cdots \subset \Lambda_h$, where Λ_i are $\mathbf{Z}$-modules corresponding to the matrices B_i. Indeed, since B_i has the same characteristic polynomial as B, we conclude that the components of the normalized Perron-Frobenius eigenvector of B_i must lie in the number field k. Therefore Λ_i is a full $\mathbf{Z}$-module in the field k. It is well known, that set $\{\Lambda\}_{i=1}^h$ of such modules can be ordered by inclusion $\Lambda_1 \subset \cdots \subset \Lambda_h$, where Λ_h is the maximal $\mathbf{Z}$-module.

REMARK 3.6. The inclusion of scales $\Sigma_1 \subset \cdots \subset \Sigma_h$ follows from the inclusion $\Lambda_1 \subset \cdots \subset \Lambda_h$ and the inequality $0 \leq \Lambda_i \leq 1$.

(iii) Let us prove that the matrix L_v given by formula (2.7) does not depend on $\{B_i \mid 1 \leq i \leq h\}$. Indeed, the a_k in (2.7) are coefficients of the characteristic polynomial of the matrix $B^{\pi(p)}$, where $\pi(p)$ is a positive integer. Notice that the spectrum $Spec\ (B_i) = \{\lambda_1^i, \ldots, \lambda_{2n}^i\}$ does not depend on the matrix B_i, since $Char\ (B_i)$ does not vary. Because $Spec\ \left(B_i^{\pi(p)}\right) = \{(\lambda_1^i)^{\pi(p)}, \ldots, (\lambda_{2n}^i)^{\pi(p)}\}$, we conclude that the spectrum of the matrix $B_i^{\pi(p)}$ is independent of B_i. Therefore the polynomial $Char\ \left(B_i^{\pi(p)}\right)$ and the coefficients a_k in (2.7) are the same for all matrices B_i. It follows from (2.7), that L_v is independent of the choice of matrix $\{B_i \mid 1 \leq i \leq h\}$.

(iv) Let us calculate the kernel of homomorphism (3.8). Without loss of generality, we assume $B \cong B_h$. By remark 3.6, we have the inclusion of torsion groups:

$$\Sigma_1 \subset \cdots \subset \Sigma_{h-1} \subset \Sigma\left(\mathcal{A}_{\Theta(k)}\right). \tag{3.9}$$

Let $\mathbf{e} \in \prod_v K_0\left(\mathcal{A}_{\Theta(k)} \rtimes_{L_v} \mathbf{Z}\right)$ be the trivial element of torsion group at the RHS of (3.8). From item (iii), it is known that the preimage of $\mathbf{e}$ under the homomorphism (3.8) consists of h distinct elements of the group $\Sigma\left(\mathcal{A}_{\Theta(k)}\right)$. Indeed, each Σ_i in (3.9) has a unique such element lying in $\Sigma_i \backslash \Sigma_{i-1}$, since otherwise the corresponding abelian variety would have two different reductions modulo p. We conclude therefore, that the kernel of homomorphism (3.8) is an abelian group of order h.

(v) Let us calculate the Shafarevich-Tate group $Ш(\mathscr{A}_K)$. It follows from item (iv), that $h = |Cl\ (\Lambda)|$, where $Cl\ (\Lambda)$ is the ideal class group of Λ, see [Latimer & MacDuffee 1933] [**LM**]. It is easy to see, that the kernel of (3.8) is isomorphic to the group $Cl\ (\Lambda)$. To express $Ш(\mathscr{A}_K)$ in terms of the group $Cl\ (\Lambda)$, recall that the K-homology is dual to the K-theory [Blackadar 1986] [**B**, Section 16.3]. Roughly speaking, the cocycles in K-theory are represented by the vector bundles. Atiyah proposed using elliptic operators to represent the K-homology cycles. An elliptic operator can be twisted by a vector bundle, and the Fredholm index of the twisted operator defines a pairing between the K-homology and the K-theory. In particular, $K^0(\mathcal{A}_{\Theta(k)}) \cong K_0(\mathcal{A}_{\Theta(k)})$, where $K^0(\mathcal{A}_{\Theta(k)})$ is the zero K-homology group of the the algebra $\mathcal{A}_{\Theta(k)}$. By analogy with (3.4), one gets $\Sigma_0(\mathcal{A}_{\Theta(k)}), \Sigma^0(\mathcal{A}_{\Theta(k)}) \hookrightarrow \mathbf{Q}/\mathbf{Z}$ and, therefore, a pair of embeddings:

$$Cl\ (\Lambda_0),\ Cl\ (\Lambda^0) \hookrightarrow \mathbf{Q}/\mathbf{Z}. \tag{3.10}$$

Since $Cl\ (\Lambda_0) \cong Cl\ (\Lambda^0)$ are finite abelian groups, the Atiyah pairing gives rise to a bilinear form $Q_{\mathbf{F}_p}(x, y)$ over the finite field $\mathbf{F}_p$ for each prime p dividing $|Cl\ (\Lambda)|$. The (3.10) is a group homomorphism, if and only if, the $Q_{\mathbf{F}_p}(x, y)$ is an alternating form, i.e. $\{Q_{\mathbf{F}_p}(x, x) = 0 \mid \forall x \in Cl\ (\Lambda_0) \oplus Cl\ (\Lambda^0)\}$. Recall that the alternating bilinear forms $Q_{\mathbf{F}_p}(x, y)$ exist, if and only if, $p \neq 2$. Namely, the form $Q_{\mathbf{F}_2}(x, y)$ is always symmetric, i.e. $Q_{\mathbf{F}_2}(x, x) \neq 0$ unless $x = 0$. Thus there are no Atiyah pairing in the case $p = 2$. The rest of the proof follows the Basis Theorem for finite abelian groups. Lemma 3.5 is proved. □

REMARK 3.7. Both cases of lemma 3.5 are realized by the concrete abelian varieties, see [Rubin 1989] [**R**, Examples (B) and (C)] and [Poonen & Stoll 1999] [**PS**, Proposition 27], respectively.

Theorem 1.1 follows from lemma 3.5.

4. Complex multiplication

The $Ш(\mathscr{A}_K)$ can be calculated using theorem 1.1, if the functor F is given explicitly. To illustrate the idea, we consider the abelian varieties with complex multiplication.

4.1. Abelian varieties. Denote by $\mathbf{K}$ a CM-field, i.e. the totally imaginary quadratic extension of a totally real number field $\mathbf{k}$. The abelian variety $\mathscr{A}_{CM}$ is said to have complex multiplication, if the endomorphism ring of $\mathscr{A}_{CM}$ contains the field $\mathbf{K}$, i.e. $\mathbf{K} \subset End\ \mathscr{A}_{CM} \otimes \mathbf{Q}$. The $deg\ (\mathbf{K} \mid \mathbf{Q}) = 2n$, where n is the complex dimension of the $\mathscr{A}_{CM}$. For a canonical basis in $\mathbf{K}$ the lifting of the Frobenius endomorphism has the matrix form:

$$Fr_v = \begin{pmatrix} a_1 & 1 & \dots & 0 & 0 \\ -a_2 & 0 & 1 & 0 & 0 \\ \vdots & \vdots & \ddots & \vdots & \vdots \\ a_{2n-1} & 0 & \dots & 0 & 1 \\ -p & 0 & \dots & 0 & 0 \end{pmatrix}. \tag{4.1}$$

The functor F is acting by the formula $Fr_v \mapsto L_v$, where L_v is given by (2.7), see [**N**, p. 179] for the details. Since $L_v \in End\ \Lambda$, one can recover the ring Λ from the eigenvalues of matrix L_v. Let us consider the simplest case $n = 1$, i.e. the elliptic curves with complex multiplication.

4.2. Elliptic curves. Denote by $\mathcal{E}_{CM}$ an elliptic curve with complex multiplication by the ring $R = \mathbf{Z} + fO_{\mathbf{K}}$, where $\mathbf{K} = \mathbf{Q}(\sqrt{-D})$ is an imaginary quadratic field with the square free discriminant $D > 1$ and $f \geq 1$ is the conductor of R. The ring $\Lambda = F(R)$ is given by the formula $\Lambda = \mathbf{Z} + f'O_k$, where $k = \mathbf{Q}(\sqrt{D})$ is the real quadratic field and the conductor $f' \geq 1$ is the least integer solution of the equation $|Cl\ (\mathbf{Z} + f'O_k)| = |Cl\ (R)|$, see [**N**, Theorem 1.4.1]. Moreover, there exists a group isomorphism $Cl\ (\Lambda) \cong Cl\ (R)$. Using theorem 1.1, one gets the following corollary.

COROLLARY 4.1. $Ш(\mathcal{E}_{CM}) \cong Cl\ (R) \oplus Cl\ (R)$.

EXAMPLE 4.2. (**[Rubin 1989]** [**R**, Example (B)]) Let $\mathcal{E}_{CM}$ be the Fermat cubic $x^3 + y^3 = z^3$. The $\mathcal{E}_{CM}$ has complex multiplication by the ring $R \cong \mathbf{Z} + O_{\mathbf{K}}$, where $\mathbf{K} \cong \mathbf{Q}(\sqrt{-3})$. It is well known, that in this case $Cl\ (R)$ is trivial. We conclude from corollary 4.1, that the $Ш(\mathcal{E}_{CM})$ is a trivial group. An alternative proof of this fact is based on the exact calculation of the p-part of $Ш(\mathcal{E}_{CM})$ and can be found in [Rubin 1989] [**R**, Theorem 1 and Example (B)].

EXAMPLE 4.3. (**[Rubin 1989]** [**R**, Example (C)]) Let $\mathcal{E}_{CM}$ be the modular curve $X_0(49)$ given by the equation $y^2 + xy = x^3 - x^2 - 2x - 1$. The $\mathcal{E}_{CM}$ has complex multiplication by the ring $R \cong \mathbf{Z} + O_{\mathbf{K}}$, where $\mathbf{K} \cong \mathbf{Q}(\sqrt{-7})$. It is well known, that the group $Cl\ (R)$ is trivial. Using corollary 4.1, one concludes that the $Ш(\mathcal{E}_{CM})$ is a trivial group. A different proof of this result can be found in [Rubin 1989] [**R**, Theorem 1 and Example (C)].

REMARK 4.4. Theorem 1.1 is true for the simple abelian varieties $\mathcal{A}_K$. If $\mathcal{A}_K$ is not simple, then the functor F splits and one gets different formulas for the group $Ш(\mathcal{A}_K)$, see the corresponding calculations in the excellent paper [Stein 2004] [**S**, Theorem 3.1].

References

[B] B. Blackadar, *K-theory for operator algebras*, Mathematical Sciences Research Institute Publications, vol. 5, Springer-Verlag, New York, 1986, DOI 10.1007/978-1-4613-9572-0. MR859867

[BS] A. Borel and J.-P. Serre, *Théorèmes de finitude en cohomologie galoisienne* (French), Comment. Math. Helv. **39** (1964), 111–164, DOI 10.1007/BF02566948. MR181643

[C] J. W. S. Cassels, *Diophantine equations with special reference to elliptic curves*, J. London Math. Soc. **41** (1966), 193–291, DOI 10.1112/jlms/s1-41.1.193. MR199150

[LM] C. G. Latimer and C. C. MacDuffee, *A correspondence between classes of ideals and classes of matrices*, Ann. of Math. (2) **34** (1933), no. 2, 313–316, DOI 10.2307/1968204. MR1503108

[M1] H. Minkowski, *Geometrie der Zahlen* (German), Bibliotheca Mathematica Teubneriana, Band 40, Johnson Reprint Corp., New York-London, 1968. MR249269

[M2] D. Mumford, *Tata lectures on theta. I*, Progress in Mathematics, vol. 28, Birkhäuser Boston, Inc., Boston, MA, 1983, DOI 10.1007/978-1-4899-2843-6. MR688651

[N] I. V. Nikolaev, *Noncommutative geometry—a functorial approach*, 2nd ed., De Gruyter Studies in Mathematics, vol. 66, De Gruyter, Berlin, 2022, DOI 10.1515/9783110788709. MR4486440

[PS] B. Poonen and M. Stoll, *The Cassels-Tate pairing on polarized abelian varieties*, Ann. of Math. (2) **150** (1999), no. 3, 1109–1149, DOI 10.2307/121064. MR1740984

[R] K. Rubin, *Tate-Shafarevich groups of elliptic curves with complex multiplication*, Algebraic number theory, Adv. Stud. Pure Math., vol. 17, Academic Press, Boston, MA, 1989, pp. 409–419, DOI 10.2969/aspm/01710409. MR1097625

[S] W. A. Stein, *Shafarevich-Tate groups of nonsquare order*, Modular curves and abelian varieties, Progr. Math., vol. 224, Birkhäuser, Basel, 2004, pp. 277–289. MR2058655

[T] J. Tate, *WC-groups over $\mathfrak{p}$-adic fields [MR0105420 (21 #4162)]*, Séminaire Bourbaki, Vol. 4, Soc. Math. France, Paris, 1995, pp. Exp. No. 156, 265–277. MR1610926

Department of Mathematics and Computer Science, St. John's University, 8000 Utopia Parkway, New York, New York 11439

Email address: igor.v.nikolaev@gmail.com

SELECTED PUBLISHED TITLES IN THIS SERIES

798 **Marat V. Markin, Igor V. Nikolaev, and Carsten Trunk, Editors,** Advances in Functional Analysis and Operator Theory, 2024

794 **Jean-Pierre Magnot, Editor,** Recent Advances in Diffeologies and Their Applications, 2024

793 **Abba Gumel, Editor,** Mathematical and Computational Modeling of Phenomena Arising in Population Biology and Nonlinear Oscillations, 2024

792 **Shaoming Guo, Zane Kun Li, and Brian Street, Editors,** Recent Developments in Harmonic Analysis and its Applications, 2024

791 **Mee Seong Im, Bach Nguyen, and Arik Wilbert, Editors,** Algebraic and Topological Aspects of Representation Theory, 2024

790 **Peter Crooks and Alexandru I. Suciu, Editors,** Compactifications, Configurations, and Cohomology, 2023

789 **I. S. Krasil′shchik, A. B. Sossinsky, and A. M. Verbovetsky, Editors,** The Diverse World of PDEs, 2023

788 **I. S. Krasil′shchik, A. B. Sossinsky, and A. M. Verbovetsky, Editors,** The Diverse World of PDEs, 2023

787 **Hemen Dutta, Editor,** Mathematical Modelling, 2023

786 **Hemen Dutta, Editor,** Mathematical Modelling, 2023

785 **A. Leroy and S. K. Jain, Editors,** Algebra and Coding Theory, 2023

784 **Dinh-Liem Nguyen, Loc Hoang Nguyen, and Thi-Phong Nguyen, Editors,** Advances in Inverse Problems for Partial Differential Equations, 2023

783 **Alla Detinko, Michael Kapovich, Alex Kontorovich, Peter Sarnak, and Richard Schwartz, Editors,** Computational Aspects of Discrete Subgroups of Lie Groups, 2023

782 **Galina Filipuk, Alberto Lastra, and Sławomir Michalik, Editors,** Recent Trends in Formal and Analytic Solutions of Diff. Equations, 2023

781 **Emanuel Indrei, Diego Marcon, and Levon Nurbekyan, Editors,** Geometric and Functional Inequalities and Recent Topics in Nonlinear PDEs, 2023

780 **Erik Koelink, Stefan Kolb, Nicolai Reshetikhin, and Bart Vlaar, Editors,** Hypergeometry, Integrability and Lie Theory, 2022

779 **Samuele Anni, Valentijn Karemaker, and Elisa Lorenzo García, Editors,** Arithmetic, Geometry, Cryptography, and Coding Theory 2021, 2022

778 **Carlos Galindo, Alejandro Melle Hernández, Julio José Moyano-Fernández, and Wilson A. Zúñiga-Galindo, Editors,** p-Adic Analysis, Arithmetic and Singularities, 2022

777 **Bang-Yen Chen, Nicholas D. Brubaker, Takashi Sakai, Bogdan D. Suceavă, Makiko Sumi Tanaka, Hiroshi Tamaru, and Mihaela B. Vajiac, Editors,** Differential Geometry and Global Analysis, 2022

776 **Aaron Wootton, S. Allen Broughton, and Jennifer Paulhus, Editors,** Automorphisms of Riemann Surfaces, Subgroups of Mapping Class Groups and Related Topics, 2022

775 **Fernando Galaz-García, Cecilia González-Tokman, and Juan Carlos Pardo Millán, Editors,** Mexican Mathematicians in the World, 2021

774 **Randall J. Swift, Alan Krinik, Jennifer M. Switkes, and Jason H. Park, Editors,** Stochastic Processes and Functional Analysis, 2021

773 **Nicholas R. Baeth, Thiago H. Freitas, Graham J. Leuschke, and Victor H. Jorge Pérez, Editors,** Commutative Algebra, 2021

772 **Anatoly M. Vershik, Victor M. Buchstaber, and Andrey V. Malyutin, Editors,** Topology, Geometry, and Dynamics, 2021

For a complete list of titles in this series, visit the
AMS Bookstore at **www.ams.org/bookstore/conmseries/**.